PATENTS

Patents

Economics, Policy and Measurement

F.M. Scherer

Aetna Professor Emeritus, John F. Kennedy School of Government, Harvard University and Lecturer, Woodrow Wilson School of Public and International Affairs, Princeton University, USA

Edward Elgar

Cheltenham, UK • Northampton, MA, USA

Published by
Edward Elgar Publishing Limited
Glensanda House
Montpellier Parade
Cheltenham
Glos GL50 1UA
UK

Edward Elgar Publishing, Inc.
136 West Street
Suite 202
Northampton
Massachusetts 01060
USA

A catalogue record for this book
is available from the British Library

ISBN 1 84542 481 6

Printed and bound in Great Britain by MPG Books Ltd, Bodmin, Cornwall.

Contents

PART III WHAT DO THE MEASURES MEASURE?

PART IV THE ROAD AHEAD

Acknowledgements

The author and publishers wish to thank the following who have kindly given permission for the use of copyright material:

Association of American Medical Colleges for 'The Economics of Human Gene Patents', *Academic Medicine*, **77**, December 2002, 1348–67.

American Economic Association for 'Nordhaus' Theory of Optimal Patent Life: A Geometric Reinterpretation', *American Economic Review*, **62** (3), June 1972, 422–7 and 'Firm Size, Market Structure, Opportunity, and the Output of Patented Inventions', *American Economic Review*, **55** (5), Part 1, December 1965, 1097–125.

Blackwell Publishing for 'A Note on Global Welfare in Pharmaceutical Patenting', *The World Economy*, **27**, July 2004, 1127–42 and 'Demand-Pull and Technological Invention: Schmookler Revisited', *Journal of Industrial Economics*, **30** (3), March 1982, 225–37.

Elsevier Ltd for 'Comment on Edmund Kitch', in John Palmer (ed.), *Research in Law and Economics*, **8**, 1986, 51–8 and 'The Propensity to Patent', *International Journal of Industrial Organization*, **1** (1), 1983, 107–28 and 'Inter-Industry Technology Flows in the United States', *Research Policy*, **11**, 1982, 227–45.

Houghton-Mifflin for 'The Economics of the Patent System', in F.M. Scherer, *Industrial Market Structure and Economic Performance*, 2nd edition, Rand McNally, 1980, 439–58.

Kluwer Academic Publishers NL for 'Exploring the Tail of Patented Invention Value Distributions' (with Dietmar Harhoff and Katrin Vopel), in Ove Granstrand (ed.), *Economics, Law and Intellectual Property*, Kluwer, 2003, 279–309.

MIT Press Journals for 'Citation Frequency and the Value of Patented Inventions' (with Dietmar Harhoff, Francis Narin and Katrin Vopel), *Review of Economics and Statistics*, **81** (3), August 1999, 511–15.

OUP for 'The Innovation Lottery', in Rochelle Dreyfuss, Diane Zimmerman and Harry First (eds), *Expanding the Boundaries of Intellectual Property*, Oxford University Press, 2001, 3–21 and 'Post-TRIPS Options for Access to Patented Medicines in Developing Nations' (with Jayashree Watal), *Journal of International Economic Law*, **5** (4), December 2002, 913–39.

Pergamon Journals Inc. for 'The Office of Technology Assessment and Forecast Industry Concordance as a Means of Identifying Industry Technology Origins', *World Patent Information*, **4** (1), 1982, 12–17.

Taylor and Francis Ltd for 'Technology Flows Matrix Estimation Revisited', *Economic Systems Research*, **15** (3), September 2003, 327–49.

University of Chicago Press for 'Corporate Inventive Output, Profits, and Growth', *Journal of Political Economy*, **73** (3), June 1965, 290–97.

Vanderbilt University Law School for 'The Pharmaceutical Industry and World Intellectual Property Standards', *Vanderbilt Law Review*, **53** (6), November 2000, 2245–54.

Wiley – VCH Verlag GmbH for 'Economic Effects of Strengthening Pharmaceutical Patent Protection in Italy' (with Sandy Weisburst), *International Review of Industrial Property and Copyright Law*, **26** (6), 1995, 1009–24.

Every effort has been made to trace all the copyright holders but if any have been inadvertently overlooked the publishers will be pleased to make the necessary arrangements at the first opportunity.

Introduction

This volume reproduces selected articles I have written on various aspects of the patent system over the course of four decades.

My interest in the patent system came from the conjunction of three forces: a youthful fascination with technology that led me to enter undergraduate studies at the University of Michigan with the intention (abandoned after a year-long course) of majoring in chemistry; my reading of Joseph A. Schumpeter's *Capitalism, Socialism, and Democracy* in Shorey Peterson's economics honors seminar during the spring semester of 1954; and my decision to write on the atomic energy patent laws for the requisite honors thesis.

As the seminar was progressing, the US Congress was debating a new law whose purpose was to move atomic energy from a strictly military endeavor into the commercial realm, with electric power generation as the first of many foreseen applications. Under the US Atomic Energy Act of 1946, private patents in specific nuclear energy areas were prohibited. To encourage the development of a new commercial industry, a new law was being written, which among other things would permit patenting, but the relevant committees were struggling with the question of whether private patents would be subject to compulsory licensing. My paper, 'The Atomic Energy Patent Laws and Economic Progress', analyzed the incentive effects of compulsory licensing provisions, concluding inter alia that they were likely to have their main adverse effect on small firms' R&D investment incentives. The law eventually passed, the Atomic Energy Act of 1954, reserved compulsory licensing for inventions made under government contract and those covering 'special nuclear material' in which the government had a direct interest.

My honors thesis experience must have sensitized me to new developments in the patent field. A triggering event came in the settlements during January 1956 of two major antitrust cases, one requiring AT&T and the other IBM, to license the approximately 9,000 patents in their portfolios, in most instances without receiving royalties in return. There had been earlier compulsory licensing precedents in US antitrust settlements, some surveyed in my senior honors thesis. But the judgments of 1956 were of unusual scope, involving companies pressing the frontiers of technology. In an editorial 27 January 1956, the *Wall Street Journal* warned:

> So it may turn out that these are dangerous victories the Government boasts about. The settlements in these cases indicate a belief that everybody's patents should be everybody else's. But this is a philosophy that strikes at incentive; new ideas and new inventions may be lost. Such Government victories may turn out to be far more costly for the nation than for the companies.

Twenty months later, I was a second-year MBA student at the Harvard Business School, enrolled in the 'Manufacturing' course taught by Georges Doriot. Doriot had been a general responsible for US Army research and development during World War II and in 1957 was president of the first high-technology venture capital firm,

American Research and Development Corporation. Among the requirements in his Manufacturing course was that we form groups of around nine students to conduct two hands-on projects in local companies and also to collaborate in writing a so-called 'topic report'. Our group decided for our topic report to address the question of how compulsory patent licensing under antitrust decrees affected R&D investment incentives. The idea was mine, and I was chosen as the project leader. During our Christmas vacation we fanned out to interview 22 companies, many of which had entered compulsory licensing decrees, and we later received mail questionnaires from an additional 69 companies. We also conducted statistical analyses of patenting by a broader universe of US corporations. Our findings were self-published in a monograph, *Patents and the Corporation*.[1] The first edition sold out quickly, so I rewrote parts for a revised second edition, adding new insights I had gleaned from several innovation case studies as a research assistant at the Business School.

What we learned from our interview and survey work was quite contrary to our expectations. The compulsory licensing decrees had for the most part very little adverse impact on the R&D investments of the firms that had experienced them and also of companies that might consider themselves vulnerable in the future. And indeed, the expectation of patent protection seldom played a significant role in the companies' R&D decisions. Rather:[2]

> The survey results showed clearly that competitive exigencies were far more important among established corporations as reasons for research and development than considerations of potential patent protection. Once a company has reached a point where it can regard itself as a going concern, the need to remain on an even plane with, or to surpass, competitors is a far more active incentive to do research and development than the lure of possible exclusive rewards.

Both our interviews and our statistical analysis showed, however, that compulsory licensing decrees did have some negative influence on patenting, the more so, the more directly impacted companies were. The evident reason was that companies were being more selective in applying for new patents and relying more on secrecy, especially for process inventions that were not available to others for physical inspection.

The findings we reached in *Patents and the Corporation* have since been confirmed and amplified through more sophisticated and wider-ranging surveys conducted at Yale University, Carnegie–Mellon University, the University of Pennsylvania and the University of Cambridge.[3]

Another formative experience underlay the development of my views on how the patent system works in the real world. Between 1966 and 1969, I was principal economic adviser to the Committee on Government Patent Policy (CGPP), a multi-agency task force created to study in depth how the varying disposition of patent rights to inventions made under US government contracts affected contract participation and the diffusion of patented inventions into the commercial sector. The committee's extensive findings were published in a series of volumes by the Harbridge House consulting firm. My direct contributions were made in a series of memoranda, but I summarized what I had learned from the CGPP study and other work in a 1977 monograph, *The Economic Effects of Compulsory Patent Licensing*.[4] That monograph, once virtually inaccessible, has been reprinted in a previous Edward

Elgar Publishing compendium, so no excerpts are presented here.[5] A more succinct summary of my views on how the patent system functions and malfunctions is presented in Article 1 here, which is taken from the second edition of my textbook on industrial organization.[6]

Prior to the 1969 publication of a book by William Nordhaus,[7] most of the theorizing about how the patent system balances incentives for investment in invention against the desire for maximum diffusion of inventions was murkily qualitative. Nordhaus' analysis established a new, more rigorous paradigm. Article 2 reprinted here provides my first (but not final) reaction. It generalizes Nordhaus' special mathematical assumption about the relationship between R&D spending and the extent of invention and shows how, in the more general case, one needs to consider not only the continuous relationship between patent life and the extent of invention, but also whether patent protection makes it possible to satisfy a 'benefits greater than cost' test that might otherwise not be met.

Article 3 is a brief comment on a paper by Edmund Kitch, the originator of the 'prospect theory' of patent rights. In the paper on which I commented, Kitch argues that the Xerox Corporation had no real monopoly power despite its strong patent position. I argue to the contrary, and indeed, observe that Xerox's patent protection was so strong that Xerox represented a rare historical case consistent with Kitch's prospect theory. It should be noted here that the prospect theory proposes, and the Nordhaus theory assumes, quite similar protected positions for a patent holder – a case that, I argue in Article 8 later, implies quite different welfare tradeoffs than does an alternative rent-seeking view of patents. Through my analysis of Xerox's monopoly power I provide hints as to how one derives the demand curve for a market-blocking patent.

Whether or to what extent 'inventions' involving the human genome should receive patent protection has been a controversial policy question in both the United States (which has chosen a relatively tolerant stance) and Europe (which has discouraged gene patents). Article 4 reviews the arguments, first with an overview of patent logic generally and then with a focus on the special features of genome patenting – notably, the key position of startup companies in pushing the technology forward and the strong sequential impact one generation of inventions has on the possibility of making next-stage inventions. It also reports on an empirical survey of the kinds of gene sequence patents issued in the US and the types of firms that received them. It recommends measures to ensure that prior patent positions do not close off subsequent lines of discovery.

During the 1980s, a handful of industries strongly dependent upon patent and copyright protection began intense lobbying efforts to induce third-world nations to emulate the most industrialized nations in the extent of protection they provided to so-called 'intellectual property'. They were led by the pharmaceutical industry, in which patents provide particularly strong and effective protection. Their efforts culminated during the early 1990s in an agreement that for nations to join the newly-created World Trade Organization, they had to provide a high level of patent protection, among other things, for pharmaceutical products, which many less-developed nations had excluded from patentability. As the debate evolved, I grew increasingly concerned that the world's poor nations were being blackmailed to adopt policies that were not in the best interest of their citizens. The question became a major item on my research agenda, leading to a series of papers. Article 5 provides my historical overview of the pharmaceutical patent controversy.

One argument advanced by the patent proponents is that the grant of patent rights would stimulate greater inventive effort within nations offering patents for the first time. That such a tendency would exist can hardly be denied, but how strong would the behavioral change be? On this there was little relevant evidence. Fortuitously, a natural experiment occurred in Italy, which in 1978 began granting pharmaceutical product patents after a long period of denying them. During the period of denial the world's leading generic drug supply industry took root in Italy. Fortunately, a Harvard undergraduate student, Sandy Weisburst, had the ability to read Italian and took an interest in the question as a senior thesis topic. Article 6 summarizes the results of his research. In the first decade after Italy began awarding pharmaceutical product patents, drug R&D and the marketing of new drug products scarcely increased relative to world trends, but Italy's balance of trade from generic drug exports dropped sharply. The experiment revealed that making the transition to high-technology product innovation is clearly not assured simply through a change in patent policies. To my dismay, the publishers of the journal in which our joint article appeared changed the designated order of authors; Weisburst should be considered the senior author.[8]

With the inclusion of the TRIPS (trade-related aspects of intellectual property rights) articles in the World Trade Organization treaty, how those provisions would be interpreted became a priority question, especially in view of the AIDS pandemic escalating in the third world. Article 7 is distilled from a longer paper commissioned by the World Health Organization and circulated at the WHO–WTO Høsbjør, Norway, conference in April 2001. It was written with Jayashree Watal, who served previously as a member of the Indian delegation at the WTO treaty talks. We attempted to extend the frontiers in several controversial areas. We reviewed the logic of patent protection for pharmaceutical products and examined how compulsory licensing, authorized under two articles of the TRIPS agreement, might work. Our analysis identified a lacuna – that production under compulsory license was required to be preponderantly for home consumption, but small, least-developed nations were unlikely to have the domestic capability to produce their own generic drugs under compulsory license. That lacuna was remedied through mutual agreement in August 2003 in preparation for the (otherwise unsuccessful) Cancun Round of trade negotiations. We also showed, as several other economists had done earlier, that when pharmaceutical manufacturers enjoy patent protection in all the world's nations, the best pricing strategy would be for them to engage in Ramsey-type price discrimination, charging relatively high prices in the richest nations and low prices in the poorest nations. From an analysis of AIDS anti-retroviral drug pricing in 15 nations for the years 1995–9, we found only a weak tendency toward such pricing. We identified reasons why and proposed changes to encourage the optimal policy. In fact, the price front broke in 2001, triggered by offers from Indian generic drug producers, who were not yet bound by first-world patent laws, and encouraged by the consensus reached at Høsbjør. Thus, in the years that followed, anti-retrovirals were available to third-world nations at prices much lower than in the first world. The final section of the article explores a loophole in US tax law to show how under certain conditions pharmaceutical makers can actually increase their after-tax profits by giving away drugs to charitable health-care organizations.

During the approximately eight years I struggled with the problems posed under TRIPS for the supply of pharmaceuticals to low-income nations, it was clear that a theory of optimal patent protection richer than the one originated by William Nordhaus was needed. Article 8 is the fruit of that struggle. It combines a generalization of the Nordhaus invention inducement theory with a rigorous approach to measuring the benefits of various patent regimes in rich and poor nations, assuming the Ramsey pricing recommended in Article 7. It adds a component that is seldom utilized in economists' benefit–cost analyses, but which I considered vital to addressing the value-laden question of third- vs first-world equity in patent questions – weighting benefits by the ratio of the marginal utility of income for poor consumers relative to the marginal utility for rich consumers. It shows that for plausible values of that ratio, global welfare is higher if low-income nations are not required to provide strong patent protection on pharmaceutical products invented in the first world. An important distinction is made between what might, given Article 3, be called the Nordhaus–Kitch prospect view of invention inducement mechanisms and rent-seeking models, which differ by assuming that R&D outlays rise to exhaust discounted quasi-rent potentials. The case for third-world free-riding on first-world inventions is shown to be stronger in the rent-seeking case, which, empirical evidence suggests, appears to reflect real-world tendencies more closely.

Patents are of interest to economists not only because of their incentive role in stimulating technological innovation, but also because they provide a potential lever for the difficult task of measuring the rate and direction of inventive activity. During the 1960s Jacob Schmookler and I collaborated in extending the art of using patent statistics as instruments of measurement.[9] Articles 9 through 17 stemmed from that collaboration, at first directly and then, after Schmookler's untimely death in 1967, by inspiration.

I had already used patent data for *Patents and the Corporation* to measure statistically how compulsory licensing affected corporate patenting. Further exploratory work was carried out in an econometrics paper for Hendrik Houthakker at Harvard in 1962. The first project on which I embarked as a new assistant professor at Princeton in 1963 was to compile what at the time was considered a large database on patenting and, to a more limited extent, R&D employment by 448 large American manufacturing corporations. Article 9 attempted a definitive test of various structural hypotheses derived from the work of Joseph A. Schumpeter, notably, that large corporations provide more fertile soil for research, development and technological innovation than small firms, and that innovative activity is more intense in industries dominated by only a few firms (that is, tight oligopolies) than in atomistically structured industries. The analysis supported for the most part a rejection of the Schumpeterian hypotheses, recognizing to be sure possible exceptions, given the substantial diversity of industrial circumstances. I presented the results inter alia in September 1965 at a conference in Fontainebleau, France, where French finance minister (later president) Valery Giscard d'Estaing and others, persuaded by Schumpeterian theorizing, argued that Europeans could best catch up to the US juggernaut technologically by creating large national 'champion' enterprises dominating the industries they occupied. I argued then (joined by several other Americans attending the conference) that they were mistaken. The hindsight of 40 years and numerous European industrial policy failures supports my skepticism much more than their optimism.

In Article 10, I extended the 448-company database to analyze how inventive activity, measured by patents, affected corporations' sales growth and profitability. This was one of those hypotheses everyone knew to be true, but which had at the time very little systematic statistical support. The results confirmed the hypotheses, but showed also that the business cycle had an important confounding influence. The linkages I found have been confirmed in numerous subsequent studies.

In our occasional meetings during the 1960s, Jacob Schmookler and I debated whether technological changes were triggered more by 'demand-pull' influences or advances in the science and technology base, which I named 'technology-push' inducements. In Chapter VIII of *Invention and Economic Growth*, Schmookler proposed an ingenious synthesis of the contending hypotheses: technological advances were responsive to the pull of demand, but the industrial locus from which the advances originated followed patterns of specialization, with industries enjoying the richest science and technology bases contributing a disproportionate share of advances, which they sold across industry boundaries to meet less well-endowed industries' demands. He conceptualized this idea as a kind of input–output matrix, but did not live long enough to implement the concept empirically. Having worked during the mid-1970s at the US Federal Trade Commission to collect an extraordinarily detailed set of data, including R&D expenditure data, broken down by narrowly-defined line of business, I took up the challenge in the late 1970s where Schmookler left off. To advance toward the goal, I had a team of four unusually bright Northwestern University students analyze 15,112 US invention patents obtained by 443 corporations, ascertaining from the printed specifications for each patent and information on company product lines by geographic location the industry in which the invention originated as well as the industry(ies) expected to derive benefit from the invention. The patent data were then linked to R&D data in the Federal Trade Commission files. Article 11 provides the main descriptive result of this grueling data collection effort. It shows rich flows of technology from originating industries to using industries and reveals inter alia why agriculture, mining and many service industries have enjoyed rapid rates of productivity growth even though they performed at the time (and most still do perform) very little research and development. The productivity consequences were analyzed more fully in separate articles not reprinted here.[10]

Article 12 makes further use of the industry-linked patent data to retest for a much broader array of industries. Schmookler's finding that inventive activity, measured by patents assigned to the industries using the inventions, was responsive to the pull of demand. Schmookler's basic hypothesis is confirmed, although not as strongly as in *Invention and Economic Growth*, and less so for industrial materials inventions than for the capital goods inventions his original analysis emphasized. One of the most striking results came from distinguishing between capital goods inventions made within the industry using them and those made by other industries and sold in inter-industry transactions. The correlation with demand variables was if anything stronger for the inter-industry transactions than for inventions made in response to demand within the inventing firm's own sphere of operation. Markets for technology transmit their incentive signals remarkably well!

The linked patent and R&D data were used again in Article 13 to determine differences in industries' 'propensity to patent', which I defined as differences in the number of patents received per million dollars of R&D expenditures. Fairly wide differences are observed – from 0.45 patents per million dollars in motor vehicles to 3.20 patents in stone, clay and glass products. Various qualitative and quantitative hypotheses explaining the reasons for these differences are explored. The data are also used to achieve new insights into the Schumpeterian firm size hypotheses tested earlier in Article 9. At the individual line of business level, the probability that at least one patent will be obtained rises nonlinearly, the higher R&D outlays are. Counts of the number of patents received per line of business rise most often proportionately with R&D outlays, and next most frequently with a diminishing returns tendency. Again, the Schumpeterian firm size hypotheses are not supported.[11]

Linking 15,112 patents to the lines of business that originated them and then tracing their effects through a chain of inter-industry sales was a prodigiously labor-intensive operation. The question nagged at me, was there an easier way? Article 14 revisits the topic, constructing a matrix of inter-industry technology flows under the assumption that inventions are traced more easily to the industries in which they originate, after which they can be flowed out to using industries in rough proportion to the patterns of inter-industry sales recorded in input–output matrices periodically published by the US Department of Commerce (and by other governments' statistical agencies). The answer is, similar results can be obtained, although there are some differences, and difficult methodological questions must be resolved and new data classifications must be undertaken to do the job satisfactorily. Because it is generally similar to the matrix included with Article 11, the detailed inter-industry technology flows matrix originally published with Article 14 is omitted in the version reprinted here.

In using patents to measure the amount, origins and uses of inventive activity, one is well-advised to be sensitive to the limitations of one's yardstick. Articles 15 to 17 reflect this concern.

Rather than constructing a matrix of technology flows through the laborious examination of individual patents, as my team did for Article 11, it was suggested during the early 1980s that a concordance from US Patent Office classes to industries, devised by Patent and Trademark Office staff, be employed. At an early stage of our linking project we attempted to use patent classifications, but found them ill-suited to the task, since many emphasize generic technological characteristics (for example, 'article dispensing', 'furnaces' and 'measuring and testing') that might be conceived or used in any industry. To illustrate this point more generally, a selection of 100 patents from our sample of 15,112 was linked to industries through the Patent Office concordance and, alternatively, through our reading of patent specifications and information on the products manufactured in establishments where the inventors worked. Disagreement rates of up to 49 percent, mostly attributable to problems in the Patent Office concordance, were reported in Article 15. My experience since then suggests that the International Patent Classification is more useful than the US classification for industry linkings, since it is organized mainly by industries rather than narrow technological features. Even then, it is hard to tell for inventions sold across industry lines whether an origin industry or a using industry has been identified.

In my research during the early 1960s for Article 9, I happened upon a small sample for which George Washington University researchers had identified through survey methods the profitability of individual patents. I also attended a seminar at which Benoit Mandelbrot discussed the extraordinary statistical properties of highly skew distributions of the Pareto form, that is, with a few outlying extreme values. When Pareto α coefficients are below 1.0, such distributions are characterized by asymptotically infinite means and variances. A crude test with the George Washington University data suggested that the profitability distribution might be Pareto with α less than 0.5, leading me to conclude that 'patent statistics are likely to measure run-of-the-mill industrial inventive output much more accurately than they reflect the occasional strategic inventions which open up new markets and new technologies'. Those inventions, I continued, must remain the domain of economic historians.

This worry remained latent for a quarter century until an opportunity arose during the late 1990s to take advantage, with the collaboration of Dietmar Harhoff, of data generated during 1977, the last year in which patent protection could be obtained in Germany only through domestic patents without the complication of newly-authorized Europatents. We selected what were likely to be the most valuable patents as those running to full term because their holders had paid annual renewal fees totalling DM16,075 (about $7,500 at 1986 exchange rates). We then sought to ascertain through telephone and face-to-face interviews with patent holders the economic value of each such patent. A smaller parallel sample was drawn for the US. Using these peculiarly rich data, we found in Article 16 that the distribution was indeed highly skew, as expected, but that it was more likely to conform to a tolerably well-behaved log normal form rather than the Pareto form. We also found that patent value estimates made using our direct methods were several hundred times higher than those derived by other scholars through inferences based only upon patent renewal rates. The main reason for the difference is that renewal rate studies assume that through non-renewal the patent holder forgoes only the right to exclude others from using the invention but does not surrender its own right to use the invention, whereas our counter-factual question assumed that the patent holder would surrender the right to use its own invention. With our US patent sample we were able to estimate that the aggregate profitability of the selected inventions exceeded the total amount spent on research and development by the patent-holding companies in 1976, the most likely year of invention.

Article 17 carries our analysis of the patent value data further. In it, we discovered that the most valuable patents tended to be cited most frequently in subsequently issued patents. Patents not renewed to full term were in turn cited less frequently on average than those for which renewal fees were paid over the maximum patent life of 18 years. A richer analysis of the German data (not reprinted here) uncovered additional ascertainable correlates of patent value.[12] Notably, those German patents that were targeted for and survived external 'opposition' proceedings proved to be especially valuable.

The implications of highly skew payoff distributions are profound. Basically, the law of large numbers works poorly, or with extreme Pareto distributions, not at all.

This means that portfolio strategies lose their effectiveness in ensuring that if one invests in a large enough number of projects, returns on investment will converge toward stable averages.[13] Article 18 carries the findings from our skewness research in new, more speculative directions. Is it possible that skewness – that is, the fact that relatively few investments yield most of the returns – is actually a positive motivating force, rather than a factor discouraging investment through risk aversion? The selection provides a theoretical explanation why those who take high-technology plunges might be skewness lovers, advancing complementary evidence from the analysis of horse race betting. Further investigation of the behavioral implications is very much to be desired.

Notes

1. The authors, listed in approximate order of their contributions to the book, were F.M. Scherer, Sigmund Herzstein, Alex W. Dreyfoos, William G. Whitney, Otto J. Bachmann, Cyril P. Pesek, Charles J. Scott, Thomas G. Kelly and James J. Galvin; 1958, *Patents and the Corporation: a Report on Industrial Technology under Changing Public Policy*, Boston, MA.
2. Ibid., p. 149 of the revised edition.
3. See Richard C. Levin et al., 'Appropriating the Returns from Industrial Research and Development', *Brookings Papers on Economic Activity* 3, 1987, 783–820; Wesley M. Cohen, 'Patents and Appropriation: Concerns and Evidence', *Journal of Technology Transfer*, **30**, January 2005, 57–71; Edwin Mansfield, 'Patents and Innovation: An Empirical Study', *Management Science*, **32**, 1986, 173–81; and C.T. Taylor and Z. Aubrey Silberston, *The Economic Impact of the Patent System: A Study of the British Experience*, Cambridge: Cambridge University Press, 1973. An important anticipatory publication was Arnold Plant, 'The Economic Theory Concerning Patents for Inventions', *Economica*, **1** new series , February 1934, pp. 30–51.
4. New York University Graduate School of Business Administration Monograph Series in Finance and Economics, 1977.
5. Ruth Towse and Rudi Holzlhauer, *The Economics of Intellectual Property*, International Library of Critical Writings in Economics, Cheltenham, UK and Northampton, MA, USA: Edward Elgar, 2002, vol. II, pp. 315–400.
6. In the third edition, the chapter on patents was combined with other material and is therefore not used here.
7. *Invention, Growth, and Welfare: A Theoretical Treatment of Technological Change*, Cambridge, MA: MIT Press, 1969.
8. After serving a term as a Supreme Court clerk, Weisburst is now an associate with the Mayer, Brown, Rowe and Maw law firm.
9. Schmookler's magnum opus was *Invention and Economic Growth*, Cambridge, MA: Harvard University Press, 1966.
10. F.M. Scherer, 'Inter-Industry Technology Flows and Productivity Growth', *Review of Economics and Statistics*, **64**, November 1982, 627–34; and 'R&D and Declining Productivity Growth', *American Economic Review*, **73**, May 1983, 215–18.
11. For additional results, see F.M. Scherer, 'Technological Change and the Modern Corporation', in Betty Bock et al. (eds), *The Impact of the Modern Corporation*, New York: Columbia University Press, 1984, 286–93.
12. Dietmar Harhoff, F.M. Scherer and Katrin Vopel, 'Citations, Family Size, Opposition and the Value of Patent Rights', *Research Policy*, **32**, 2003, 1343–63.
13. See F.M. Scherer and Dietmar Harhoff, 'Technology Policy for a World of Skew-Distributed Outcomes', *Research Policy*, **29**, 2000, 559–66.

PART I

ECONOMIC ANALYSIS AND POLICY

[1]

The economics of the patent system

Few economic institutions have stirred as much controversy for such a long time as the patent system. Debate over the granting of patent monopolies on inventions has continued ever since the practice was formalized by the Republic of Venice in 1474.[1] In the United States, an interminable procession of blue-ribbon commissions and special legislative committees has been constituted, but from their labors no major substantive reforms have followed. Every attempt to change the system has been drowned in a sea of argument and special pleading. Here we shall explore the logic and paradoxes of the system, seeking to shed light where too frequently there has been only heat.[2]

Some background on the system

An invention patent is an exclusive right to one's invention, including the derivative right to prevent others from using it. Patents are issued in the United States under the broad authority of Article I, Section 8 of the Constitution, which gives Congress the power "to promote the progress of science and useful arts, by securing for limited times to authors and inventors the exclusive right to their respective writings and discoveries." The term of a U.S. patent grant is 17 years from the date of issue.

Under the applicable U.S. laws, as amended in 1952, a patent can be issued to cover "any new and useful process, machine, manufacture, or composition of matter, or any new and useful improvement thereof."[3] Before a patent can be granted, the application is examined in the Patent Office to ensure that the invention is new and nonobvious, that it was not commercialized or known to the public for more than a year before the date of application, and that it has practical utility.

The test for utility has in most instances been perfunctory, requiring nothing more than the barest suggestion that some practical use might ensue. However, more stringent standards have been applied to chemical inventions. For example, in 1966 the U.S. Supreme Court held that the test could not be satisfied for a chemical com-

1. For a survey of the early history of patent grants and the controversy that attended them, see Fritz Machlup, *An Economic Review of the Patent System*, Study no. 15 of U.S., Congress, Senate, Subcommittee on Patents, Trademarks, and Copyrights, 85th Cong., 2nd sess., 1958, pp. 1–5, 22–44.

2. There is an enormous literature on patent systems in the United States and abroad. The most comprehensive compact source is the series of studies commissioned by the Senate Subcommittee on Patents, Trademarks, and Copyrights during the late 1950s. Contemporary U.S. developments are followed in two journals, the *Journal of the Patent Office Society* and *Idea*.

3. 35 U.S. Code 101.

pound simply by showing that molecularly related compounds inhibited tumors in mice.[4]

A more critical hurdle in determining patentability is often the test for nonobviousness. The basic rule is that an invention is not patentable if "the subject matter as a whole would have been obvious at the time the invention was made to a person having ordinary skill in the art."[5] More stringent tests have been applied in certain Supreme Court decisions, e.g., that an invention must reveal "the flash of creative genius," and not merely crystallize gradually through trial and error.[6] In an apparent effort to neutralize such precedents, Congress specified in 1952 that "patentability shall not be negatived by the manner in which the invention was made."[7] Nevertheless, in 1966 the Supreme Court declared that it would continue to impose strict criteria of nonobviousness in contested cases and noted disparagingly the "notorious difference between the standards applied by the Patent Office and by the courts."[8]

The criteria implemented by the Patent Office have not in fact been very restrictive, since many gadgets and minute improvements receive patent protection. Picking up at random an issue of the weekly *Official Gazette,* one can find patents covering a hand-operated nutcracker (No. 4,009,651), a bird perch (No. 4,009,686), and a coin-holding pocket comb (No. 4,009,725), along with such impressive contributions as a method for removing radioactive contaminants from nuclear reactor coolants (No. 4,010,068). Roughly three-fourths of the applications filed with the U.S. Patent Office eventually lead to the issuance of invention patents, and during the 1970s the number of patents issued (excluding design and plant patents) ranged between 70,000 and 81,000 per year. Most covered inventions of slight technological and economic significance, but in any given year there are likely to be a thousand or so moderately to extremely important inventions patented.

In principle, a patent goes to the person making the invention, but scientists, engineers, and technicians employed by corporations normally assign all rights in their work-related inventions to their employer. Of the U.S. invention patents issued between 1971 and 1975, 51 percent were assigned to domestic corporations, 23 percent to foreign corporations and governments, 2 percent to the U.S. federal government, and 23 percent were issued to individual inventors. As one might suppose, the role of the unaffiliated inventor has been declining both relatively and absolutely over time, while corporate patenting has been on the ascendancy. The share of all U.S. patents issued to individual inventors was 81 percent in 1901, 72 percent in 1921, 42 percent in 1940, and 25 percent during the 1960s. Individual inventors received their largest absolute number of patents (30,332) in 1925. From 1971 through 1975, they averaged 17,607 patents per year.

The purposes of the patent grant

Governments have chosen to grant exclusive patent rights on inventions for three main reasons: to promote invention, to encourage the development and commercial utilization of inventions, and to encourage inventors to disclose their inventions to the public.

Early writers on the patent system seldom defined the term invention carefully, and as a result they were vague about how they perceived the role patents played in fostering invention. This has led to considerable confusion, for many scholars have assumed that the patent laws were enacted primarily to promote invention, construed narrowly to include only conception and crude proof testing. Some then infer that the traditional justification has lost most of its relevance to the modern industrial world, where inventive and developmental activities are often divorced from one another organizationally and where development outlays constitute more than three-fourths of all industrial R & D expenditures. Critics seize upon this interpretation by insisting that the system should be abolished; defenders argue in return that the system's rationale has changed over time.

It is questionable whether such a sharp historical dichotomy is empirically tenable. In England, from which U.S. patent precedents stem, patent rights were granted both before and after the 1624 Statute of Monopolies to protect the domestic exploitation of inventions made abroad as well as to encourage invention. One of the most famous English patents covered James Watt's separate condenser principle for steam engines. In extending that patent's life, Parliament in 1775 recognized that the central invention had already been made, and that what it was doing was encouraging further work by Watt to perfect the engine and to "render the same of the highest utility to the publick of which it is capable." Its decree implied furthermore that this was not an unusual case, calling attention to "the many difficulties which always arise in the execution of such large and complex machines . . . and the long time requisite to make the necessary trials."[9] From this and other cases, it would appear that when early patent grants were made ostensibly to encourage invention, what was meant was the complete spectrum of invention, investment, and development—that is, the whole process of innovating.

At any rate, encouraging developmental investment is clearly not an inherently inappropriate function of the patent grant. It might even carry more weight as a public policy objective than merely fostering invention. As Judge Jerome Frank observed in an *obiter dictum* on the conventional rationale:

> The controversy between the defenders and assailants of our patent system may be about a false issue—the stimulus to invention. The real issue may be the stimulus to investment. On that assumption, a statutory revision of our patent system should not be too drastic. We should not throw out the baby with the bathwater.[10]

Yet to state the issue in this way is not to resolve it. On whether patent protection is actually necessary and desirable to stimulate investment in development, we must for the moment retain an open mind.

The third purpose of the patent system—encouraging disclosure—can be dealt with more briefly. Supposedly, the prospect of patent protection leads inventors to make public what they otherwise would keep secret. This view has numerous critics, who argue that inventors will conceal whatever they can, with or without a patent system, and that only the inventions that would be found out anyway will be patented.[11] Inventions affecting internal production processes are especially susceptible to secrecy while product inventions are not, since a product can be purchased and inspected by any would-be imitator. The assertion that patent rights make no difference at all in the choice between secrecy and disclosure is nevertheless an oversimplification. Relying on secrecy is always risky. Business espionage is common, and engineers migrate frequently from one job to another, carrying knowledge of trade secrets with them.[12] If there were no patent system, producers would have no alternative to accepting the risk of trying to maintain secrecy. But patent protection *is* an alternative, and it can be sufficiently attractive to tip the decision toward disclosure in borderline cases. This interpretation draws support from a statistical study of corporate patenting, which showed that firms deprived of patent rights through antitrust judgments reduced their patenting relative to that of unaffected firms.[13] Thus, the patent system probably does encourage more disclosure than there otherwise would be. How great the difference is,

4. *Brenner* v. *Manson*, 383 U.S. 519 (1966).
5. 35 U.S. Code 103, in effect codifying the rule stated in *Hotchkiss* v. *Greenwood*, 11 Howard's Reports 248, 267 (1851).
6. *Cuno Engineering Corp.* v. *Automatic Devices Corp.*, 314 U.S. 84, 91 (1941).
7. 35 U.S.C. 103.
8. *Graham et al.* v. *John Deere Co. et al.*, 383 U.S. 1, 18 (1966).
9. 15 George III, c. 61 (1775), cited in F. M. Scherer, "Invention and Innovation in the Watt-Boulton Steam-Engine Venture," *Technology and Culture* 6 (Spring 1965): 184, 187.
10. *Picard* v. *United Aircraft Corp.*, 128 F. 2d 632, 643 (1942).
11. Cf. Machlup, *An Economic Review of the Patent System*, p. 76; and Alfred E. Kahn, "The Role of Patents," in J. P. Miller, ed., *Competition, Cartels and Their Regulation* (Amsterdam: North-Holland, 1962), p. 317.
12. On the legal protection accorded trade secrets, see *Kewanee Oil Co.* v. *Bicron Corp. et al.*, 416 U.S. 470 (1974).
13. F. M. Scherer, S. E. Herzstein, A. W. Dreyfoos et al., *Patents and the Corporation*, 2nd ed. (Boston: Galvin, 1959), pp. 137–46, 153–55. Further support was drawn from interviews and a questionnaire survey.

and how it affects the rate at which new technology diffuses throughout industry, is uncertain.

The costs and benefits of the patent system

Stimulating the invention and development of new products and processes is without doubt the most important benefit expected of the patent system. For it society pays a price: the monopoly power conferred by patent grants. In simplest terms, the overriding issue of patent policy is whether the benefits of the system outweigh the costs. Or on a more sophisticated plane, the problem is to design a system—e.g., by adjusting the length or strength of patent grants—that will yield the maximum surplus of benefits over costs.

Inventions and innovations bestow benefits upon society. How beneficial they are depends upon how fully they are utilized. Here one of the patent system's many paradoxes appears. Under the system, inventors are given the right to control and restrict utilization of their inventions, so outputs may be lower and prices higher than they would be if the inventions were utilized under purely competitive conditions. Normally, patent holders can choose between alternative methods of controlling utilization. They can reserve exploitation of the invention exclusively to themselves, calling upon the courts to enjoin anyone who attempts to infringe upon that right. In this way, the profit-maximizing price can be set directly. Or they can license as few or as many firms as they please to exploit the invention, charging royalties for the privilege. Through astute determination of the royalty rate, the patentees can in theory achieve the same price-quantity outcome and profits as they could retaining exclusive exploitation, other things (such as the costs of internal versus licensed production) being equal.[14] In many nations, patent holders can also solidify their control over licensees by prescribing prices at which the product can be sold, imposing output quotas, and limiting licensees to particular markets or fields of use. Such restrictions have come under increasing attack in the United States, but they have not yet been ruled illegal when enforced unilaterally by the patentee.[15]

That patent owners exercise their power to set prices exploiting whatever monopoly power their patents confer does not mean that society is denied the benefits stemming from invention and innovation. On the contrary, it gains at the very least from the resources that cost-saving innovations release for alternative uses, less the research and development cost of achieving that saving. In addition, consumers other than the patentee benefit directly in two ways. First, after the patent has expired, the patent holder should in principle have no further power to restrict output, competitive pricing will prevail, and consumers will reap the full benefits of the invention. Ideally, the life of a patent should be no longer than it needs to be to encourage the optimal amount of invention, so that monopolistic restrictions are terminated as soon as possible.[16] Second, consumers may also realize immediate gains even when innovations are exploited monopolistically. Several theoretical cases must be distinguished:

Case 1 A new and superior consumer product is introduced. Although the proof is too elaborate to reproduce here, it can be shown that the innovation necessarily increases consumers' surplus unless the innovator is able to practice perfect first-degree price discrimination.[17] Consumers therefore benefit directly even when the product is priced monopolistically. Whether overall social welfare is enhanced depends upon the innovation's cost, the impact of the innovation on substitute products, and other considerations analyzed in Chapter 14.[18]

Case 2 A new and more efficient production process is introduced by a firm already exercising monopoly power. Here the effect is to shift the monopolist's marginal cost curve downward, inducing a decrease in price if the marginal revenue curve is continuous. Consumers enjoy lower prices and higher consumers' surplus. An exception occurs if the kinked demand curve mentality prevails.

Case 3 A substantially more efficient production process is introduced under patent protection into a previously competitive industry. This is illustrated by the cost curve C_2 in Figure 16.1, where C_1 was the preinnovation cost function and OP_1 the preinnovation price. The firm controlling the new process will find it worthwhile to monopolize the industry, computing its marginal revenue *MR* and setting a price OP_2 that drives existing producers out of business. Or alternatively, it will license the invention at a per-unit royalty equal to the difference between the cost of production with C_2 and price OP_2. In either case, consumers benefit from the lower price. This result is more likely, the greater the reduction in cost is and the more elastic demand is at outputs exceeding the preinnovation output, *ceteris paribus.*[19]

Case 4 A slightly more efficient production process is introduced under patent protection into a competitive industry. This is illustrated by cost curve C_3 in Figure 16.1. If the patent holder could monopolize the industry without restraint by virtue of its patent, it would like to set price OP_3. But it cannot do this because of competition from the preinnovation process C_1. It must therefore either set a price slightly below OP_1 and drive others out, or license others at a per-unit royalty slightly less than the difference between C_3 and C_1. Here the reduction in price will be insignificantly small, and consumers at large will benefit little from the innovation until its patent protection has expired and competitive pricing commences. Then the innovator's profits are redistributed to consumers and, in addition, the dead-weight welfare loss triangle *EFG* is transformed into consumers' surplus.

In three cases out of the four, consumers gain immediately to some extent from the introduction of a patented invention, though they enjoy the full price reduction benefits only after its patent protection has expired. And except when innovators' profits come largely from cannibalization of the profits that would otherwise be enjoyed by the producers of substitute products, it is likely that society as a whole (i.e., including both consumers and producers) gains from inventions and innovations induced or hastened by the grant of patent rights.

Alternative incentives to invent and innovate

There is a rub, however. Patents are granted upon all inventions meeting the established standards of novelty and utility. But the exclusionary power conferred by patents may not be necessary to induce invention and/or innovation in all cases. There are alternative inducements that could render the patent incentive redundant. If an invention would be conceived and developed without patent protection, no social benefit can be attributed to that particular patent grant. Yet the inventor is likely to take advantage of the opportunity to secure patent protection, and so the restrictive aspects of the system take their toll. On such inventions, the social benefits from patent rights are zero but the social costs may be positive. In assessing the patent system's overall value, we must weigh the net benefits associated with inventions that would not have been availa-

14. This point is stressed, and a partial proof is provided, in John S. McGee, "Patent Exploitation: Some Economic and Legal Problems," *Journal of Law and Economics* 9 (October 1966): 135-62.

15. The governing precedent is *U.S.* v. *General Electric Co.*, 272 U.S. 476 (1926). Two attempts to overturn the doctrine were held back by a 4-4 split among the participating Supreme Court justices. *U.S.* v. *Line Material Co. et al.*, 333 U.S. 287, 315 (1948); and *U.S.* v. *Huck Mfg. Co. et al.*, 382 U.S. 197 (1965).

16. For a path-breaking treatment of this problem, see William D. Nordhaus, *Invention, Growth, and Welfare* (Cambridge, Mass.: M.I.T. Press, 1969), pp. 70-90. See also Carole Kitti, "Patent Policy and the Optimal Timing of Innovations" (Ph.D. diss., University of Chicago, 1973).

17. See Dan Usher, "The Welfare Economics of Invention," *Economica* 31 (August 1964): 279-87.

18. See Chapter 14, pp. 395-98 *supra.*

19. See also McGee, "Patent Exploitation," pp. 143-44; Kenneth J. Arrow, "Economic Welfare and the Allocation of Resources for Invention," in the National Bureau of Economic Research conference report, *The Rate and Direction of Inventive Activity* (Princeton: Princeton University Press, 1962), pp. 619-22; and S. C. Hu, "On the Incentive To Invent: A Clarificatory Note," *Journal of Law and Economics* 16 (April 1973): 169-77.

ble without patent protection against the net social losses associated with patented inventions that would be introduced even if no patent rights were offered. This is a difficult and perhaps impossible task. Our insight into the problem can nevertheless be sharpened by trying to determine what kinds of inventions fall into each of these two categories.

If pure and perfect competition in the strictest sense prevailed continuously—i.e., if sellers and buyers were numerous, products homogeneous, resources highly mobile, entry easy, and knowledge perfect—incentives for invention and innovation would be fatally defective without a patent system or some equivalent substitute. Developing an invention to the point of commercial applicability is costly. The risks of development, though often exaggerated, are also appreciable. To be willing to bear these costs and risks, potential innovators must have some hope of being able to sell their product at a price exceeding the cost of production so that they can recoup their development costs plus a premium for risk. But if imitators can swarm in to copy the invention as soon as it has been introduced, postinnovation prices will fall rapidly to the level of production cost, wiping out supranormal profits for innovators and imitators alike. One would have to be a fool or a philanthropist to invest heavily in invention and development when the imitative steamroller is expected to eliminate divergences between price

F 16.1 Optimal output with patented cost-reducing inventions

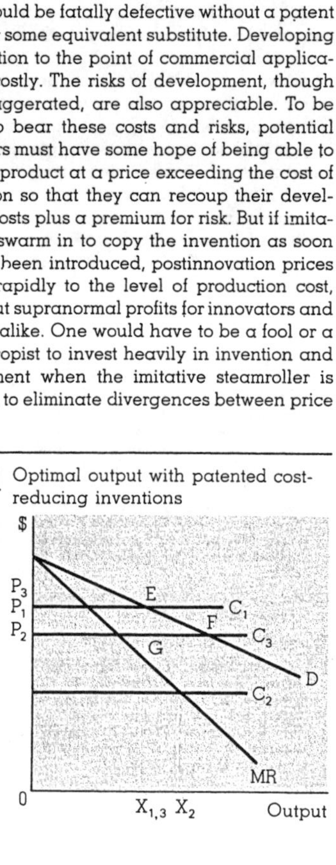

and production cost so quickly and relentlessly. The fundamental argument for a patent system is that patent protection permits the innovator (or inventor) to retard competitive imitation, and hence to anticipate earning supranormal profits if its contribution proves technically and commercially successful.

However, real-world markets are almost never purely and perfectly competitive. Therefore, the crucial question is, Does imitation actually eliminate innovative profits so rapidly? It may not because of three phenomena: natural imitation lags, the advantages of competitive product leadership, and the existence of nonpatent barriers to the emergence of a competitive market structure.

There are several reasons why imitation naturally lags behind the profitable introduction of new products and processes, often by a substantial interval. Secrecy is one. Even when patent protection is sought in exchange for disclosure, the details of an invention may with luck be kept secret until the patent is issued—in the United States, nearly two years on the average after an application is filed.[20] During this interim, the invention's sponsor may be able to secure a significant head start preparing its production facilities and marketing channels. According to a 1957 statistical survey, roughly 90 percent of the sampled inventions used commercially were first put into use before the covering patent was issued and thus before patent disclosure hoisted a telltale flag to potential imitators.[21]

Second, knowledge is almost never perfect. It takes time for entrepreneurs to learn about a new and promising invention, even after its existence has been publicized, and it takes even longer for them to decide it is worth imitating. Some are quicker than others in this regard, but imitation must be widespread in order to eliminate the innovator's profits, and for nontrivial inventions the diffusion process often proceeds slowly. For example, the approximate intervals elapsing between first use in the United States and the date when 60 percent of all relevant producers had imitated for seven innovations studied by Edwin Mansfield were as follows:[22]

Packaging beer in tin cans	1 year
High speed beer bottle filler	7 years

Continuous wide strip steel mill	9 years
By-product coke oven for steel mills	18 years
Continuous annealing of tin-plated steel	20 years
Continuous coal mining machinery	4 years
Diesel locomotives	11 years

In a statistical analysis of these and other innovations, Mansfield found that the speed of imitation was positively correlated with the profitability of the new process or product and inversely related to the magnitude of the investment required. For product innovations in particular, this implies among other things that the speed of imitation depends upon the innovator's pricing policy. Companies pricing their new products to make a quick killing will encourage rapid imitation, while those pursuing a limit pricing strategy will experience slow imitation. Depending upon the circumstances, either strategy may suffice to yield substantial profits.

Third, merely knowing that an attractive new product or process exists and examining the patent specifications is often an inadequate basis for imitating successfully. Know-how that can come only from carrying out one's own R & D effort, receiving generous technical assistance from the innovator, or hiring away several of the innovator's key engineers may also be required. The importance of know-how as a barrier to rapid imitation varies widely from field to field. A new garment design or a children's toy can be copied in hours, while it may take years to develop independently the know-how required to produce rayon or high-quality color film.[23] Moreover, there is probably a positive correlation between the complexity of the original development job and the difficulty of acquiring sufficient know-how to imitate it. As a result, those developments that entail the highest risks tend to be the least rapidly imitated.[24] For all these reasons, innovators often enjoy freedom from competitive imitation for a sufficiently long time to recoup their original investment manyfold even without patent protection.

Competition, product differentiation, and natural lags interact to form a second set of incentives for investment in research and innovation. Frequently a company's image is enhanced by being first on the market with a new product, and through this product differentiation advantage it may be able to maintain a favorable price differential or retain a sufficiently large share of the market to earn supranormal profits for some time.[25] The expectation of gaining such a position is a tempting carrot for investment in R & D. There is also a complementary stick. Firms that lag in the innovative race may find their differentiation advantages and market positions eroding, and to avoid this fate they invest in defensive research and development. Competition to be

20. In many foreign jurisdictions publication is more rapid, sometimes following shortly after application and before the patent issues. In the 1950s and 1960s, U.S. application issuance lags averaged three to four years.

21. Barkev Sanders, "Speedy Entry of Patented Inventions into Commercial Use," *Patent, Trademark, and Copyright Journal* 6 (Spring 1962): 87–116. With the acceleration of processing by the U.S. Patent Office more recently, the percentage used before patent issue has undoubtedly fallen.

22. Estimated (with assistance from the author) from Edwin Mansfield, *Industrial Research and Technological Innovation: An Econometric Analysis* (New York: Norton, 1968), pp. 134–35. Mansfield's figures cover any use, however small, by industry members. Much longer average imitation lags are found when the percentage of industry output produced with a new process is analyzed. See Bela Gold, W. S. Peirce, and Gerhard Rosegger, "Diffusion of Major Technological Innovations in U.S. Iron and Steel Manufacturing," *Journal of Industrial Economics* 18 (July 1970): 219–22.

23. See Jesse W. Markham, *Competition in the Rayon Industry* (Cambridge, Mass.: Harvard University Press, 1952), pp. 20–24, 56–57; and George E. Frost and S. Chesterfield Oppenheim, "A Study of the Professional Color Motion Picture Antitrust Decrees and Their Effects," *Patent, Trademark, and Copyright Journal* 4 (Summer 1960): 126–27.

24. An exception might occur when protracted research is required to identify a workable solution, but once that solution is known, it becomes easy to find slightly different but equally effective inventions. Trial-and-error search for chemical compounds with desired new therapeutic effects, where there was no theory to guide the search toward specific molecular configurations, was an example before changes in the U.S. drug laws imposed heavy clinical test burdens on imitators as well as innovators.

25. See especially Ronald Bond and David Lean, *Sales, Promotion, and Product Differentiation in Two Prescription Drug Markets*, Federal Trade Commission staff report (Washington, D. C.: Government Printing Office, February 1977), Chapters 3–6.

first or at least not to be left behind thus acquires a self-sustaining momentum that needs little or no extra boost from patent rights. That this is commonly the case in American industry is suggested by the results of a questionnaire and interview survey covering 91 large corporations holding approximately 30 percent of all U.S. corporate patents in 1956. When asked what factors played an important role in their R & D investment decisions, only 7 companies selected "patent protection to be secured" as first or second most important of five factors. The majority indicated that patent protection was least important, emphasizing instead the necessity of maintaining competitive leadership or remaining competitively viable.[26]

Here, however, we must be wary of oversimplifying complex relationships. The patent system may affect the competitive struggle for innovative superiority in subtle but important ways. If there were no patents, the advantages of being first and the disadvantages of being a poor second might be less pronounced, though not vanishingly small. It is also conceivable that the prospect of gaining patent protection serves as a primer in igniting an innovation rivalry which then becomes self-sustaining. That is, some firm decides to pioneer a new field of technology because it believes *inter alia* that patent barriers to rapid imitation will help make the venture profitable. Others react after a lag with countervailing programs to protect their flanks. The threat each firm's program poses to the market positions of rivals becomes in and of itself a sufficient incentive to induce continued R & D investment. Whether there are many cases of innovative rivalries that would not have been triggered without an initial patent stimulus, or that would have begun much later, is hard to tell. The development of television is one plausible candidate; the development of xerography another; and the race to introduce new products in the pharmaceuticals industry following World War II may be a third.[27] But innovative rivalry can and does get under way for reasons unrelated to the expectation of patent rights in many instances, and once it does, it generates its own incentives for continuation.

Another patent-related reason why companies conduct R & D is to guard against being foreclosed from some field of technology by another concern exercising parallel patent rights. By making and patenting appropriate inventions, a firm stakes out its own independent claim in the field and obtains a bargaining counter to deal with others holding complementary inventions. However, few patents are sufficiently basic and broad to 'fence in' a field altogether. Most can be invented around, and an appreciable amount of industrial R & D is motivated by a desire to avoid infringing or paying royalties upon other firms' patented inventions. The fact that patent rights stimulate this search, and that sometimes the search yields an unambiguously superior new product or process, is frequently cited as a major benefit of the patent system. Yet the sword cuts both ways, for the resources devoted to circumventing patents might, if inventions were made freely available, be allocated instead to alternative activities with higher incremental social payoffs. For instance, Comanor found that while a number of important pharmaceutical improvements were discovered in this manner, competitive research also led to a great many products with therapeutic effects quite similar to drugs already available. He concluded that society would lose relatively little if fewer resources were devoted to duplicative "molecule manipulation" R & D.[28] S. G. Sturmey observed that British radio manufacturers wasted "a lot of ingenuity" during the 1920s devising circuit arrangements that reduced royalties paid to the Marconi Company, even though radio tube technology advanced in the process.[29] On the other hand, the pace of advance in petroleum cracking technology was almost certainly accelerated by the vigorous efforts of companies to invent around rival processes.[30] All we have to go by at present are examples and counterexamples, and so it is impossible to determine whether, on balance, the benefits of R & D motivated by the hope of circumventing patents exceed the costs of duplication.[31]

The third major reason why firms may find investment in innovation profitable despite the absence of patent protection is the existence of nonpatent barriers to the emergence of a competitive market structure. Here we tie into the analysis of the previous chapter. For all but those innovations that define a completely new field, the most likely early imitators are companies already operating in the industry to which the innovation pertains. Lack of production facilities, managerial experience, and distribution channels impedes the entry of outsiders. When in addition the market is moderately or heavily concentrated, postimitation pricing discipline is apt to remain fairly firm. Even for innovations requiring substantial development outlays there may be sufficient room for all industry members to imitate without preventing the leader from recovering its development investment. It follows obversely that patent protection will be most important as an inducement to progress when there are large numbers of probable imitators or when development costs are expected to be high in relation to the stream of quasi-rents the innovation could support under favorable (e.g., oligopolistic) pricing conditions.[32]

We find then that business firms may invest in innovation without patent protection if natural imitation lags are substantial, if there are major competitive differentiation advantages from being first in the market with a new product, and/or if the relevant market is oligopolistic. All three characteristics are widespread in a modern industrialized economy. Any one of the three can provide a sufficient incentive if the invention is economically important—that is, if it has the potential of yielding marketwide cost savings or premium profits large in relation to the innovator's development costs. In such instances, the prospect of a relatively short lead over imitators—e.g., as little as two or three years—may be attractive enough to induce investment. To cite an extreme example, the Reynolds International Pen Co. earned back its original investment a hundredfold during the 18 months it enjoyed an almost exclusive position, before a host of firms imitated its ball point pen.[33] It is only when the barriers to widespread and rapid imitation are weak, or when the advantages of competitive leadership are modest, or when the profit potential of the innovation is small, or when there is some adverse combination of the three, that patent protection becomes an important incremental stimulus.

A similar conclusion emerges when one attempts to determine, assuming that no other incentives for innovation exist, how long the life of a patent must be to induce the amount of cost-saving innovation that maximizes the surplus of discounted consumers' plus producer's surplus less R & D costs. William Nordhaus has shown that the more important a process innovation

26. Scherer, Herzstein, Dreyfoos et al., *Patents and the Corporation*, pp. 107, 118, 149.

27. See W. R. Maclaurin, "Patents and Technical Progress: A Study of Television," *Journal of Political Economy* 58 (April 1950): 152; and William S. Comanor, "Research and Competitive Product Differentiation in the Pharmaceutical Industry," *Economica* 31 (November 1964): 372-84. On the other hand, in *Sales, Promotion, and Product Differentiation*, p. 45, Bond and Lean show that innovative rivalry in the anti-anginal drug field emerged without significant patent protection.

28. Comanor, "Research and Competitive Product Differentiation," pp. 381-84.

29. S. G. Sturmey, *The Economic Development of Radio* (London: Duckworth, 1958), p. 223.

30. See McGee, "Patent Exploitation," p. 151; John L. Enos, "Invention and Innovation in the Petroleum Refining Industry," in *The Rate and Direction of Inventive Activity*, pp. 302-303; Roger L. Beck, "Patents, Property Rights, and Social Welfare," *Southern Economic Journal* 43 (October 1976): 1050-54; and Daniel L. Landau, "Patents and Over-Investment in Process Inventions? Comment," *Southern Economic Journal* 45 (July 1978): 285-88, with a reply by Beck, pp. 289-92.
In xerographic copying and float glass production, competitive research was apparently inhibited by decisions of the dominant patent holders to license at least some of their patents on favorable terms. See G. H. Wierzynski, "The Eccentric Lords of Float Glass," *Fortune*, July 1968, p. 123; and Erwin Blackstone, "The Copying Machine Industry: A Case Study" (Ph.D. diss., University of Michigan, 1968, pp. 98-112).
For a different view of the "inventing around" problem that seems little influenced by any concern for reality, see Edmund W. Kitch, "The Nature and Function of the Patent System," *Journal of Law and Economics* 20 (October 1977): 265-90. Compare the *Report to the Federal Trade Commission on Federal Energy Land Policy*, Senate Committee on Interior and Insular Affairs print (Washington, D.C.: Government Printing Office, 1976), pp. 657-74.

31. See also C. T. Taylor and Z. A. Silberston, *The Economic Impact of the Patent System* (Cambridge, England: Cambridge University Press, 1973), pp. 199-201, who found in a survey of British companies relatively little wastefully duplicative R & D spurred by the desire to invent around competitors' patents.

32. Along analogous lines, Jesse Markham suggests that the patent system has its strongest incentive value when all other means of obtaining a monopoly are foreclosed by law, since then innovation will be the main open route to supranormal profits. "The Joint Effect of Antitrust and Patent Laws Upon Innovation," *American Economic Review* 56 (May 1966): 293.

33. See Thomas Whiteside, "Where Are They Now?," *New Yorker*, 17 February 1951, pp. 39-58.

is—that is, the more it reduces unit production costs—the shorter the socially optimal patent life will be.[34] This is so for two reasons. An R & D effort that facilitates large cost savings before running into the stage of severely diminishing returns pays for itself quickly. And, secondly, the incremental resource misallocation losses attributable to extended patent lives soon begin to outweigh the added induced cost-reduction gains. Making arbitrary but plausible estimates for the parameters of his model, Nordhaus discovered that the optimal life of a patent for process inventions reducing unit production costs by roughly 10 percent ranges between three and seven years for end product demand elasticities in the neighborhood of 0.7 to 4.0. Only when research yields cost savings grudgingly and in modest amounts—e.g., 1 percent of unit production costs—and when demand is of unit elasticity or less are patent lives of 15 to 20 years optimal.

This perspective on incentives for innovation has profound implications for judging the patent system's value. In most situations, to repeat, natural lags and other advantages permit sponsors to tap an innovation's full profit potential without experiencing substantial competition for at least a couple of years, if not more. For the most part, it would appear, only those innovations that offer potentially exploitable benefits small in relation to development costs need the extra lure of patent protection before entrepreneurs will choose to plunge. Important contributions, on the other hand, should pay for themselves in short order. We are led then to ask, Would society really lose much at the margin if, by abolishing the patent system, it sacrifices mainly innovations with low benefit/cost ratios? The answer implied by the analysis thus far is negative. A negative answer is also suggested by the detailed interviews Taylor and Silberston conducted with 27 British companies operating in research-oriented industries. They found that without effective patent protection, R & D expenditures would be reduced negligibly in the electrical industries, by 5 percent on average in the machinery and mechanical components industries, by 5 percent also in basic chemicals, by 25 percent in such specialty chemicals as pesticides and adhesives, but by 64 percent in pharmaceuticals (where imitation in the absence of patents was expected to be especially swift and effective).[35]

Despite this sanguine prognosis (except for drugs), it is conceivable that without a patent system certain spectacular technical contributions—those effecting a genuine revolution in production or consumption patterns—might be lost or (more plausibly) seriously delayed because their support lends itself poorly to rational benefit/cost calculation. Such innovations may lie off the beaten paths of industrial technology, where no firm or group of companies has a natural advantage; and the innovator may be forced to develop completely new marketing channels and production facilities to exploit them. They may entail greater technological and market uncertainties, higher development costs, and longer inception-to-commercialization lags than the vast bulk of all industrial innovation. Entrepreneurs may be willing to accept their challenge only under highly favorable circumstances—notably, when it is anticipated that if success is achieved, it can be exploited to the fullest, among other things through the exercise of exclusive patent rights.

That such cases exist is virtually certain. Black-and-white television and the early development of xerographic concepts are probable examples.[36] Both would undoubtedly have come to fruition eventually, but patent protection evidently speeded up the process. The most important single gap in our knowledge of patent system economics is how many inventions and innovations fit this pattern, and how long their introduction would have been delayed if there were no patent system. It is the kind of gap that could be narrowed by spending on research sums small in relation to the cost of maintaining the patent system or the social losses incurred when an innovation as important as xerography is delayed by only a single year. The author's best guess is that such cases are rare, occurring perhaps a few times per decade. But this is no more than a stab in the dark.

A related question is what kind of reward structure best encourages work in such unexplored, uncertain new technologies. A structure that satisfies orthodox notions of equity, proportioning gains to the investments made, might be the wrong answer. Under the patent system, enormous gains out of all proportion to original investments are occasionally realized. Edwin Land of Polaroid camera fame amassed a fortune estimated at $500 million from his numerous inventive contributions; the market value of securities received by Chester Carlson for his xerography patent rights had mounted to $160 million in 1968; Pilkington Brothers, Ltd., was expected to collect $100 million in royalties alone for the use of its float glass patents; and Hoffmann-La Roche of Switzerland must have realized profits in the multibillion dollar range from the worldwide sales of its Valium and Librium tranquilizers.[37] Occasional success stories like these may be what best energizes technical risk bearing, as Schumpeter observed more generally of the capitalist reward system:

> Spectacular prizes much greater than would have been necessary to call forth the particular effort are thrown to a small minority of winners, thus propelling much more efficaciously than a more equal and more "just" distribution would, the activity of that large majority of businessmen who receive in return very modest compensation or nothing or less than nothing, and yet do their utmost because they have the big prizes before their eyes and overrate their chances of doing equally well.[38]

Whether the largest prizes need to be as large as some of those cited here is doubtful, but too close a proportioning of gains to costs is sure to discourage investment in the most uncertain technical ventures.

Within this overall incentive framework, is the patent system of more value in stimulating invention and innovation by large firms, small firms, or by independent inventors? In general, it appears to be the small firm and the independent inventor who are affected the most. Small companies commonly lack the distribution channels and market acceptance of their larger rivals. As a result, their rate of penetration into new markets through innovation is likely to be slower and the profit they realize in the first years after innovation lower.[39] Also, their market positions are more vulnerable when large, well-entrenched rivals retaliate with imitative products. The independent inventor occupies an even less tenable position. Patent protection should in principle help offset these disadvantages, permitting the small entrepreneur to compete in the innovation game on a more equal footing with the giants, who enjoy adequate (or perhaps even excessive) protection against premature imitation by virtue of their established market positions.

Nevertheless, the real-world patent system does not always operate in this ideal fashion. Size confers an advantage not only in promoting new products, but also in waging harassment campaigns. The grant of rights conferred by a patent is not a certain thing. The validity of a patent can be challenged in court, and among the scores of lawsuits fought to completion each year, more than 60 percent have ended in invalidation decisions.[40] As one jurist remarked, "A patent is merely a license to bring a lawsuit."[41] In such contests the odds are weighted in favor of the large corporation, which has superior legal

34. *Invention, Growth, and Welfare*, pp. 76–82. See also F. M. Scherer, "Nordhaus' Theory of Optimal Patent Life: A Geometric Reinterpretation," and the reply by Nordhaus, *American Economic Review* 62 (June 1972): 422–31.

35. Taylor and Silberston, *The Economic Impact of the Patent System*, pp. 194–208, especially p. 199. The figure for pharmaceuticals may be biased upward because two of the three drug makers interviewed were newcomers to the industry who might not have entered without patent protection. See also note 27 *supra*.

36. Cf. Maclaurin, "Patents and Technical Progress"; and Blackstone, "The Copying Machine Industry," pp. 7–23, 75–80.

37. See Arthur M. Louis, "America's Centimillionaires," *Fortune*, May 1968, pp. 152–53; Wierzynski, "The Eccentric Lords," p. 91; Robert Ball, "The Secret Life of Hoffmann-LaRoche," *Fortune*, August 1971, p. 134; and F. M. Scherer, *The Economic Effects of Compulsory Patent Licensing* (New York University Monograph Series in Finance and Economics, 1977), pp. 43–47.

38. Joseph A. Schumpeter, *Capitalism, Socialism, and Democracy*, 3rd ed. (New York: Harper, 1950), pp. 73–74.

39. F. M. Scherer, "Research and Development Resource Allocation Under Rivalry," *Quarterly Journal of Economics* 81 (August 1967): 385, 388–89.

40. See Charles L. Trozzo and Carole E. Kitti, *The Effects of Patent and Antitrust Laws, Regulations, and Practices on Innovation* (Washington, D.C.: Institute for Defense Analyses Paper P-1075, February 1976), Chapter 9; and Lawrence Baum, "The Federal Courts and Patent Validity," *Journal of the Patent Office Society* 56 (December 1974): 758–87.

41. *Guide* v. *Desperak et al.*, 144 F. Supp. 182, 186 (1956).

and financial resources. Therefore, small enterprises whose patents are challenged by powerful rivals on either good or spurious grounds often choose to settle out of court, giving up their exclusive position and licensing the challenger to avoid the cost and uncertainty of protracted litigation. Or they may sell out altogether to the challenger. The ability of well-heeled companies to harass smaller patent holders in this way has been curbed considerably by unsympathetic antitrust judgments. But to the extent that patent validity contests persist, apprehension about them diminishes the incentive value of the patent system for small firms and independent inventors. Yet on balance, the system probably does more to stimulate invention and innovation by small entrepreneurs than by the large corporation, and this is a telling argument in its favor.[42]

One final comment on the benefits of a patent system is appropriate. Our discussion thus far has been from the perspective of industrialized nations. But what role should patent grants play in less developed lands? Newly emerging nations seldom sustain much of a domestic R & D effort, partly because they lack the necessary human resources and partly because they can tap into the reservoir of technology available in other nations.

An indication of this dependence upon external technology sources is given by statistics on the percentage of patents granted to foreigners by various nations during 1970 through 1972. In Argentina, foreigners received 76 percent of all patents issued; for Venezuela, the figure was 94 percent; for Iran, 92 percent; and for Portugal, 94 percent. This can be compared with 29 percent in the United States, 31 percent in Japan, and 53 percent in West Germany.[43]

Since the amount of domestic inventive activity is modest in underdeveloped countries, one might expect the benefits of a patent system to be particularly small relative to the social costs of granting foreigners patent protection. Indeed, it is somewhat surprising that such countries offer conventional patent rights. Professor Machlup attributes many emerging nations' membership in the International Union for the Protection of Industrial Property to pressures from industrialized nations and to prestige motives, suggesting that newly independent nations seem irrationally eager to "have the honor of paying higher prices for imported products."[44] An alternative rationalization is that foreign firms are more willing to establish a base of operations in underdeveloped lands or license their know-how to domestic firms if they can protect their position through patents.[45] But this is not completely convincing, since know-how is often sold in packages independent of patent licenses, and since emerging nation markets are so imperfect and imitation is so sluggish that an efficient producer should usually be in a favorable position to realize normal and quite possibly even supranormal profits without the fringe benefit of patent protection.[46]

The social costs of the patent system

While encouraging some inventions and innovations that would not otherwise be made and hastening the introduction of others, the patent system simultaneously generates social costs. The most obvious cost is the resource misallocation attributable to monopolistic pricing of patented inventions that would have been available without patent protection. How much power over price a patent confers varies widely from case to case, depending upon the availability of substitutes and consumers' knowledge of their availability. Some of the most extreme cases concern the pricing of patented pharmaceutical items, for which demand is typically quite inelastic over a considerable price range. From 1956 through the mid-1960s, the Pfizer Company and its four licensees sold the antibiotic tetracycline to druggists at a wholesale price of $30.60 per bottle of 100 capsules. Total sales at wholesale to drug stores exceeded $1 billion during this period. Production costs ranged between $1.60 and $3.80 per bottle; and when doubts about the validity of Pfizer's patent began to mount, several unlicensed firms began producing and selling tetracycline at approximately $2.50 per bottle wholesale. Many similar cases of price-cost margins on the order of 90 percent for patented drug products have been identified.[47] They are, to repeat, extreme, but it is not at all rare for patent holders to charge prices substantially in excess of full production and distribution costs and to earn handsome profits as a result of their protected market position.

Furthermore, companies have managed through various legitimate and shady practices to extend and pyramid the monopoly power derived from their patents.[48] One way they do this is by fencing in a field of technology. As we have seen earlier, it is usually possible to invent around a single patent, and so the amount of monopoly power conferred by any but the most basic and sweeping patents is constrained, perhaps severely, by the possibility of substitution. However, by accumulating an extensive portfolio of patents, producers can solidify their monopolistic domination of a field. When du Pont scientists "invented" nylon, for instance, they did not rest content with patenting the basic superpolymer's composition and processes for producing it. They systematically investigated the whole array of molecular variations with properties potentially similar to nylon, blanketing their findings with hundreds of patent applications to prevent other firms from developing an effective substitute. Similar tactics have been pursued by du Pont and many other corporations in such fields as cellophane, plastics, synthetic leather, synthetic rubber, photo supplies, tranquilizers, radio, television, shoe machinery, data processing equipment, electric lamps, telephone equipment, copying processes, and can-closing machinery, to name just a few. When one company comes to dominate a field by accumulating a massive patent portfolio, it not only prevents rivals from operating except on its acquiescence, but also becomes the logical buyer for related new concepts patented by independent inventors. Thus, at some point the process of patent accumulation may take on a "snowball effect."[49]

Under current U.S. judicial interpretations, this is perfectly legal when it is achieved through one's own internal research and development efforts unaccompanied by abusive practices. As the Supreme Court commented in one leading case, "The mere accumulation of patents, no matter how many, is not in and of itself illegal."[50] However, companies run the risk of antitrust violation if they build up an impregnable patent portfolio by systematically buying out rival patents, especially when coercive tactics are employed to soften up the sellers.

A dynamic ramification of the pyramiding phenomenon is the extension of one's monopoly power over time by prolonging the effective life of basic patents and by amassing improvement patents once basic patents expire. Since the grant of U.S. patent rights continues for 17 years from the date of issue, applicants have been known to delay a patent's ultimate expiration date by dragging their heels on Patent Office procedural matters while the application is pending. No doubt more important is the prolongation of control achieved through improvement patenting. Electric lighting is the classic illustra-

42. See also John Jewkes, David Sawers, and Richard Stillerman, *The Sources of Invention* (New York: St. Martin's, 1959), p. 251, for a similar conclusion.

43. *Industrial Property*, Statistical Annexes, December 1971, December 1972, and December 1973.

44. Fritz Machlup, "Patents," *International Encyclopedia of the Social Sciences*, vol. 11 (London: Macmillan, 1968), p. 465.

45. See Raymond Vernon, *The International Patent System and Foreign Policy*, Study no. 5 of U.S., Congress, Senate, Subcommittee on Patents, Trademarks, and Copyrights, 85th Cong., 2nd sess., 1958; Edith Penrose, "International Patenting and the Less Developed Countries," *Economic Journal* 83 (September 1973): 768–86; Douglas Greer, "The Case Against the Patent System in Less-Developed Countries," *Journal of International Law and Economics* 8 (December 1973): 223–66; and the articles by S. J. Patel, Pedro Roffe, and Peter O'Brien in *World Development* 2 (September 1974): 223–66.

46. On the costs of transferring know-how abroad, see David J. Teece, "Technology Transfer by Multinational Firms: The Resource Cost of Transferring Technological Know-how," *Economic Journal* 87 (June 1977): 241–61.

47. See Henry Steele, "Monopoly and Competition in the Ethical Drugs Market," *Journal of Law and Economics* 5 (October 1962): 131–63; *idem*, "Patent Restrictions and Price Competition in the Ethical Drugs Industry," *Journal of Industrial Economics* 12 (July 1964):198–223; and (on Valium and Librium) Scherer, *The Economic Effects of Compulsory Patent Licensing*, pp. 44–45.

48. For a comprehensive catalogue, see Floyd L. Vaughan, *The United States Patent System* (Norman: University of Oklahoma Press, 1956).

49. See Carl Kaysen, *United States* v. *United Shoe Machinery Corporation* (Cambridge, Mass.: Harvard University Press, 1956), p. 90; and Carl Kaysen and Donald F. Turner, *Antitrust Policy* (Cambridge, Mass.: Harvard University Press, 1959), p. 166.

50. *Automatic Radio Mfg. Co.* v. *Hazeltine Research, Inc.*, 339 U.S. 827, 834 (1950).

tion. General Electric virtually regimented the domestic incandescent lamp industry from 1892 through the 1930s, first by acquiring the basic Edison patents; then through patents on the argon-filled lamp and tungsten filaments; and finally through patents on such features as tipless bulbs, internally frosted bulbs, and nonsag filaments.[51]

A further charge against the patent system is that it permits the suppression of inventions from which the public might otherwise benefit. Nearly everyone has heard the recurrent rumor that some shadowy power in the automobile or petroleum industry has obtained and suppressed patents on a carburetor that would let full-sized autos travel 50 miles per gallon of gasoline. Most such rumors, including the present example, prove on investigation to have little or no substance.[52] Yet some valid complaints remain. When corporations assemble huge patent portfolios to bolster their market positions, they inevitably include many inventions they do not intend to use directly, patenting them only to prevent others from using them. If decision makers choose rationally, the inventions left unused will normally be those that are economically inferior to, or at least not superior to, those that are used.[53] Whether this should be called suppression is an empty semantic question. Despite their inferiority, such inventions might well be commercialized profitably by other firms if the superior invention is priced monopolistically and if the inferior inventions are at least better than the freely available technology. It is also possible that patent holders make mistakes, utilizing inventions inferior to others in their portfolios because they misjudge costs or demand. Abstracting from such situations, there are very few convincingly documented cases of the deliberate suppression of clearly superior inventions. The least ambiguous instances typically involve products which for various reasons could not be priced as profitably as available inferior variants, or inventions whose use could upset the status quo in a delicately coordinated price-fixing scheme.[54]

As noted previously, companies may spend substantial sums inventing around the patent positions of rivals. This cost is offset to an unknown extent by the benefits of superior inventions resulting from the circumventing effort. One firm's patent position may also block another producer from introducing improvements complementary to the original invention. An early example was James Watt's steam engine patent. Access to it was essential if one were to develop high-pressure engines, in which Watt saw little value. Watt's refusal to grant licenses impeded the work of Jonathan Hornblower, Richard Trevithick, and others on high-pressure engines until the patent expired in 1800, and this may in turn have had some small effect in delaying the introduction of steam locomotives and steamboats.[55] Similar blockages occur frequently in modern times—e.g., when the Pfizer Company found that it had to produce aureomycin, patented by the American Cyanamid Company, as an intermediate step in making tetracycline. In more cases than not, such problems are resolved through purchase of the improvement patents or through a cross-licensing agreement under which each party permits the other to utilize its patents, usually with provisions for the payment of royalties proportionate to the parties' respective patent contributions.

Cross-licensing is a constructive means of avoiding stalemates between complementary patent portfolios. However, it also has considerable potential for abuse. Some of the most egregious price-fixing schemes in American economic history were erected on a foundation of agreements to cross-license complementary and competing patents. Industries cartelized at one time or another in this way include electric lights, glass bottles, parking meters, eyeglasses, magnesium, synthetic rubber, titanium paint pigments, radio broadcasting equipment, motion picture production, gypsum board, hardboard, machine tools, bathtubs, and a host of other electrical and chemical products.[56] Typically, such arrangements have been implemented by adding to the patent exchange agreement provisions specifying prices, market quotas, membership in the industry, and other aspects of conduct and structure. All attempts reciprocally (i.e., not unilaterally by a single patentee) to extend patent licensing agreements into the realm of price and output determination have been ruled illegal by the American courts, but restrictions of this sort continue to be legal in most other nations. Where they persist, the restrictive effects of a patent system are magnified.

A more subtle practice in the grey area of U.S. law concerns the settlement of patent ownership and validity disputes.[57] It is not unusual for two or more corporations to have conflicting claims to the priority of their employees in making an invention, or they may possess evidence that an invention did not satisfy the criteria for patentability, unknown to the responsible Patent Office examiner. They can fight out their case before the Patent Office and in the courts, but this is costly and it also risks invalidation of the patent, so that no one is able to bar entry into the field. Fearing this, they may simply agree to live and let live, permitting the most advantageously situated firm to receive the patent without opposition and then to license the potential challengers. If there is an explicit agreement that no other producers are to be licensed, a violation of the antitrust laws exists. But it is difficult to prove this in court, and in any event it may be unnecessary for the companies involved to reach an outright agreement. Recognizing where its interests lie, the patent recipient merely elects unilaterally to license only rivals able to endanger its patent position, restricting the size of the 'club' to sufficiently few sellers that awareness of mutual interdependence makes tacitly collusive pricing likely. Thus, results similar to what would obtain under overt price fixing are secured. Still it is possible to be tripped up playing this game of legal brinkmanship, as the makers of tetracycline discovered in a series of antitrust cases in which they ultimately achieved substantial acquittal, but only after paying damages of some $120 million.[58]

We observed earlier that the patent system permits powerful corporations to harass financially weak firms and independent inventors through protracted litigation. This causes obvious inequities, and the legal costs can be significant. Between 1900 and 1941, 684 radio patents were entangled in a total of 1,567 infringement suits.[59] Inventors like Lee de Forest and Edwin Armstrong were forced to sell out their rights in key patents because, as Armstrong later lamented, he was "in danger of being litigated to death."[60] A single lawsuit over petroleum cracking patents lasted 15 years, piling up court costs and legal fees exceeding $3 million.[61]

An additional source of inequity arises from the priority system, under which (in the United States) the first person to invent a device or process receives full patent rights, while those who finish second in the race get nothing.[62] Invention

51. See Arthur A. Bright, *The Electric Lamp Industry* (New York: Macmillan, 1949), pp. 84–104, 235-302; and G. W. Stocking and M. W. Watkins, *Cartels in Action* (New York: Twentieth Century Fund, 1946), pp. 304–12.

52. But see "The Culprit: Carburetion," *Business Week*, 7 April 1973, p. 5, which provokes continuing apprehension.

53. See Machlup, *An Economic Review of the Patent System*, pp. 12, 41; McGee, "Patent Exploitation," pp. 145–48; and Beck, "Patents, Property Rights, and Social Welfare," pp. 1047–55.

54. One of the best-documented cases is New Jersey Standard Oil's suppression of the Santopour pour-point depressant for lubricating oils. See Stocking and Watkins, *Cartels in Action*, pp. 497–98. For a more extensive list, see Vaughan, *The United States Patent System*, pp. 231–38; and for a list that errs on the side of including doubtful cases, U.S., Senate, Committee on the Judiciary, Subcommittee on Antitrust and Monopoly hearings, *Economic Concentration*, Part 6, 90th Cong., 1st sess., 1967, pp. 3271–74. On Romanoff's apparent suppression of low-cost synthetic caviar, see "Marketing Observer," *Business Week*, 28 June 1976, p. 51.

55. See Charles Singer, E. J. Holmyard et al., *A History of Technology*, vol. IV, "The Industrial Revolution" (London: Oxford University Press, 1958), pp. 188–97.

56. For a summary of the leading cases, see Vaughan, *The United States Patent System*, pp. 105–67. For an attempt to develop criteria for identifying cross-licensing agreements of a primarily restrictive character, see George L. Priest, "Cartels and Patent License Agreements," *Journal of Law and Economics* 20 (October 1977): 309–78.

57. Vaughan, *The United States Patent System*, pp. 203–10.

58. *Federal Trade Commission* v. *American Cyanamid Co. et al.*, 363 F. 2d 756 (1966), 72 F.T.C. 623 (1967); *U.S.* v. *Charles Pfizer and Co., Inc., et al.*, 281 F. Supp. 837 (1968), reversed 426 F. 2d 32 (1970), 404 U.S. 548 (1972), and 367 F. Supp. 91 (1973); and "5 Drug Makers Will Settle Price Suits for 120 Million," *New York Times*, 7 February 1969, p. 1. See also *U.S.* v. *Singer Manufacturing Co. et al.*, 374 U.S. 174 (1973).

59 W. R. Maclaurin, *Invention and Innovation in the Radio Industry* (New York: Macmillan, 1949), p. 273.

60. Maclaurin, *Invention and Innovation*, pp. 256–57.

61. McGee, "Patent Exploitation," p. 153.

62. Section 102 (g) of the Patent Code mentions three distinct and potentially conflicting criteria for deciding priority contests: "In determining priority of invention there shall be considered not only the respective dates of conception and reduction to practice of the invention, but also the reasonable diligence of one who was first to conceive and last to reduce to practice." Many other nations such as West Germany, Japan, and France award priority to the first person who files a patent application.

is responsive to changes in the state of knowledge and demand; and since new opportunities may become evident to many persons at about the same time, it is quite common for two or more inventors to arrive at the same result independently and almost simultaneously. When a priority contest subsequently materializes in the Patent Office, the odds favor the inventor backed by ample resources.[63]

Finally, we must take into account the direct costs of administering a patent system. The budget of the U.S. Patent Office for fiscal year 1978 was $90 million, including salaries for 2,945 authorized employees plus sundry other expenses. The American Patent Bar Association has nearly 3,000 members, most of whom work outside the Patent Office. Assuming salary, clerical support, transportation, and office maintenance costs of roughly $80,000 per practicing patent attorney, we estimate total annual patent system administration costs of roughly $330 million at 1978 prices.

The costs and benefits on balance

Only in this last category are we able to hazard anything like a numerical estimate of the patent system's costs. The monopolistic restrictions facilitated by the system add costs that are no doubt higher but unmeasurable. On the opposite side of the balance, the system allows society to enjoy the benefits of some inventions that would not otherwise be made or that would become available at a later date without the patent incentive. These benefits are probably confined largely to two categories of inventions: those whose economic value is modest in relation to development costs, and those that represent unusually bold, risky departures from known technology. The social gain forgone through losing an invention in the first category because no patent protection is offered would by definition be slight. How the ledger would stand when all such gains are accumulated—on hundreds or perhaps even thousands of borderline inventions per year—is hard to say. The author's best guess is that the total would still not be enormous. Inventions in the second category represent a horse of a different color. They are few and far between, but even a few can make a big difference in the efficiency of production or the quality of life. For instance, the introduction of xerographic copying processes permitted business enterprises and government agencies to realize savings of at least a quarter billion dollars per year at 1967 levels of utilization.[64] Without a patent system, the American economy would probably have been forced to wait at least several years longer for xerography. Recognition that a few such cases exist, offsetting the patent system's substantial costs, is what deters governments from scrapping the system altogether. As Jewkes, Sawers, and Stillerman observed:

> It is almost impossible to conceive of any existing social institution so faulty in so many ways. It survives only because there seems to be nothing better.[65]

Proposals for reform

This is perhaps too complacent a stance. Even if we concur that the basic system should be preserved, we can try to improve it by making major or peripheral adjustments. Several substantive proposals for reform deserve consideration.

One possibility is to reduce the life of a patent. Obviously, any decision on the duration of the patent grant must be a compromise.[66] Some inventions will be made even if the term of protection is a year, others only if the term runs for 30 years. An ideal patent system would hand-tailor the life of each patent to the peculiar circumstances of the invention it covers, but this is administratively infeasible. Debate centers, therefore, on the optimal average life. For a special case Nordhaus has shown that terms longer than 15 years are optimal only for inventions yielding very modest cost savings or facing unusually inelastic demand, but his analysis ignores such elements as uncertainty, risk aversion, and alternative incentives to invent and innovate.[67] Making horseback assumptions about relevant but highly aggregated variables, Professor Machlup concluded that the marginal benefits of the patent system fall short of the marginal costs for terms longer than 15 years.[68] Here again, however, the impact of longer protection on rare but spectacularly novel and ambitious developments is ignored. In the absence of more conclusive support for a shorter term, policymakers have been reluctant to move away from the status quo.

A different way of coping with the optimal life problem is to draw a dichotomy, as the Germans do, granting full-term patents on significant inventions but only petty patents (*Gebrauchsmuster*), with a term of three years, on minor inventions and improvements.[69] This proposal has considerable appeal, especially as a means of preventing the extension of monopoly power for decades on end, as in the incandescent lamp industry. Of course, if companies seek full-term rather than petty patent protection and succeed, the most a petty patent system will accomplish is the weeding out of nuisance patents with little entry-blocking power.

There are also alternative or complementary ways to prevent patents on minor inventions from clogging the arteries of technological advance. More rigorous standards might be imposed in deciding whether an invention is patentable, so that trivial contributions receive no protection. The American courts have waged a continuing campaign on this front without striking success. And at the administrative level, the proportion of patent applications approved by the Patent Office increased from roughly 60 percent in the 1950s to more than 70 percent in the 1970s, suggesting a possible reduction of standards. Second, to discourage purely defensive patenting, provision can be made for the mere registration of inventions, after which the inventions are rendered unpatentable unless another inventor proves priority within a limited period of time. Such a procedure was instituted by the U.S. Patent Office in 1968. It appears to have met with little success, for in contrast to the hundred-thousand-odd regular patent applications filed each year, there have only been one to two hundred defensive registrations. Third, patents in force could be subjected to an annual tax or renewal fee and cancelled in the event of nonpayment, so that patent holders are forced each year to reassess whether it is worthwhile maintaining their exclusive rights. This is done in many countries outside the United States. In Germany, the annual renewal fee in 1976 was DM 50 (about $21) for the third and fourth years, escalating to DM 1,700 ($720) for the terminal eighteenth year. In order to retain exclusive rights over the full 18-year term, a German patent holder must pay a cumulative total exceeding DM 10,000. Under this scheme fewer than 5 percent of all German patents remain in force for the full term, and the average life of a patent is less than eight years.[70] The system is highly effective in weeding out patents with nothing but minor nuisance value. However, neither a defensive registration system nor a renewal fee system can encourage voluntary abandonment of patents that protect important inventions in use or play a key role in fencing off some area of technology. Thus, they would not do much to reduce the amount of monopoly power based upon patents.

In many nations no patent protection is given for drug entities, other chemical compounds, and foodstuffs. This is ostensibly done to protect the public from monopolistic exploitation on the pur-

63. For a possible exception to the rule, in which an erstwhile graduate student inventor appears to have won rights to basic laser patents following an 18-year legal struggle against Bell Telephone Laboratories and others, see "A Laser Patent That Upsets the Industry," *Business Week*, 24 October 1977, pp. 122–23. Compare "Who Invented the Laser?," *Business Week*, 27 November 1965, pp. 132–37; and *Gould* v. *Schawlow and Townes*, 363 F. 2d 908 (1966).

64. Estimated from cost and demand data in Blackstone, "The Copying Machine Industry."

65. Stillerman, *The Sources of Invention*, p. 253.

66. The 17-year term of United States patents was a compromise between following the British precedent of a 14-year term (originally, time to train two sets of apprentices) and permitting special exceptions to allow a 21-year term.

67. Nordhaus, *Invention, Growth, and Welfare*, pp. 81–82.

68. Machlup, *An Economic Review of the Patent System*, pp. 66–73; and Machlup, "Patents," p. 471.

69. See Alfred F. Crotti, "The German *Gebrauchsmuster*," *Journal of the Patent Office Society* 39 (August 1957): 566ff. Other nations with similar systems include Japan, Italy, Spain, and Venezuela.

70. See P. J. Federico, "Renewal Fees and Other Patent Fees in Foreign Countries," Study no. 17 of U.S., Congress, Senate, Subcommittee on Patents, Trademarks, and Copyrights, 85th Cong., 2nd sess., 1958; and Thomas Dernburg and Norman Gharrity, "A Statistical Analysis of Patent Renewal Data for Three Countries," *Patent, Trademark, and Copyright Journal* 5 (Winter 1961–62): 340–60.

chase of vital staples. The notion has an element of paradox. If the patent system is justified at all as a means of encouraging technological progress, one might think the case for patent rights should be at its strongest in fields like medicine, where the public stands to gain the most from innovation. One can logically favor both the patent system generally and medical exemptions from it only if there are peculiar extenuating circumstances—e.g., if pharmaceutical research would be almost as vigorous without patents as with them because it is concentrated in universities and government laboratories, or for smaller or less developed countries that have little hope of stimulating significant domestic research but can benefit from the unimpeded importation of foreign medical technology.[71]

A more fundamental reform—the introduction of general compulsory licensing provisions—has been proposed numerous times by commissions and congressional committees, but every attempt to alter the U.S. law in this direction has been beaten down as a result of determined opposition from industrial groups and the patent bar.[72] Compulsory licensing is an accepted feature of patent laws abroad.[73] Typically, it can be invoked when a patent recipient fails to utilize an invention in the domestic market within a specified period of time, when licensing is essential to bring a complementary invention into use, or when the patent owner abuses a patent position—e.g., by restricting supply excessively. If a potential licensee has been refused a license but can prove that one of the applicable criteria is met, the courts will compel the patent holder to issue the license, intervening further to set a "reasonable royalty" when the parties are unable to come to mutually satisfactory terms. Curiously, these sanctions are not invoked very frequently. In West Germany and Japan, for example, a number of compulsory license applications were filed under post-World War II patent laws, but as of 1975 no licenses had been mandated.[74] Sixteen applications were filed between 1959 and 1968 under the British provisions governing general compulsory licensing, but only two license grants were compelled.[75] The number of applications and grants might be small because inventions worth the cost of a legal battle are too valuable not to be utilized, because the criteria that must be satisfied before licensing is ordered are stringent, and/or because the very existence of compulsory licensing procedures encourages patent holders and would-be licensees to settle out of court. That stringency of the criteria is an important factor is suggested by the British and Canadian experience with compulsory licensing of pharmaceutical patents, for which special and more permissive legal provisions have been adopted. Between 1950 and 1971, there were 56 drug patent compulsory license applications in the U.K., 20 culminating in Comptroller of Patents decisions favoring the applicant.[76] Similarly, 15 licenses were granted during the 1960s under the Canadian drug patent provisions.[77] There is also reason to believe that compulsory licensing provisions do encourage voluntary agreements to license.[78]

Although the U.S. Congress has not seen fit to enact general compulsory licensing laws, compulsory licensing has been specified as a remedy in more than 125 antitrust cases, making available tens of thousands of patents at "reasonable" royalties or (in a few instances) royalty free.[79] In this respect the United States has outdistanced its European neighbors, whose antitrust and patent bureaucracies have for the most part been chary about intervening in patent matters. By vigorously prosecuting predatory practices and attempts to extend monopoly power beyond the limited bounds of specific patent grants, the U.S. antitrust authorities have managed materially to limit the social costs of the patent system.[80] This has been achieved without doing serious damage on the benefits side of the equation, for most of the corporations subjected to compulsory licensing orders had alternative incentives to innovate by virtue of natural lags, competitive differentiation, and established market positions. To them the prospect of profits gained by abusing the patent grant apparently had little incremental incentive value, for the effect of compulsory licensing on their R & D efforts appears to have been minute. Thus, an analysis of 1975 research and development expenditures data revealed that 42 companies subjected to compulsory licensing decrees of appreciable scope spent no less on R & D relative to sales than nonimpacted corporations of comparable size and industrial orientation.[81] The only observable adverse consequences entailed a likely propensity for decree-impacted firms to patent fewer of their inventions, resorting to secrecy more often.[82]

All in all, the substantial amount of evidence now available suggests that compulsory patent licensing, judiciously confined to cases in which patent-based monopoly power has been abused or extended far beyond levels needed to provide adequate incentive, would have little or no adverse impact on the rate of technological progress and would on occasion mitigate significant monopoly burdens.[83]

Finally, if one despairs of reforming the patent system through halfway measures, there remains the possibility of abolishing it. This is not an alternative favored by the author, but that should scarcely bar considering it seriously. We have no reason to fear that such a step would bring all inventive and innovative activity to a halt, since there are many alternative incentives. Yet some contributions would undoubtedly be lost, unless substitute incentives were created. There are two main alternatives.

First, the government might assume an increasingly active role in sponsoring and subsidizing innovation where private initiative proves inadequate. As Chapter 15 brought out, the U.S. federal government already supplies funds to support 37 percent of all industrial R & D activity plus considerable work in government and university laboratories. Most of this support is directed toward meeting unique governmental needs, as in the defense and space fields, but approximately 30 percent of federal expenditures go for R & D in such civilian areas as agriculture, food products, mining techniques, energy technology, water desalination, civil aviation, and the full spectrum of medical and medicinal technologies. Extension of the government's role would therefore not be a radical departure. Still, it has certain drawbacks. A centralized R & D resource allocation process could hardly avoid overlooking promising new opportunities, backing the wrong concept, neglecting unknown young people at the peak of their creative powers, and in general bogging down in red tape. A steady diet of research at the government trough can also

71. On India, see Sanjaya Lall, "The International Pharmaceutical Industry and Less-Developed Countries, with Special Reference to India," *Oxford Bulletin of Economics and Statistics* 36 (August 1974): 143-69.

72. See Catherine S. Corry, *Compulsory Licensing of Patents—A Legislative History*, Study no. 12 of U.S., Congress, Senate, Subcommittee on Patents, Trademarks, and Copyrights, 85th Cong., 2nd sess., 1958.

73. See Fredrik Neumeyer, *Compulsory Licensing of Patents Under Some Non-American Systems*, Study no. 19 of U.S., Congress, Senate, Subcommittee on Patents, Trademarks, and Copyrights, 85th Cong., 2nd sess., 1959; and P. J. Federico, "Compulsory Licensing in Other Countries," *Law and Contemporary Problems* 13 (Spring 1948): 295-309.

74. See David J. Henry, "Multi-National Practice in Determining Provisions in Compulsory Patent Licenses," *Journal of International Law and Economics* 11 (March 1977): 325-51.

75. Committee to Examine the Patent System and Patent Law, *The British Patent System* (London: HMSO, July 1970), p. 214.

76. *The British Patent System*, p. 214; and Scherer, *The Economic Effects of Compulsory Patent Licensing*, pp. 40-43.

77. O. J. Firestone, *Economic Implications of Patents* (Ottawa: University of Ottawa Press, 1971), p. 38. For an analysis of royalty setting in several cases, see Scherer, *The Economic Effects of Compulsory Patent Licensing*, pp. 43-50.

78. See Taylor and Silberston, *The Economic Impact of the Patent System*, pp. 16, 111-40.

79. U.S., Congress, Senate, Committee on the Judiciary, Subcommittee on Patents, Trademarks, and Copyrights, staff report, *Compulsory Patent Licensing Under Antitrust Judgments*, 90th Cong., 1st sess., 1960; and Scherer, *The Economic Effects of Compulsory Patent Licensing*, pp. 62-66.

80. For a quite different view, see Ward S. Bowman, Jr., *Patent and Antitrust Law* (Chicago: University of Chicago Press, 1973). Bowman argues that many of the so-called patent abuses subjected to antitrust attack were in fact desirable because they enhanced the patent-holding firm's profits and hence added incentive for investment in innovation. He largely ignores competitive stimuli and other nonpatent inducements to such investment and hence bases his case upon an unrealistically narrow view of the innovative process. Bowman is on sounder ground in arguing that certain patent "abuses"—e.g., discriminatory royalty rates—contribute to economic efficiency.

81. Scherer, *The Economic Effects of Compulsory Patent Licensing*, pp. 67-75.

82. Scherer, *The Economic Effects*, pp. 64-67; and Scherer et al., *Patents and the Corporation*, pp. 137-46, 153-54.

83. Cf. Taylor and Silberston, *The Economic Impact of the Patent System*, who conclude at p. 350 that "the balance of economic advantage for the [United Kingdom] lies narrowly with the existing patent system" rather than a worldwide system of compulsory licensing at "reasonable" royalties for all patents, without any required showing of abuse or excessive monopoly power.

dull incentives for efficiency, encouraging those who are subsidized to substitute money for ingenuity. Therefore, while government sponsorship will continue to be a valuable supplement, it holds little promise as a complete alternative to incentive systems that rely as much as possible on private initiative.

The other possibility is to institute a system of governmental awards or bonuses for individuals and firms making significant technological contributions. Such a system was proposed by James Madison at the Constitutional Convention of 1787 as an explicit alternative to the patent system.[84] Something like it was written into the U.S. Atomic Energy Act of 1946, establishing a Patent Compensation Board to confer monetary awards upon individuals making inventions related to military uses of atomic energy, which under the law could not be patented. The award approach is also a standard method of motivating invention by individuals in the Soviet Union, where inventors can apply for certificates of invention. When a certificated invention has been introduced into industrial practice, the inventor is entitled to a 2 percent share of the cost savings during the first five years of use up to a ceiling of 20,000 rubles.[85] Despite their evident attractions, award systems of this genre suffer from three main drawbacks. First, estimating the value of inventive contributions is a difficult task, and any bureaucratic council entrusted with the job is bound to make mistakes and perpetrate inequities. When inequity is inevitable, one might prefer that it be the result of a more impersonal income distribution mechanism. Second, as the approach has been traditionally interpreted, it pertains only to inventions in the narrow sense. If no provision were made to reward developmental contributions, and if in addition alternative incentive structures were deficient, there might be some danger, to reiterate Judge Frank's caveat, of throwing out the baby with the bathwater. It is worth noting that a chronic complaint of Soviet inventors is that all too frequently their inventions are not accepted or put into practice by unsympathetic managers.

Third, there is an inherent conservative bias in the prizes granted by administrative and quasi-judicial bodies. Munificence is a rare committee virtue. For example, the Atomic Energy Commission's Patent Compensation Board awarded the assignee of Enrico Fermi's basic patent on the production of radioactive isotopes—forerunner of the methods for producing plutonium—a sum of $300,000 when the patent was dedicated to the public. In an analogous case, the federal government agreed to pay $1 million as compensation for utilizing Robert H. Goddard's basic liquid rocket engine patents. During the life of the Goddard patents, U.S. expenditures on liquid-propelled rockets amounted to roughly $10 billion. Compared to the value of the inventions and the profits that might have been earned if exclusive patent rights could have been enforced, the Fermi and Goddard awards were miserly. They were certainly not what Schumpeter had in mind in describing "spectacular prizes . . . thrown to a small minority of winners." It is doubtful whether a generalized reward system administered in this conservative tradition would motivate as much risk bearing as the patent system presently does. Still, the crucial policy question is how serious the loss of inventions and innovations owing to lessened risk bearing would be. And that, unfortunately, is a subject on which much more remains to be learned.

84. See Machlup, *An Economic Review of the Patent System*, pp. 15-16.

85. See Francis Hughes, "Soviet Invention Awards," *Economic Journal* 55 (June-September 1945): 291-97; *idem*, "Incentive for Soviet Initiative," *Economic Journal* 56 (September 1946): 415-25; E. Artemiev, "The New Soviet Law on Inventive Activity," *Industrial Property* (July 1974): 320-25; and (for details of the reward determination system), the "Laws and Treaties" section of *Industrial Property* (February 1977).

[2]

Nordhaus' Theory of Optimal Patent Life: A Geometric Reinterpretation

By F. M. SCHERER*

For more than a century the patent system has been studied with remarkable care by economists.[1] Yet only recently, in a contribution by William Nordhaus, has formal economic theory been brought to bear successfully on the central policy issue of the patent system—*how much protection* should be accorded inventors and innovators. This article extends Nordhaus' pioneering work and corrects what in some cases is a significant interpretational error. Nordhaus' original presentation was largely algebraic, but certain problems he left unsolved can be tackled more directly through the geometric approach taken here. This mode of attack has the fringe benefit of making what in the original paper was a rather forbidding tangle of mathematical notation more comprehensible intuitively.

I. The Basic Model

I begin by observing, as Nordhaus did, that inventions and innovations are not free goods. To make and introduce an invention which reduces unit production costs, research and development (*RD*) outlays must be incurred. For any given production task there exists at some moment in time an "invention possibility function" (*IPF*) which relates the percentage unit production cost reduction *B* achieved (the "output" of an inventive effort) to the expenditure on *RD*. The more research input, the greater will be the cost saving, *ceteris paribus*. For mathematical convenience Nordhaus considers only the very simple invention possibility function $B=\beta RD^{\alpha}$, which with $\alpha<1$ implies continuously diminishing marginal returns to inventive effort. However, it seems more plausible (and as we shall see, more flexible)

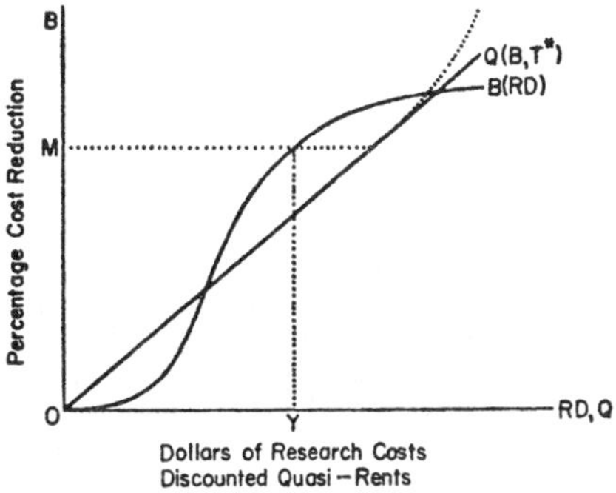

FIGURE 1

to assume an inflected function like $B(RD)$ in Figure 1, where at first there are increasing returns to the research effort, after which diminishing returns set in.

The benefits to the firm depend in a slightly more complex manner upon the cost savings facilitated. Following Kenneth Arrow, Nordhaus assumes that production is initially carried out under competitive conditions at constant unit cost and price $0C_0$, as shown in the conventional supply and demand diagram Figure 2.[2] The firm which secures exclusive patent rights on an invention reducing unit costs to $0C_1$ can either drive other firms out of business, producing the whole former output $0X_0$ and commanding a monopoly rent of C_0EAC_1 per year, or it can license the patent to existing producers, charging a royalty which extracts the same surplus C_0EAC_1 from them. Note that even though the patent confers some monopoly power, it does not permit the patent holder to charge a price above the cost $0C_0$ associ-

* Professor of economics, University of Michigan. I am grateful to William Nordhaus and the editor for helpful criticisms.

[1] See the survey article by Fritz Machlup.

[2] As long as the new process is merely used internally and not licensed to outsiders, the analysis applies equally well to the case of a monopolist or group of colluding oligopolists pricing to deter all entry in a range of relatively inelastic demand. See Scherer, pp. 219–21.

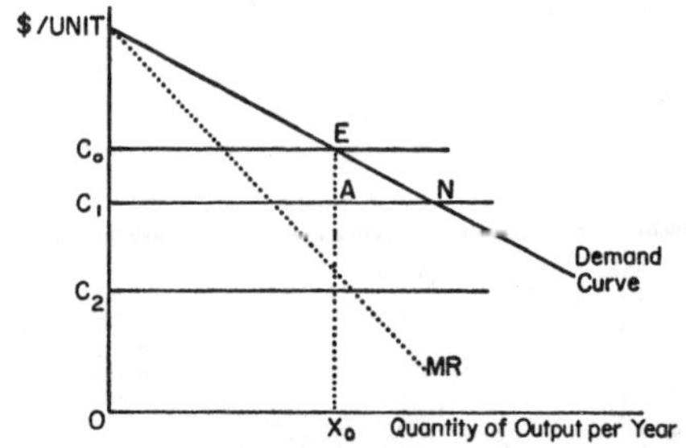

FIGURE 2

ated with the (now inferior) competitive process. Because of this, and if demand is not very elastic in the neighborhood of the competitive price, the optimal postinvention price and quantity under a patent monopoly will be identical to those in the preinvention equilibrium. However, if the invention reduces cost by so much (i.e., to $0C_2$) that the new long-run cost curve cuts the monopolist's marginal revenue curve MR to the right of the old competitive output $0X_0$, the patent holder will find it advantageous to see that price is reduced below $0C_0$ and that output is expanded beyond $0X_0$. Nordhaus assigns the name "run-of-the-mill" to inventions which reduce cost insufficiently to induce a price reduction and output expansion, as with process $0C_1$ in Figure 2. Following Nordhaus, I shall focus here on the run-of-the-mill case, touching only peripherally on the case of "drastic" price-reducing process inventions like those leading to unit cost $0C_2$ in Figure 2.

Since there is no output expansion effect with run-of-the-mill inventions, the annual monopoly rent to the patent holder is a linear function of the percentage cost reduction B. For a given patent life $T = T^*$, the patent monopolist's total discounted quasi-rent function $Q(B, T^*)$ can therefore be shown as a straight line in the coordinates of Figure 1.[3] (If the cost reduction effort is pursued into the stage where the "drastic" invention case holds, the quasi-rent function begins to curve upward, as shown by the dotted offshoot from $Q(B, T^*)$ in Figure 1.) Given the market demand and competitive supply conditions, the invention possibility function $B(RD)$, and some patent life T^*, the firm maximizes its profits by extending its RD expenditures to that level where the horizontal distance between the "cost of invention" function $B(RD)$ and the quasi-rent function $Q(B, T^*)$ is a maximum—i.e., at RD outlay $0Y$ and percentage cost reduction $0M$ in Figure 1.

The patent grant's duration T is usually fixed by the government. This means that T is a parameter to the inventor-innovator (though some exceptions exist). But to the government the patent life is a policy variable—normally decided upon for broad classes of inventions, although case-by-case determination is not inconceivable. As the government increases the patent grant's life under given technological and market conditions, the number of years over which the patent recipient can command monopoly rents rises and the quasi-rent function $Q(B, T)$ shifts to the right. For simplicity, I (and Nordhaus) assume that the patent holder reaps the full monopoly potential of its invention while the patent is in force and that competitive imitation wipes out supranormal rents completely at the patent's expiration date. Thus, we tentatively ignore imitation lags, reputation effects, and the obsolescence which occurs due to exogenous technological change and competitive "inventing around" one's patent. Given these assumptions and constant market demand, the existence of a positive discount rate ensures that the rightward shift of $Q(B, T)$ with increases in T exhibits diminishing returns, since the gains from incremental patent life extensions are discounted more and more heavily. Figure 3 illustrates the relationships between run-of-the-mill invention quasi-rent functions for various patent lives, assuming constant demand and continuous compounding at a 12 percent discount rate. As the period of patent protection rises, the equal-slope, maximum profit point on the invention possibility function shifts to the right, as indicated by arrows designating the sev-

[3] Where r is the discount rate, $Q(B, T^*) = B(0C_0)(1-e^{-rT^*})/r$. This formulation is used by Nordhaus after setting $0C_0 = 1$.

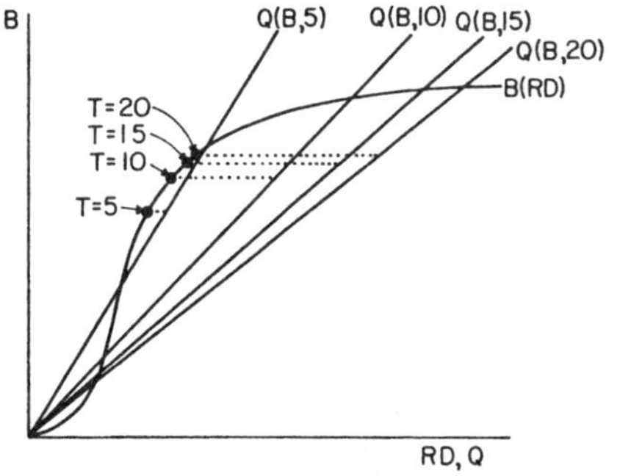

FIGURE 3

eral patent-holder's optima in Figure 3. The longer the patent's life, the further a profit-maximizing firm will carry its cost-reducing *RD* expenditures.

How far should this cost-reducing effort proceed if society's interest is to be served? To find out, Nordhaus uses a standard technique of welfare economics. Returning to Figure 2, we observe that while a run-of-the-mill invention leading to unit cost $0C_1$ enjoys patent protection, the patent owner realizes each year a producer's surplus C_0EAC_1 equal to the cost saving on the preinvention output. When the patent expires, competition drives the price down to $0C_1$, output is increased, the producer's surplus is wiped out, and society gains a new consumers' surplus C_0ENC_1. Ignoring the redistributive implications of this change (or assuming constant and equal marginal utilities of income between the patent holder and society in general), the "price" society pays to induce a reduction in unit costs from $0C_0$ to $0C_1$ is essentially the sacrifice of the "welfare triangle" $EAN(=C_0ENC_1-C_0EAC_1)$ from the time the invention is introduced until the date of patent expiration, plus the inventor's *RD* costs. To find the socially optimal patent life, one must balance the marginal deferrals of this welfare triangle surplus *EAN* and the (rising) *RD* costs against the increasing amount of cost reduction (and hence increases in producer's and consumers' surplus) stimulated by longer patent lives.

Here geometry proves a blunter tool than algebra. Nordhaus determines the socially optimal patent life by maximizing with respect to T the sum of the discounted producer's surpluses C_0EAC_1 from the start of the patent's life (t=0) to T, plus the discounted consumers' surplus C_0ENC_1 from year T to infinity, less the cost of research and development incurred; all subject to an equality constraint reflecting the inventor-innovator firm's profit-maximizing behavior in setting the slope of its *IPF* equal to the slope of its quasi-rent function $Q(B, \mathrm{T})$. The mathematics are fairly intricate and will not be reproduced. Here it suffices to characterize the heart of the determination. As the amount of induced cost reduction *B* rises due to a longer patent life, society must wait longer and longer to appropriate the welfare triangle *EAN*, though with linear demand functions the area of the triangle increases as the square of *B*, *ceteris paribus* (see Scherer, p. 402). But for each additional year's wait it motivates less and less incremental cost reduction because the gain to the patent holder increases at best only linearly with increases in *B*,[4] because later years' monopoly rents are discounted more heavily than those in the early years, and because additional percentage points of cost reduction are achieved only at rising research cost. Sooner or later, these diminishing-return effects overpower society's interest in stimulating additional cost reduction by extending the patent life. Therefore, in all but some special limiting cases there exists a finite socially optimal patent life.

II. Comparative Statics Results

From the constrained welfare-maximizing conditions reflecting this balancing, Nordhaus derives three important comparative statics results. All can be comprehended intuitively from the geometric presentation.

First, the larger is the arc elasticity of demand in the neighborhood of the preinvention and postinvention competitive equilibria, the shorter is the socially optimal patent life. This is so because, at least for run-of-

[4] And in the "drastic" invention case, at less than a linear rate. See the dotted segment of $Q(B, \mathrm{T}^*)$ in Figure 1.

the-mill inventions, nothing varies with elasticity but the base (and hence area) of the welfare triangle *EAN*. As price elasticity increases, the area of the welfare triangle increases proportionately, making society less and less willing to postpone its capture in exchange for a given incrementally induced cost reduction.

Second, the "easier" it is to achieve a given cost reduction—that is, the steeper the *IPF*, and hence the larger the equilibrium induced level of cost reduction *B*, *ceteris paribus*—the shorter the socially optimal patent life will be. This is so because the area of the awaited welfare triangle rises quadratically with increases in *B*, as noted earlier, whereas the patent holder's monopoly rent C_0EAC_1 rises only at a linear rate with *B*. Hence, when big cost reductions are likely whether the allowed patent life is modest or long, society is less willing to defer the realization of its net welfare surplus to motivate still more cost reduction than it would be if the cost savings under comparable patent life conditions and research investments were modest.

Third, the optimal patent life is shorter, the sharper the curvature of the invention possibility function in the neighborhood of the optimal *RD* expenditure. This result is best illustrated with a new diagram. Panels (a), (b), and (c) of Figure 4 each have identical run-of-the-mill invention quasi-rent functions corresponding to patent lives of eight and seventeen years. The invention possibility functions, however, display increasingly sharp curvatures, analogous to declining elasticities of substitution in orthodox production function theory. The sharper the curvature, the smaller is the difference in the amount of cost reduction induced by a given increase in patent life. And as the cost reduction effect attenuates, *ceteris paribus*, society's welfare gain from deferring competitive imitation falls correspondingly, so shorter patent lives prove optimal.

From Nordhaus' equations 5.13 and 5.14, p. 78, it can be deduced that in the limiting case of a rectangular or "stair-step" *IPF*, as illustrated in Figure 4 (c), the socially optimal patent life is zero. Here, however, the theory as developed thus far goes astray by focusing exclusively on the first-order marginal conditions for a patent-holder's local profit maximum, ignoring the necessary *total condition* that expected profits be positive.[5] If the patent's life is set at zero, the quasi-rent function $Q(B, T)$ coincides with the vertical axis of Figure 4(c), lying at all points but the origin to the left of the *IPF*. Faced with a situation in which expected quasi-rents are less than *RD* outlays at any positive *RD* level, no rational firm will invest, even though the consumers' (and producer's) surpluses potentially available exceed prospective *RD* costs. The investment incentive

[5] To deal with this problem algebraically, it is necessary to impose upon the social welfare function being maximized an inequality constraint reflecting the necessity of positive profit expectations, along with Nordhaus' equality constraint reflecting the inventor's marginal optimizing conditions.

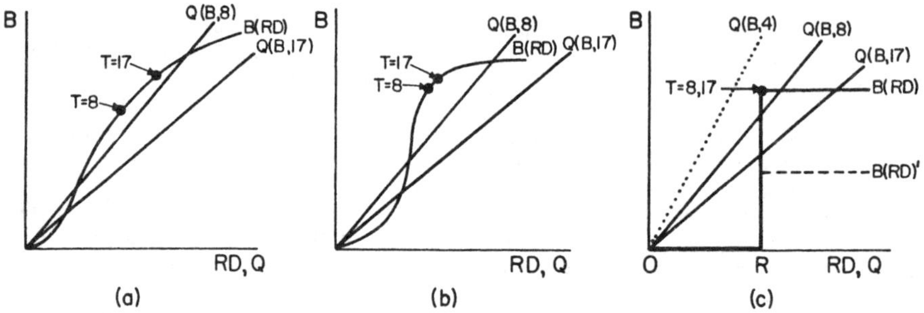

FIGURE 4

is likewise stultified by short patent lives—i.e., with the assumed four-year life indicated by a dotted line in Figure 4(c). And although following Nordhaus' welfare-maximizing marginal rule *necessarily* implies failure to satisfy the positive-profits condition only in this limiting stair-step *IPF* case, the incompatibility can also arise in other cases—notably, when the cost of achieving a given value of B is relatively high and when the invention possibility function's curvature is relatively sharp in the neighborhood of the inventing firm's equilibrium.

We see then that the patent grant really plays two investment-inducing roles, which in a different context I have called the "stimulus effect" and the "Lebensraum effect" roles (see Scherer, p. 369). In its stimulus effect role, emphasized by Nordhaus, an optimal patent policy sees to it that the monopoly rent lure induces *RD* investment just sufficient to equate the marginal social gain from *further* cost reduction with marginal social cost. In its Lebensraum effect role, the patent grant must persuade investors that competitive imitation will be deferred sufficiently long to make discounted quasi-rents exceed *RD* outlays for at least *some* positive *RD* investment level. Neither aspect can be ignored in designing an optimal patent policy.

Fortuitously, the comparative static relationships derived through Nordhaus' stimulus effect analysis and those associated with the Lebensraum effect are identical for what, from a policy standpoint, is undoubtedly the most important exogenous variable—the "ease" of invention. This is demonstrated in Figure 4(c) by considering two different stair-step invention possibility functions $B(RD)$ and $B(RD)'$. With "hard invention" *IPF* $B(RD)'$ (broken line), the cost saving realized by spending $0R$ dollars on research is considerably less than with "easy invention" *IPF* $B(RD)$. The Lebensraum condition for $B(RD)$ is satisfied with a patent life shorter than eight years, but for $B(RD)'$ even a seventeen-year life is insufficient. Quite generally, the higher is the private benefit/*RD* cost ratio for a potential invention over some assumed period of exploitation, the shorter the allowed patent life can be while still satisfying the Lebensraum effect condition that expected profits be positive. This conclusion is completely consistent with Nordhaus' finding, considering marginal stimulus effects alone, that easy inventions—those yielding big cost savings in relation to the research resources invested—warrant shorter patent protection than hard inventions.

III. Some Implications

One immediate policy implication is that if law makers were to shorten the standard term of process patent grants, what would be lost would be primarily those inventions with relatively low benefit-cost ratios—those which in any event are not likely to have a great impact on social welfare. That this conclusion is based upon an extremely naive model does not necessarily nullify its thrust. Indeed, elaboration to account for a principal oversimplification strengthens the case. Contrary to what we have assumed thus far, patent protection is not the only barrier to competitive imitation. Natural inertia, secrecy, and the need to do some *RD* on one's own before mastering a new process all contribute to imitation lags in an atomistically structured industry.[6] When market concentration is high and nonpatent barriers to new entry are present, postinnovation pricing discipline is likely to be sufficiently firm to let an innovating oligopolist recover its *RD* investment even if other immediate rivals imitate quickly. When any or all of these circumstances exist, the Lebensraum condition will tend to be satisfied with short or even zero patent lives for potential process inventions promising high benefit-cost ratios. A uniform policy of long-lived patent grants confers excessive private rewards in these cases, compensated to some unknown extent by the social benefits realized from low benefit-cost projects which otherwise would not have been undertaken and by stimulus effects at the margin of projects which would have been undertaken even with short patent

[6] These points are developed more fully in Scherer, pp. 384–390, which also deals with many of the issues raised in Nordhaus' reply to this paper.

lives. And although the analysis thus far has been developed only for process inventions, it is obvious that for product inventions with high private benefit-cost ratios, the Lebensraum condition will be satisfied at shorter patent lives than for those with low benefit-cost ratios.

A "best of both worlds" policy recognizing these relationships would tailor the life of each patent to the economic characteristics of its underlying invention. This might be achieved through a flexible system of compulsory licensing, under which the patent recipient bears the burden of showing why his patent should not expire or be licensed at modest royalties to all applicants three or five years after its issue. This burden would be sustained if the patentee demonstrated that his invention fell into one or more of the categories in which longer protection is needed to satisfy the Lebensraum condition—e.g., because the market is small relative to the costs of research, or because the cost savings achieved were modest in relation to research costs. When a patent-holding corporation possesses a substantial share of the relevant market and well-established marketing channels, on the other hand, there would be a presumption in favor of early patent licensing or expiration on the assumption that positive innovation profits could normally be attained without the added inducement of strong patent protection.

There is, however, one case in which the implications of the modified Nordhaus theory must be regarded with special caution. Risk and uncertainty—surely significant elements in research and development decision making—have been ignored altogether. For most *RD* projects, Edwin Mansfield's studies suggest, this is probably not a fatal oversight. The uncertainties are sufficiently manageable that they can be taken into account by letting the invention possibility and quasi-rent functions be expressed in terms of expected values or certainty equivalents modified by an appropriate risk premium. Still it is at least conceivable that certain inventions with very high "best guess" benefit-cost ratios require unusually bold, farsighted, time-consuming departures from orthodox technology, with extraordinary attendant uncertainties and risks. In these cases, strong patent protection offering the prospect of exceptional rewards contingent upon technical and commercial success may be necessary to induce investment. Such cases, which are probably rare (i.e., embracing not more than a dozen or so major inventions per year), cannot safely be squeezed into the mold of an orthodox profit-maximizing model like the one presented here. Nevertheless, the fact that they are exceptions makes it possible to devise a policy dealing with them as such while treating the vast majority of all patented inventions under rules which assume some fairly close approximation to profit-maximizing behavior. In particular, the presumption in favor of early compulsory licensing or short patent lives for inventions with high *ex post* private benefit-cost ratios could be waived upon a showing that the patent recipient exhibited exceptional creativity or undertook unusual technical and/or commercial risks in the invention's development.

Such a policy—flexible, but rooted in the logic of economic theory—would probably be a significant improvement over the present U.S. patent system, recognized by friends and critics alike to be arbitrary and inefficient. It warrants serious consideration.

REFERENCES

K. J. Arrow, "Economic Welfare and the Allocation of Resources for Invention," in *The Rate and Direction of Inventive Activity*, Nat. Bur. Econ. Res. conference report, Princeton 1962, 619–25.

F. Machlup, *An Economic Review of the Patent System*, Study No. 15, U.S. Senate, Committee on the Judiciary, Subcommittee on Patents, Trademarks, and Copyrights, 85th Cong., 2d sess., Washington 1958.

E. Mansfield, *Industrial Research and Technological Innovation*, New York 1968, ch. 3.

W. D. Nordhaus, *Invention, Growth, and Welfare; A Theoretical Treatment of Technological Change*, Cambridge, Mass. 1969, ch. 5.

F. M. Scherer, *Industrial Market Structure and Economic Performance*, Chicago 1970.

[3]
COMMENT ON EDMUND KITCH

F. M. Scherer

Professor Kitch argues that patents rarely lead to monopolies in the sense that their owner faces, in the relevant behavioral range, a downward-sloping demand curve. That exclusive patent rights and monopoly power in Kitch's sense need not coincide has been well accepted since 1962, when Arrow published his analysis of what Nordhaus (1969, p. 72) later named "run-of-the-mill" process inventions. It is also plausible that most patents, even those covering new products, to which Arrow's analysis applies less clearly, confer at best only trivial amounts of monopoly power. My own reworking of George Washington University scholars' survey data revealed that the frequency distribution of profit returns attributed to individual company patents was so highly skewed that, asymptotically, it had neither finite mean nor variance (Scherer, 1965, p. 1098). Thus, a few "outliers" accounted for nearly all of the profits. In the sample of 74 patents, only six had profits (or rents) exceeding $1 million.

Where Kitch evokes cries of outrage (as he may have intended) is in his attempt to use the Xerox history as an example of a monopoly-free patent position. I would go so far as to suggest that he has written the wrong

Research in Law and Economics, Volume 8, pages 51–58.

ISBN: 0–89232–654–9

paper. In his introduction, he correctly notes my skepticism about the realism of his earlier theory of patents as "prospects." Had he sought evidence supporting his theory, he could have found no better case study than xerography, in which some very basic patents did stake out for more than two decades the exclusive right to "prospect" in a field of technology. It is a rare case, I might add. But Xerox as the classic example of monopoly-free patents? Neither theory nor evidence will support the leap.

Kitch goes astray for three reasons that, alas, are all too typical of legal scholars brandishing Chicago-acquired "price theory." First, he consistently rejected my urging, in response to a preconference draft of the paper, that he become familiar in depth with the facts of the Xerox story. Second, his Chicago training prepared him only to apply static apparatus where dynamics must rule the day. And third, his command of the static tools leaves a fair amount to be desired.

Let me start with the most obvious manifestation of the third problem, for it underlies some of what will follow. As Chicagoan John McGee (1966) recognized long ago, the dilemma posed in connection with Kitch's figures 1-3 is no dilemma at all. Assuming fixed proportions, the derived demand curve for the patent is correctly measured in terms of units of output produced using the patent. It has the shape of the shaded rectangle in Kitch's Figure 1 plus the right-abutting triangle, not the shape proposed in his Figure 2. Contrary to Kitch's assertion, a violation of the fixed proportions assumption creates no theoretical difficulty. To be sure, only the fixed proportions case is taught in introductory "price theory" courses. But a course only slightly more advanced would reveal that if there is a well-defined mapping of cost advantages from using the patented process, the appropriate demand curve can be derived. In Kitch's two-machine case, costs alone are irrelevant; what matters is the difference between costs and the revenues obtainable from customers, ranked in descending order of the excess of reservation prices over costs. In the increasing marginal production cost case, the derived demand curve will begin sloping downward immediately instead of at output Qo.

I raise this point not simply to be critical, but because the ability to derive appropriate demand curves is useful in attempting to estimate the bounds of Xerox's monopoly power. To that task we now turn.

Kitch claims two insights as to why Xerox may have had no monopoly power: because its pricing freedom was inhibited in the early years by "newly obsolete technology," and because with the expiration of its key patents (clearly plural and not single-dated) drawing near, it had to "occupy" the untapped market at prices approximating marginal cost.

First, the early years. What was that "newly obsolete technology"? At first, it was diffusion transfer, accounting for about 47 percent of photocopying activity in 1960, and Thermofax, with about 45 percent of the activity.[1] In the 1960s, the average total cost, including machine amortization, maintenance, and paper, but excluding clerical labor, of a Xerox copy on machines used for more than 250 copies per day was less than 1 cent. The variable materials cost of diffusion transfer copies on already existing machines was in the range of 8-10 cents, and Thermofax copies were about twice as expensive at the margin as Xerox copies. Diffusion transfer copies took about two minutes to make, or about 5 cents per copy at clerical wages of $1.50 per hour. Xerox was much faster. For multiple copies, it was also about four times as fast as Thermofax. The quality of Xerox copies was greatly superior to both alternatives. (At the conference I passed around a faded 1960 Thermofax copy as evidence.) In other words, in terms of machine and supplies costs, clerical labor costs, and quality, operating a *new* Xerox machine priced at its own full cost easily dominated over continuing to operate an *existing* diffusion transfer or Thermofax copier. So where's the constraint?

To be sure, Xerox had to compete with carbon paper, which may even today have lower full costs in very low-volume applications. The quality difference is plain. Less obvious is the difference highlighted in a 1950s song, "Secretary's Lament":

> Carbon paper
> Stains my hands.
> All I touch is
> Remington-Rands
> But never a man.

I have yet to hear the 1960s analogue:

> I'm a sorrowful loner
> All splattered with toner.

A more direct constraint on Xerox's plain-paper machine pricing, at least from 1962 on, was the existence of the coated-paper Electrofax process. My figure graphs Blackstone's data on the comparative average total costs (excluding clerical labor costs) of Xerox copies (solid line) vs. Electrofax copies (dashed line) over a spectrum of monthly machine utilization rates. With its higher fixed machine amortization costs, Xerox is the higher-cost process at monthly volumes of less than 1,852 copies (about 93 per day). Above that threshold, it dominates Electrofax, and hence could have swept Electrofax from the higher-volume market, had it so chosen.[2] Why didn't it? Why did it charge (at least until 1965) 5 cents per copy (2.5 times its full costs) in the 2,000 copies per month range and 3.5 cents per copy in higher volumes? Because it was more profitable to share the market

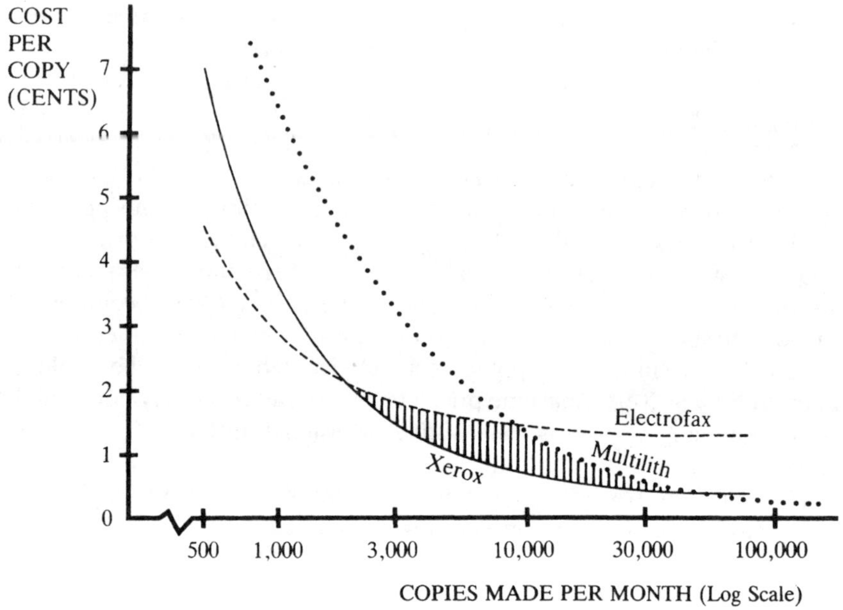

Unit Costs of Alternative Copying Processes

Source: Erwin A. Blackstone. (1968) "The Copying Machine Industry: A Case Study," Ph.D. dissertation, University of Michigan, p. 133.

with Electrofax at high prices than to dominate it at lower prices. And if making such a trade-off is not the essence of monopoly power, I should be dispatched to Chicago for a refresher course.

Professor Kitch stresses the constraining role of mimeograph and multilith processes on Xerox's pricing of multiple-copy applications. He dismisses (note 10) Blackstone's detailed study of the economics of the copying machine industry because it purportedly excluded consideration of such processes as mimeograph. In fact, Blackstone did analyze the impact of those processes, observing that "Xerox has rapidly moved into the large duplicating market, estimated conservatively at . . . three and one-half times as large as that of copying" (p. 43) and that "The reason why the profit maximizing price is low (in very high volume copying) can be ascribed to the availability of duplicating" (p. 157). Blackstone did not provide comparative cost data on duplicating processes. With its wax stencils, messy ink, and inferior quality, mimeograph's variable costs, with or without quality adjustment, were probably greater than the full costs of Electrofax at low volumes and Xerox at high volumes. At intermediate volumes, mimeograph might arguably have displaced Electrofax as the next-best alternative to Xerox. As one who earned college money running a Multilith machine, I can attest that on low-volume jobs, the substantial setup and cleanup time costs made Multilith prohibitively costly. On high-volume applications, Multilith's costs were probably a bit lower than those of 1960s-vintage Xerox machines (i.e., before more recent improvements were made). Thus, Multilith's cost curve was somewhat like that shown by the dotted line in my figure.

For a static world free of informational imperfections and other frictions, assuming purely competitive pricing by the suppliers of alternative processes, and ignoring differences in copy quality, the demand curve for Xerox's patents can be derived from the information in my figure plus data on the distribution of copies made by users at all points in the monthly copy volume spectrum. One begins from what can be called Xerox's "cost advantage area"—that is, the shaded area in my figure bounded by the envelope of the minima of higher-cost alternative processes and Xerox's costs. That area is divided into a series of thin vertical segments whose height is the magnitude of Xerox's cost advantage over the next-best process. Each such segment is multiplied by the number of copies made by users whose volume requirements lie in that segment. (The complication that users' locations in the volume spectrum depend upon price must be sloughed off.) On a new graph (not explicitly shown) whose horizontal axis is total copies made per month and whose vertical axis is cents per copy, the segments are rearrayed in descending order of Xerox's cost advantage. The result is the derived demand curve for Xerox's patented technology. Given virtually any plausible shape of Xerox's cost advantage area, it must slope downward. Assuming, as befits a static analysis, Xerox's technology to be fully developed, marginal R&D cost is zero, and so one finds the optimal uniform price for the technology by computing the marginal to the derived demand curve and establishing its zero ordinate. To determine the optimal uniform price of Xerox *machine* services, one must remember that the curves in my figure are not conventional unit cost curves. Rather, each point on a curve reflects a level of customer utilization and hence a given amortization of fixed machine capital costs. To set the service price, therefore, a two-part tariff must be used (as Xerox did). A fixed charge equal to those fixed costs is levied, and the variable charge is the sum of variable costs to Xerox (e.g., for maintenance and bundled supplies) plus the optimal technology price. Given a downward-sloping patent demand curve, a downward-sloping average fixed machine cost curve, and approximately constant machine operating costs, the resulting equilibrium will lie on a downward-sloping portion of the demand curve for Xerox machine services.

However, the real world is much more complex, in part because Xerox could gain greater profits, as it did, by price-discriminating among users according to monthly volume and the number of copies made per job. With perfect (first degree) price discrimination, it could in principle capture profits equal to the entire shaded cost advantage area, multiplied by the relevant user densities. Also, rival processes were not in fact priced competitively. Partly because of this and partly because of its substantial quality advantage, Xerox was able also to tap high-reservation price demands in the lower-volume spectrum. There were also entry and exit response lags that Xerox exploited in an important way (Scherer, 1980, p. 240). These qualifications would strengthen the conclusion that Xerox faced significant customer reservation price differences and was a monopolist in Kitch's sense.

Kitch's argument to the contrary rests entirely upon the assumptions of his Figure 4. It is possible that at some "point in time" Xerox may have faced some such constellation of demand and marginal cost functions. Setting aside quibbles about the empirically implausible downward swoop of Kitch's marginal cost curve, that curve's *position* depends critically upon the capacity investment decisions made by Xerox. With Xerox's sales (and presumably its capacity) rising at a compounded rate of roughly 50 percent per year during the early 1960s, it is most unlikely that the functional relationships postulated by Kitch could have persisted more than fleetingly. Kitch provides no clue as to where the average cost curve corresponding to his constructed marginal cost curve lies. His argument cannot be interpreted without locating that short-run average cost curve and the long-run envelope to which it belongs. Blackstone estimates that the minimum efficient scale in copying machine production was at roughly 8 to 10 percent of 1968 volume. Beyond that volume, production would occur at approximately constant unit costs. If long-run unit costs were roughly constant beyond some intermediate capacity level, and a fortiori if economies of scale persisted into the natural monopoly range, a profitable firm that has expanded to Kitch's output, 0, will continue expanding into the downward-sloping demand curve range. Xerox was, of course, highly profitable. If, contrary to Kitch's argument and the empirical likelihoods, copying machine supply took place at increasing long-run unit costs, Xerox would sooner or later be impelled to contract from 0 and produce in the first downward-sloping demand region—if not by deliberate choice, then from the entry of smaller, lower-cost competitors. Thus, Kitch's Figure 4 is an illustration of disequilibrium, not equilibrium.

There remains Kitch's suggestion that Xerox pre-emptively set price low and "occupied" the residual market to keep competitors out in anticipation of patent expiration. The precise means by which this entry deterrence was to occur is left vague, except that it was necessary "to lower the price down to the competitive price." In a world of equal eventual access to technology and no product differentiation, this implies preferring the discounted present value of zero profits from now to infinity over the discounted present value of positive but declining profits. No rational firm would choose Kitch's alternative. Assuming the patent holder to enjoy no postpatent advantages other than inertia, Gaskins (1970, pp. 51–53) shows that the profit-maximizing dynamic strategy is to *raise*, not reduce, price as the patent expiration date approaches.[3] If pre-expiration price reductions are to pay, the patent holder must expect to enjoy thereby some postexpiration advantage that allows it to hold price above its costs (and those of rivals with access to the same technology). It is easy enough to visualize such scenarios, but not those from which monopoly power is absent.

To sum up, it is possible to speculate that Xerox possessed no monopoly power, just as one can speculate that 30 angels dance every Friday evening on the head of a pin. One might have thought that, since the age of reason dawned, arguments of this sort would have given way to rigorous logic and careful empirical observation. But one might expect too much.

Nothing in this critique implies my endorsement of SCM's treble damages suit against Xerox. That was the rent-seeking action of an unprogressive and impotent rival, not an attempt to invigorate competition. As such, it deserved the disapprobation of the courts and Professor Kitch. The FTC suit to which Kitch alludes only briefly was different in both motivation and effect. On it, the reader is referred to Bresnahan's (1985) thoughtful analysis.

NOTES

1. These figures, like all used in this comment, are drawn from Blackstone (1968).
2. Note that Electrofax, which first appeared on the market in 1962, was not a newly obsolete technology in the sense that its fixed costs were already sunk. Thus the comparison here is to the full costs of Electrofax copying. If the average variable costs of Electrofax are compared to the full costs of xerography, the breakeven volume shifts to roughly 4,000 copies per month.
3. The relevant theory is not presented in the Gaskins paper cited by Kitch. Kitch's implied criticism of Gaskins, however, is incorrect. Pricing to protect the value of an asset—the state variable in a control theory equation system—is the essence of dynamic optimal limit pricing. Gaskins' limit pricing model, it is worth noting, was inspired by Blackstone's empirical study of Xerox pricing. See Gaskins' dissertation, p. 107.

REFERENCES

Arrow, Kenneth J. (1962) "Economic Welfare and the Allocation of Resources for Invention," pp. 619–622 in *The Rate and Direction of Inventive Activity*, National Bureau of Economic Research conference report, Princeton, N.J., Princeton University Press.

Blackstone, Erwin A. (1968) "The Copying Machine Industry: A Case Study," Ph.D. thesis, University of Michigan, Ann Arbor.

Bresnahan, Timothy F. (May 1985) "Post-Entry Competition in the Plain Paper Copier Market," *American Economic Review*, Vol. 75, pp. 15–19.

Gaskins, Darius W. (1970) "Optimal Pricing by Dominant Firms," Ph.D. thesis, University of Michigan, Ann Arbor.

McGee, John S. (October 1966) "Patent Exploitation: Some Economic and Legal Problems," *Journal of Law and Economics*, Vol. 9, pp. 138–143.

Nordhaus, William D. (1969) *Invention, Growth, and Welfare*, Cambridge, Mass., M.I.T. Press.

Scherer, F. M. (December 1965) "Firm Size, Market Structure, Opportunity, and the Output of Patented Inventions," *American Economic Review*, Vol. 55, pp. 1097–1125.

———. (1980) *Industrial Market Structure and Economic Performance*, rev. ed., Boston, Houghton Mifflin.

[4]

The Economics of Human Gene Patents

Frederic M. Scherer, MBA, PhD

ABSTRACT

The author examines patents on DNA sequences, including data on gene sequence grants issued by the PTO during a 33-month period from 1998 to 2001. Policy supporting patents on DNA sequences and other elemental information that are far "upstream" in the product development pathway is contrasted with the economic bases and rationale for patents to pharmaceuticals, which require a protracted and expensive process of development and testing but that can be relatively cheaply and competitively imitated once they are approved and disclosed. How to allocate appropriately the economic returns among the upstream and downstream inventors is a challenging problem for economic theory, as well as for contemporary biomedical research, and is perhaps most familiarly embodied in licensing and cross-licensing disputes involving "reach-through" and "reach-back" rights. Such disputes can generate enormous transaction costs. They may become increasingly frequent and vexing with respect to the scope and overlap of patent claims on human gene sequences. On the basis of his analyses, the author argues that genome patent claims should be interpreted narrowly. He is particularly concerned with ensuring that the development of new (therapeutic) products is not blocked or retarded by a multiplicity of prior patent claims, but he is pessimistic that the diversity of participants in biotechnology will provide a "sufficient community of interest to organize comprehensive low-royalty cross-licensing" regimes. Accordingly, he suggests mandatory arbitration as one mechanism for resolving such problems.

Acad. Med. 2002;77:1348–1367.

In the millennium year 2000 two teams, one organized through the public sector and one privately financed, completed preliminary sequence maps of the human genome, encompassing an estimated three billion DNA base pairs. Current estimates, still subject to considerable uncertainty, suggest that the human genome includes 30,000 to 40,000 distinct gene sequences that may in combination express 100,000 proteins. The United States Patent and Trademark Office has been accepting patent applications for individual human genome sequences whose "utility" can be shown by links to particular genetically-based diseases, genetically expressed proteins whose function in the human body can be identified, and/or medical diagnostic methods. As of 1999, nearly 3,000 patents of this nature had been issued, and a considerable backlog of applications is said to exist.[1] Whether, or under what circumstances, patents on human genome sequences should be issued has become a highly controversial issue both in the United States and in the European Community.

There are three main ways to approach this issue—precedential, ethical, and utilitarian. The precedential approach asks whether the intent of Congress in enacting patent laws under Article I, Section 8, of the U.S. Constitution was to allow patents for the identification of phenomena such as gene sequences already existing in nature. The ethical approach asks whether any individual or organization should have the exclusive right, limited to be sure in time, to control the commercial uses of phenomena that are as fundamental to human life as gene sequences. These are questions on which economists, who are more like plumbers than philosophers, have no comparative advantage. This paper, therefore, emphasizes a utilitarian approach, asking what benefits might arise from granting human gene sequence patents under diverse circumstances, what the costs might be, how the costs and benefits balance out, and how the benefits and costs are distributed among the diverse affected parties. The question is an extraordinarily difficult one, and readers

Dr. Scherer is Aetna Professor Emeritus at the John F. Kennedy School of Government, Harvard University, Boston, Massachusetts, and lecturer, Princeton University, Princeton, New Jersey.

Correspondence should be addressed care of: Stephen Heinig, Association of American Medical Colleges, 2450 N Street, NW, Washington, DC 20037.

should not expect a precise quantification of benefits and costs. The best that can be done is to survey what we know about the extent to which patent protection plays a role in accelerating (or impeding) the progress of science and technology, especially in the domain of human health, identify the relevant benefit and cost proclivities, and show how they are interrelated.

This paper begins by laying out what is known about the role patents play in providing incentives for research and development investments. Three narrower patent effect domains are explored further: pharmaceuticals, the commercialization of inventions made under government contracts, and biotechnology (which is distinctive in the large number of startup companies performing research and development). Evidence is marshaled on the kinds of organizations, public and private, performing basic research in biology, the characteristics of U.S. patents that embody DNA sequence claims, and the organizational origins of surveyed DNA sequence patents. From that empirical base, the paper focuses conceptually on the tradeoffs faced in awarding patents at diverse stages of a multi-stage discovery and development sequence for a representative therapeutic product. Finally, the empirical and conceptual threads are pulled together in an attempt to draw broad implications for public policy.

Before we begin, it is important to recognize that genomic research leads to improvements in health therapy through a complex multi-stage process. One first stage is the identification of a genome sequence. A second stage may follow, in which the functioning of the sequence in expressing proteins or (when there are mutations) causing or curing health problems is discovered. These two stages may be reversed; e.g., scientists may identify a health problem or a protein and then work backward to find the specific DNA sequence that is associated with it. The next stage entails conceiving a practical embodiment of this knowledge in the form of synthesized proteins, protein fragments, or other chemical molecules affecting them, diagnostic tests, or the like. Once a specific therapeutic modality or diagnostic approach is defined, clinical tests must be carried out to ensure that it actually works in human subjects. Regulatory hurdles must be cleared before the sequence of discoveries, inventions, and tests can be applied to improve human health and welfare. The multi-stage character of this process is a recurring theme in what follows.

THE PATENT GRANT AND ITS ALTERNATIVES

A fundamental rationale for patent grants is to impede the emergence of competitors to some newly-invented product or process, thereby allowing the inventor to realize, or at least expect to realize, profits that will on average repay the cost of its inventive efforts. The underlying theory is illustrated in Figure 1. Suppose a new product is invented and developed to the point of commercialization. The market demand for the new product is characterized by the solid line in Figure 1 labeled "Initial Demand Curve." The demand curve for any product is a schedule of units that might be purchased in a given time period, i.e., per year, depending upon the price charged, arrayed in descending order of consumers' ability and willingness to pay. Thus, some consumer would be willing to pay at most $19.36 for a single unit (where the demand curve meets the vertical axis); the 100 thousandth unit would find a consumer willing to pay no more than $14; and the 400 thousandth unit a consumer willing to pay just a bit less than $4. Readying a new product for the market also entails working out production methods, which are assumed to entail a cost of $4 per unit produced, shown by the horizontal cost curve marked "Cost per Unit" in Figure 1. At first the firm marketing this new product is the only seller of its product—in economic terms, a monopolist. It can choose the price that maximizes its profits —in Figure 1, $10.95 per unit, which will lead to the purchase of 160,000 units per year. Selling 160,000 units at a price of $10.95 per unit, and with production costs of $4 per unit, the firm will realize an annual profit of $1.112 million, shown as a diagonally shaded rectangular area in Figure 1. If all goes well, this profit will, over a sufficient period of time, repay the innovator's research and development costs plus accumulated interest to compensate for the time lag between the sinking of research and development (R&D) costs and their recoupment.

Even though the new product is priced monopolistically, consumers are also better off as a consequence of its availability. The extent to which they benefit is calculated by the difference between what they would be willing to pay, read as a descending schedule from the demand curve, and the price they must actually pay. That surplus of "willingness to

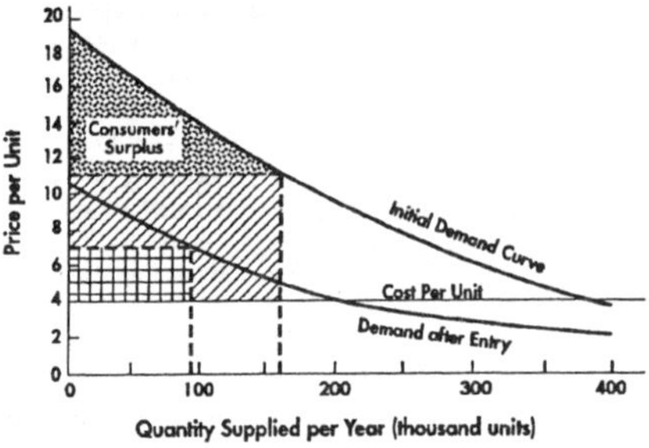

Figure 1. How competitive entry erodes the profits of an innovator.

pay" less the price actually paid is called consumers' surplus. It is shown as the nearly triangular dot-shaded area in Figure 1. As illustrated, it amounts to approximately $640,000 per year. Thus, both inventor-developers and consumers gain from the new product. In the case of life-saving medical products, consumers' willingness to pay may be very high indeed, leading to particularly large consumers' surpluses.

Sooner or later, the new product's profit potential will stimulate other firms to imitate the product and offer their own competing versions. When this happens, the demand curve remaining for the original innovator will shift downward and to the left, e.g., to the solid line labeled "Demand after Entry" in Figure 1. Making the best of its new and less advantageous situation, the innovator will reduce its price to $7.00 and sell 93,000 units per year, realizing a profit of ($7.00 − 4.00) × 93,000 = $279,000, shown by the cross-hatched rectangular area in Figure 1). If the required R&D expenditures are, for example, several million dollars, a reduced profit of this magnitude may be insufficient to recoup them. If rapid erosion of its profits by competitive imitators is foreseen, the would-be inventor–innovator may choose not to make the necessary investment in research and development, and so the product will not be brought into existence. This would be unfortunate, since, absent competitive entry, the combined benefits to participants in the economy, that is, profits plus consumers' surpluses (which amount to at least $1.75 million per year) would be more than sufficient to cover the requisite R&D investments, and so having the new product is economically worthwhile.

The role of patents is to retard competitive imitation sufficiently to let inventors and developers of potentially beneficial new products or processes expect net profits from their R&D investments. Even in the limiting case of an R&D investment so costly that it would not be made unless the innovator were guaranteed a perpetual monopoly of the new product, there is a net economic gain from having the product, since the monopoly profits needed to repay initial R&D costs are augmented by consumers' surplus; thus, total consumer plus producer benefits exceed R&D and production costs.

Alternative Barriers to Competitive Imitation

An important limitation of this rationale for the patent system is that patent protection may generate incentive "overkill," providing more protection from competitive imitation than a would-be innovator needs to justify its R&D investment. This would be so, for example, under the conditions described by Figure 1 if the innovator's head start in marketing its new product ensured that it would be free of competitive imitation for at least a year, and if the R&D costs required to make the innovation totaled (after appropriate risk-adjusted interest cost adjustments) less than the $1.112 million of profits realizable during the first year of new product sales. If the innovation would be made without patent protection, the additional monopoly profits protected by patents impose an unnecessary burden upon consumers, who, in the absence of patents, would, after competitive imitation, enjoy lower prices and larger consumers' surpluses than they would if the new product continued to be protected by patents after its R&D payback period.[2]

A head start is one "natural" barrier to rapid imitation. It is likely to protect the innovator's profits more, the longer it takes for an imitator to duplicate the innovator's product. That in turn depends in part upon how long it takes would-be imitators to recognize that there is a product worth imitating and also upon the duration and cost of R&D or other investment required to market an imitative product. Front-end imitation costs vary widely. To copy software or recorded music, the cost of imitation is close to zero; one needs nothing more than a compact disc burner. To imitate a commercial airliner design, the second mover may have to incur R&D costs close to those of the innovator, replicating detailed design drawings, prototype fabrication, and extensive prototype testing (since small design variations may render the aircraft structurally unsound or aerodynamically unstable).[3] If R&D replication costs are high and the market is too small to yield profits sufficient to cover twice-original R&D costs, imitation may not occur at all in airliner-like innovation cases.

Recognition lags may be especially long, and R&D duplication costs particularly high, for new production processes that can be kept secret from imitators, e.g., by restricting access to production sites. Secrecy during the R&D phase also enhances the innovator's head start advantage for new products, since the imitator is unlikely to begin its own R&D program until the innovator's product has been successfully launched into the market.

The advantage of a head start is magnified for products such as semiconductors and aircraft, whose production cost falls with increased experience as a result of "learning by doing." In the typical case, unit production costs fall by 20–30% with each doubling and redoubling of cumulative output up to substantial output thresholds. For a typical integrated circuit, this means that the producer who has produced 10,000 chips may incur costs of $20 per chip, while the firm with experience producing a million chips incurs a cost of only $2.25.[4] With a head start down the learning curve, the innovator can take advantage of its lower costs to charge modest but still profitable prices that render an imitator's operations unprofitable. The expectation that this could happen might sway a would-be imitator's decision toward not imitating, thereby extending the innovator's head start.

Especially for consumer goods, being the first to offer a new product also confers significant advantages in the form of difficult-to-dislodge brand loyalty and a favorable image that allows the innovator to charge premium prices even when numerous imitative substitutes are available.[5] These "pioneering brand" advantages are seen inter alia in the large price differences typically prevailing between originally branded and generic prescription drugs.

Survey Evidence

Two large-scale surveys have attempted to assess the extent to which patent protection, as compared with other barriers to competitive imitation, is important to would-be innovators hoping to capture the benefits from their technological innovations. They yield generally similar insights.

The first, the so-called Yale University survey by Richard Levin et al.,[6] elicited judgments from 650 U.S. industrial research and development laboratory managers about the relative effectiveness of several means for protecting the competitive advantages from new products and processes. The responses, on a seven-point scale, could range from 1 (not at all effective) to 7 (very effective). The average scores for various mechanisms across 130 industry groups were as follows for *new and improved products* and for *new and improved processes*, respectively:

▪ Patents to prevent duplication	4.33 and 3.52
▪ Patents to secure royalty income	3.75 and 3.31
▪ Secrecy	3.57 and 4.31
▪ Being first with an innovation	5.41 and 5.11
▪ Moving quickly down learning curves	5.09 and 5.02
▪ Superior sales or service efforts	5.59 and 4.55

Only secrecy was deemed less effective than patent protection on average in capturing the benefits from new and improved products; for new processes, patents were considered the least effective of six alternative mechanisms. The average scores varied widely, however, across industries. Among the 65 industry groups in which more than three responses were obtained, "patents to prevent duplication" had the highest average scores in agricultural chemicals (6.88), pharmaceuticals and biologicals (6.55), and industrial organic chemicals (6.05). Indeed, in those industries, patent protection was considered the most effective means of capturing the benefits from product innovations.

An even more extensive survey of 1,478 U.S. R&D laboratory managers was conducted from a Carnegie–Mellon University base by Wesley Cohen et al.[7] The authors asked their respondents to estimate inter alia the percentages of product innovations on which various mechanisms were effective in helping capture the benefits from those innovations. Across the 34 industry groups for which results have been reported, the average scores for alternative means of capturing the benefits from product innovations were as follows:

▪ Secrecy	51.0%
▪ Patents	34.8%
▪ Other legal instruments	20.7%
▪ Lead time	52.8%
▪ Complementary sales and service effort	42.7%
▪ Complementary manufacturing capability	45.6%

Thus, on average, patent protection was considered the second least effective means of capturing profits through product innovation, leading only "other legal instruments" (such as copyright and trademarks). Among the 34 industry groups, the highest average scores for product patent protection were recorded for medical equipment (54.7%) and drugs and related products (50.2%). For drugs, however, secrecy, which received much more emphasis in the Carnegie-Mellon survey than in the Yale survey for reasons that are unclear, had an even higher average score (53.6%) than patent protection, which in turn was followed closely by lead time (50.1%) and complementary manufacturing capability (49.4%). For medical equipment, the average score for drugs was exceeded by the lead time variable (58.1%). For no reported industry did patent protection hold first place among the alternatives.

A less extensive survey asked 100 U.S. industrial R&D executives what fraction of the inventions they developed during 1981 through 1983 would not have been developed had the companies been unable to obtain patent protection.[8] The R&D expenditure-weighted average for 12 industry groups was 14%. By far the largest estimated loss of developments, 60%, was projected by pharmaceutical industry executives.

The Special Role of Patents in Pharmaceuticals

The survey evidence is consistent: patents are not a very important means of capturing the benefits from innovation, and hence a stimulus to R&D investment, in most industries. But the pharmaceutical industry, whose products may be based upon human gene sequence information, is an outlier, attaching especially great importance to product patent protection. There are three probable reasons for this emphasis.

- For one, in pharmaceuticals, as in organic and agricultural chemicals, patent claims tend to define products especially precisely. If one atom is replaced by some other atom in a defined organic molecule, the product may not perform its intended function.

- Second, once a particular molecule is identified as a potentially effective therapeutic medium, it must be carried through expensive clinical trials to prove its safety and efficacy. For new chemical entities first tested in human beings in the United States between 1970 and 1982, the average clinical trial cost, including the cost of molecules dropped in failed trials, was $60 million per approved drug (assuming 1993 general price levels).[9] When clinical trial costs were capitalized at a 9% interest rate to the date at which commercial sales began, the average estimated cost rose to $93 million. For that cohort, pre-clinical discovery costs directed toward finding molecules worth testing were of roughly the same magnitude as capitalized clinical costs when failed efforts are averaged in.[10] A more recent survey of molecules introduced into clinical testing between 1983 and 1994 by main-line pharmaceutical firms yielded an estimate, including capitalized pre-clinical, clinical, and failed-molecule costs, of $802 million per approved drug (assuming price levels for the year 2000).[11] Most of the cost increase relative to prior survey estimates was attributable to more extensive and more costly clinical trials. Compared with earlier surveys, pre-clinical costs were a smaller fraction of total capitalized costs.
- Third, absent patent protection or regulatory barriers to imitation,[12] imitators might spend a very few million dollars on product formulation, process development, and clinical trials (typically on 24 human subjects) required to prove therapeutic equivalence and bring their generic substitutes onto the market in competition with the company that has incurred huge discovery and clinical testing costs. Most of the innovators' costs are incurred to provide information that a drug in fact works and is safe, information on which generic competitors can then take a free ride. Thus, the disparity between the investments of innovators and those of imitators is particularly large in pharmaceuticals—almost as large as when software pirates simply copy the diskettes of an innovator.

It is for these three reasons together that patents are accorded such high importance by pharmaceutical manufacturers. Similar reasons underlie the importance attached to agricultural chemical patents, since new products must be tested extensively for environmental safety before receiving marketing approval.

The emphasis traditional pharmaceutical manufacturers placed on patent protection had important consequences for patterns of drug development during the early 1960s.[13] Up to 1962, academic investigators performing research under grants from the National Institutes of Health (NIH) turned to private-sector pharmaceutical firms to screen thousands of newly-synthesized organic chemical molecules. If those entities showed promise of therapeutic efficacy, which occurred at frequencies above the average for company-synthesized compounds, the pharmaceutical firm secured exclusive rights to the molecule and proceeded with further pre-clinical and, when appropriate, drug development work. Beginning in 1962, however, the Department of Health, Education, and Welfare (HEW) imposed more stringent invention-reporting requirements for academic grant holders and required private-sector firms screening grant-based molecules to certify their acceptance of HEW patent policies, under which the screening company could be deprived of exclusive rights to the screened molecules and possibly also to related molecules developed in-house. When the new policy was adopted, pharmaceutical companies are said to have rejected the revised patent agreements "almost unanimously" and to have withdrawn completely from screening NIH-grant–based molecules.[14] As a consequence, academic investigators were forced to seek other ways of testing their new molecules. The best alternatives were the cancer and malaria drug screening programs operated by the NIH and Walter Reed Hospital, but they covered only a narrow spectrum of possible therapeutic uses. As a result of this stalemate, many new molecules were not screened at all or screened inadequately, making it unlikely that they would lead to new drugs and limiting researchers' publication outlets to those that did not require evidence of therapeutic merit. Collaboration between academic researchers and drug companies is said also to have ebbed. The stalemate was ended in 1968 when the Department of Health, Education, and Welfare amended its patent rules to allow the assignment of exclusive rights in grant-based patents.[15] Legislative support followed with passage of the Bayh–Dole Act in 1980. The history of the episode is a warning, however, that the emergence of new therapeutic entities could be impeded when patent-conscious private-sector firms fear that their investments in clinical testing might lack sufficient intellectual property protection.

Related Evidence

The refusal of pharmaceutical companies to screen molecules resulting from federal grant–supported research during the 1960s was one of the most extreme cases investigated by the Committee on Government Patent Policy on how patent rights affected the further development of inventions made under government contracts or grants. That committee, on which I served as economic adviser, explored on a much broader plane the links between patent rights stemming from government-supported R&D and commercial utilization. At the time of its study, some agencies of the government regularly allowed their contractors to take exclusive rights to contract- or grant-related inventions, while others pursued the opposite policy. From this "natural experiment" it was possible to analyze statistically the relationships between patent rights and commercialization for 1,720 patented inventions made by for-profit government contractors, the majority holding defense and space contracts, for which complete survey information was obtained.[16] Among those 1,720 inventions, 11.8% had been utilized in civilian market applications by the original contractor, 4.3% had not yet been utilized commercially but were expected to be, and 2.4% had been put to civilian market use by one or more licensees. An important contingent variable affecting utilization was whether the contractor had prior experience working in the civilian sector to which the invention pertained. A two-way classification revealed the following commercial utilization percentages:

- For contractors who had prior experience, utilization rates were 23.8% with exclusive rights and 13.8% without.
- For contractors who lacked prior experience, utilization rates were 6.6% with exclusive rights and 2.2% without.

Prior experience was apparently more important than exclusive patent rights in encouraging commercialization. However, both with and without prior experience, commercialization rates were higher with exclusive rights than without them. It was not possible to tell definitively from the evidence in which direction causation ran: to what extent good commercialization prospects led contractors to bargain more vigorously for exclusive patent rights, and to what extent exclusive rights facilitated investments in commercialization. From further analysis of the quantitative data and numerous case studies, I concluded that having exclusive rights on inventions made under private-sector government contracts was more apt to turn the tide toward civilian utilization

- when the firm making the invention was small both absolutely and in relation to the relevant civilian market;
- when the inventing firm had no prior commercial position and had to either break into the field itself or find a licensee;
- when substantial technical development costs and risks remained before commercialization could occur; and
- when the relevant market was small in relation to expected development costs.

The committee's work was a forerunner to passage of the Bayh–Dole and Stevenson–Wydler Acts in 1980.[17] Those acts created presumptions in favor of granting exclusive patent rights on inventions made by government contract and grant recipients and to firms carrying out the commercial development of inventions made in-house by government employees. To achieve administrative simplicity, however, they do not attempt to account for the nuances associated with the four points itemized above.

Small Firms and High-technology Startups

My inference from the committee's research was that patent rights contributed with special force to commercialization of inventions made under government grants or contracts for relatively small and/or newcomer firms. More recently, many inventions originating under government-funded research have been developed for commercial use by new high-technology ventures, typically financed in their early growth stages by organized venture capital intermediaries and then by initial public stock offerings (IPOs). With rare exceptions, such start-up firms are small in their early years. In 1996, venture capital providers disbursed an estimated $663 million to biotechnology firms and in 1997, $966 million.[18] In 1999, $1.24 billion (out of total disbursements to all industries of $50 billion) were disbursed by venture capital firms to 207 biotechnology companies.[19] In 1998, some 327 biotechnology companies had crossed the IPO threshold and had common stock listed on U.S. securities markets.[20] Their market capitalization at the time was $97 billion; their product sales $13.4 billion, and their research and development outlays $9.9 billion, or an astonishing 74% of sales. With such high outlays, aggregate biotech industry losses amounted to $5.1 billion. Indeed, according to *Business Week*, only 13 publicly listed biotech companies reported positive net profits in 1998; 22 were expected to do so by the year 2000.[21]

How important patent rights are in attracting investment to new high-technology ventures generally and, in particular to biotech ventures, is not entirely clear. From 14 case studies of technologies licensed by universities to private firms, usually small, Hsu and Bernstein reported that "many of our case study entrepreneurs, regardless of the size of the particular technology in question, would not have licensed their technologies without an exclusive license. The threat of direct competition in a niche market is usually too daunting for the licensee."[22] However, they identify exceptions, notably, the semiconductor industry, in which patent protection was considered less important. That raising venture capital is facilitated in many fields, and perhaps especially in biotechnology, when exclusive patent rights either exist or can be anticipated, appears to be the prevailing view.

The availability of venture capital has been highly cyclical. Danzon et al. show that when venture funds are forthcoming in only small amounts on capital markets, biotechnology startup firms tend to compensate by entering cooperative arrangements with and receiving funding from traditional pharmaceutical manufacturers.[23] From a survey of 118 U.S. startup companies, Gans et al. found that biotechnology companies entered into alliances—defined as either cooperative research, testing, and/or marketing agreements with larger incumbent firms, or outright mergers—far more frequently, in 55% of the cases, than did startups operating in other technological fields.[24] Sixty-five percent of the surveyed startup companies had at least one U.S. patent; the average number of patents per firm was 7.68. Gans et al. reported furthermore that firms with at least one patent to license or sell were significantly more likely to enter into alliances than firms without patents, holding other variables such as the field of technology and the source of initial financing constant. Thus, patent protection appears to facilitate access to capital not only directly but also through inter-firm alliances.

Against this evidence we must consider a conflicting strand from the Carnegie–Mellon survey of 1,478 industrial R&D executives.[25] As we have seen earlier, patent protection was not considered a very important means of securing the benefits from company product innovations in the average industry, although the pharmaceutical industry, including biopharmaceuticals, was an exception. However, Cohen et al. observe that larger firms found product patents to protect their innovations more effectively on average than did smaller companies. When they eliminated from their sample all responding firms with sales of less than $500 million, patent protection turned out to be the most important means of securing the benefits from product innovations in seven broad industrial groups (including drugs) out of 33. Their tentative rationalization of this result is that the "costs associated with patents, particularly their defense, disproportionately dissuade small firms from availing themselves of patent protection."[26] From this conclusion one might speculate that the relatively high propensity of startup firms possessing patents to enter cooperative relationships with better-established enterprises may stem not only from the need for funds, but also from perceived vulnerability vis à vis the patent portfolios of larger rivals.

For investments to develop new pharmaceutical and biological therapies, it seems clear that having patent protection is in most cases quite important. Without it, startup firms lacking robust internal cash flow are likely to have difficulty financing their research and testing programs. And for more mature enterprises, the cash flows foreseeably available to be invested inter alia in R&D would almost surely be limited, leading to lower R&D investments.[27]

Patents and Basic Biological Research

The emphasis thus far has been on the role of patent rights in encouraging the commercial development of inventions from their early inchoate stages, and in particular, in carrying out the expensive clinical trials necessary to market new therapeutic entities. However, the principal concern in this report is not whether promising molecules should receive patent protection, but whether patent rights should be extended to cover human gene sequences, which may, among other things, provide a pattern for the synthesis of useful molecules, i.e., one or two steps earlier than the drug development stage. Thus, we must explore how patent rights affect incentives to carry out research on gene sequences and their links to subsequent therapeutic entities, and how the existence of genome patents affects investment in those "downstream" entities. We begin with the first of these questions.

A map of the human genome provides scientific knowledge about the endowments of nature. Scientific knowledge is viewed by economists as the quintessential public good, that is, something whose use by one individual does not necessarily reduce the amount available for consumption by others, and whose use by anyone, once it is made known, is difficult to prevent except through special legal institutions such as patents. This view of knowledge as a public good was not original to economists. Two centuries ago, Thomas Jefferson fully anticipated the notion[28]:

> If nature has made any one thing less susceptible than all others of exclusive property, it is the action of the thinking power called an idea, which an individual may exclusively possess as long as he keeps it to himself, but the moment it is divulged, it forces itself into the possession of every one, and the receiver cannot dispossess himself of it. Its peculiar character, too, is that no one possesses the less, because every other possesses the whole of it. He who receives an idea from me, receives instruction himself without lessening mine; as he who lights his taper at mine, receives light without darkening mine. That ideas should freely spread from one to another over the globe, for the moral and mutual instruction of man, and improvement of his condition, seems to have been peculiarly and benevolently designed by nature.

From this, it is reasonable to infer, Jefferson, whose duties as Secretary of State made him the first U.S. patent examiner, would have been skeptical about awarding patents on such basic scientific discoveries as the detailed structure of the human genome.

Jefferson lived, however, in a time when science was mostly a low-cost avocation pursued by amateurs. Now science, especially biological science, is a highly-organized enterprise pursued at considerable expense by both profit-oriented and nonprofit institutions. The presumption against patenting basic information about natural phenomena might be overcome if the prospect of securing exclusive property rights in scientific discoveries for a limited period of time served as a necessary or important incentive to making investments in scientific research, and in particular, if it served to elicit discoveries that would not otherwise be made or to accelerate the pace at which scientific knowledge advances.

Scientific research directed toward advancing knowledge per se has tended to be the province of academic institutions and government laboratories. In 1980, only 2.8% of all research and development performed by U.S. industrial enterprises was reported to be basic scientific research. By 1998, the basic research share of industry's total R&D outlays had risen to 6.4%.[29] In 1980, industrial enterprises performed 13.7% of all reported basic scientific research in the United States; by 1998, their share had risen to 28.4%. Increases in biological research probably played a substantial role in this heightened industry attention to basic research. How large that role was is uncertain because of sparse data. From the available data on pharmaceutical and biotech company[30] R&D expenditures, and assuming that 20% of biotech specialists' R&D outlays were for basic research, I estimate that total industrial basic research expenditures in pharmaceutical and biological therapy areas in 1998 amounted to roughly $4 billion, a not inconsiderable amount. Better estimates will be possible when the National Science Foundation publishes industrial biotech R&D data, sought for the first time in its survey covering calendar year 2001.

Industrial enterprises pursue basic scientific research not only to secure priority in identifying promising product development candidates, but also to maintain capabilities that complement and enrich product development efforts, to attract superior talent, and to open up windows to the outside scientific world through ongoing surveillance of relevant scientific advances and direct contacts with academic investigators. Some unknown but undoubtedly considerable fraction of these motives for performing industrial basic research would exist whether or not scientific discoveries could be patented.

The other main source of support for basic research is government. In 1999, the NIH obligated an estimated $8.63 billion for basic research. Of this total, approximately $1.4 billion was for intramural research, most of the remainder going to support academic research or ($360 million) basic research in industry.[31] In the same year, 40.6% of the NIH research budget is said to have been focused on biology (as distinguished, e.g., from medical sciences and chemistry).[32] Thus, I estimate that the NIH devoted roughly a half billion dollars to intramural basic research on human biology in 1999.

In 1999, federally-funded expenditures on basic and applied research in the biological sciences at U.S. academic institutions totaled $4.23 billion.[33] An additional $1.8 billion of biological research at academic institutions was funded by non-federal sources, including state governments (e.g., from general state university support funds), university overhead, grants from industry, and grants from foundations (which probably did not exceed $30 million).[34] A large fraction of these expenditures undoubtedly entailed basic research, although the exact proportion may be unknowable, since distinctions are difficult to make. It seems clear, however, that in the biological sciences, academic basic research effort has been on the same order of magnitude as my estimate of industrial basic research effort, a marked departure from the more typical preponderance of academic over industrial basic research activities.

In the biological sciences, the fraction of all academic R&D funded by the federal government declined from 72.6% in 1979 to 67.4% in 1986 and 64.3% in 1997.[35] The proportion financed by other sources obviously had to rise to fill the gap, although no detailed breakdown is available. In 1975 grants from private industry to academic institutions lagged far behind three other sources. However, industry funding grew more rapidly than the other sources, at an average rate of 7.8% per year, so by 1998, it was almost as important as state and local governmental funds, contributing 7.2% of total academic R&D support.

For academic investigators working under government research grants, patent rights are relevant not only because they make it possible to license out commercially promising inventions for private-sector development, among other things yielding royalties and other payments shared between inventors and university budgets, but also because they encourage information sharing between academic and industry groups and financial support from industry. Since the Bayh–Dole Act was passed, a host of universities have established technology licensing offices to screen the inventions of university staff and negotiate licenses with outside firms. Between 1980 and 1998, the number of U.S. invention patents awarded per year to American universities rose from 390 to 3,151.[36] A survey of 112 business units licensing university inventions between 1993 and 1997 revealed that firms that had increased their licensing of university inventions during this period doubled their research grants to universities over the same interval.[37] Selecting from a menu of reasons to explain why they had increased their funding of university research, the surveyed companies emphasized what they perceived to be a change in universities' receptiveness to licensing and research agreements. Thus, the relatively rapid growth of industrial financial support for academic R&D during the past two decades is probably due in part to the changes in university patenting practices triggered by the Bayh–Dole Act. In this respect, an indirect link from patent rights to the support of research that yields scientific discoveries, which in turn facilitate commercial developments, is seen. However, since industrial enterprises financed 4.1% of academic R&D outlays in 1980, before the Bayh–Dole Act had its effects, the funding increment attributable as of 1998 to patent policy changes would appear at most to have been on the order of three percentage points.

RESULTS FROM A SURVEY OF GENE SEQUENCE PATENTS

In order to determine the extent and technological coverage of genome patenting, an analysis of recent patent grants was conducted. The U.S. Patent Office's computerized database was searched for all patents issued between September 1, 1998, and June 5, 2001, whose abstracts contained the terms "DNA" plus "sequence," or "nucleic acid" plus "sequence."[38] Altogether, 1,770 such patents were retrieved, 1,199 from the DNA plus "sequence" search and, after elimination of duplicates, 571 from the "nucleic acid" plus "sequence" search.[39]

To keep the analysis within feasible bounds, and because U.S. policy concerns focus primarily on the R&D incentives of domestic actors, the first step in the retrieval procedure was to separate out patents assigned to non-U.S. entities. However, some of those foreign-assignment inventions were made by U.S. inventors. Those for which at least half of the listed inventors were U.S. residents were kept within the domestic sample; those with more than half of the inventors residing abroad were considered foreign. After this division, 620 of the patents were classified to be of foreign origin and 1,150 of U.S. origin. The foreign-origin patents came from a total of 24 nations, with counts for the six leaders as follows:

Japan	151
Canada	78
France	73
Great Britain	64
Germany	49
Denmark	46

An attempt was made to classify the foreign-origin patents according to the types of organizations to which they were assigned. An estimated 65% came from companies, 18.3% from free-standing laboratories (usually government-owned and/or nonprofit), and 8.1% from universities. The university share, we shall see, is much lower than the comparable share for U.S. universities, no doubt reflecting the differential impact of the Bayh–Dole Act on U.S. universities' propensity to seek patents.

For the U.S.-origin patents, a much richer three-way classification was attempted. From inspection of the published patents, each patent was placed within one of several categories: those pertaining to the human genome, to animal genomes, plant genomes, or microorganism genomes (e.g., viruses, bacteria, and fungi); those that were methods of sequencing DNA and related genetic material; plus an "other" category that encompassed mainly uses of oligonucleotides and DNA or RNA primers (i.e., short nucleotide sequences). For patents that pertained directly or indirectly to humans, further distinctions were made among DNA sequence claims with or without expressed proteins; preventive inventions (e.g., pertaining to vaccines); diagnostic inventions; gene therapy inventions; other human treatment inventions; and a catch-all category. Since many patent specifications anticipated multiple uses, multiple codings were frequently made. From the patent assignee names, patents were classified as to whether they were assigned to a company, a university, a hospital or nonprofit research institution; or a federal, state, or local government agency; or were retained by the inventors themselves (presumably because an employer chose not to exercise its assignment rights). Within the company category, further breakdowns utilizing standard business reference works were made among *Fortune 1000* corporations, other companies large enough to be covered in *Moody's Industrials*, (mostly) smaller companies whose stocks were traded over-the-counter, companies with no known publicly-traded stocks, and foreign companies (usually, since most or all of the inventors were U.S. residents, for inventions made by their U.S. branches). Because the sample began with patents issued in late 1998, the company type distinctions were made to reflect company status as of early 1999.

Figure 2 shows the distribution of sample patents by broad use categories. Among the 1,150 U.S.-origin patents, 414 were coded as primarily for human use. The second-most-frequent-use category comprised methods of analyzing and sequencing DNA, i.e., research tools. Only two of these were found to have direct applicability to human beings. To be sure, better research tools have important indirect implications for human welfare; no attempt was made to speculate on them in the codings. Human-use implications were more frequent, but not ubiquitous, with patents on microorganisms. Other microorganism uses include industrial fermentation processes. Within the microorganism category, patents

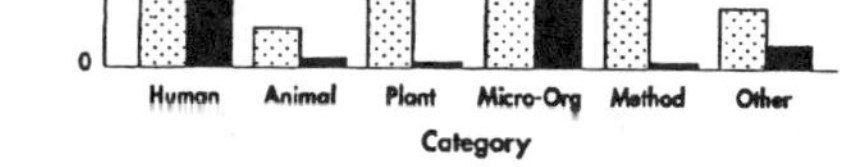

Figure 2. Distribution of patents originating in the United States, by broad categories.

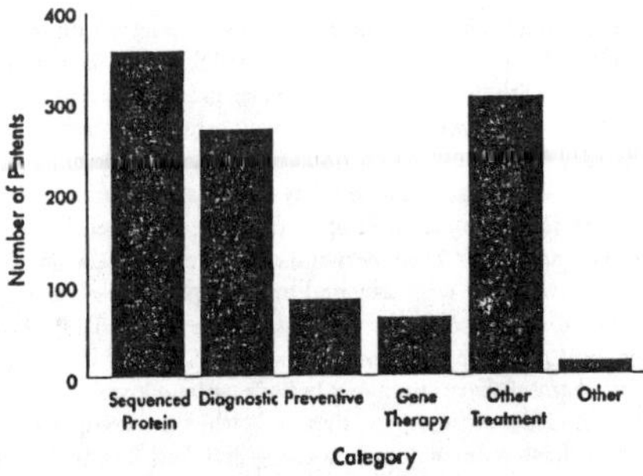

Figure 3. Distribution of human-use patents by specific use.

pertaining to virus structure had the largest fraction (60%) of anticipated human uses.

Altogether, 528 patents, or slightly less than half of the total U.S.-origin sample, were coded as having potential human uses. Figure 3 displays how these uses were distributed. Since multiple coding was allowed, 1,116 human-use applications were coded altogether for the 528 patents.

The cohort of most direct relevance to the question of whether genome sequence patents should be granted is that pertaining directly to human gene sequences, in nearly all cases, with the proteins the sequences were believed to express included as part of the patent specification. Three hundred sixty-three patents, or a bit less than a third of all sampled U.S.-origin patents, claimed human genome sequences. Among these, only six came anywhere near claiming a DNA sequence without additional expressed protein or other utility claims, but none was without some assertion of potential therapeutic or other utility. For the 363 human DNA sequence patents, 255 were coded as also having treatment applications, 212 as having diagnostic applications, 40 as having application in gene therapy, and 24 for preventive applications. Sixty-eight claimed human genome sequences, usually with expressed proteins, without any additional explicitly coded human uses.

The other most frequently claimed uses of the inventions were for treating illnesses, both through gene therapy (65 patents) and by other means, and for diagnosing genetic defects or other maladies. Preventive applications, such as the development of vaccines, were cited in a total of 84 cases.

Figure 4 shows the shares of patents originating from the various kinds of organizations to which the sample patents were assigned. The R&D advantages traditionally attributed to "big business," including "Big Pharma," are not in evidence for our sample. Companies listed among the *Fortune* 1000 for the year 1998 and/or included among the large companies covered by *Moody's Industrials* originated relatively small shares of the sample patents. The principal exception was for plant DNA patents, in which Monsanto plays a leading role. Most of the foreign companies receiving patents within the U.S.-origin group were also relatively large, or at least, large enough to maintain laboratories in the United States. Their greatest relative strength also materialized in plant DNA patents (e.g., Novartis of Switzerland), followed by patents on microorganisms (in which Novo Nordisk of Denmark is a world leader).

Universities led in the receipt of human-use patents (the

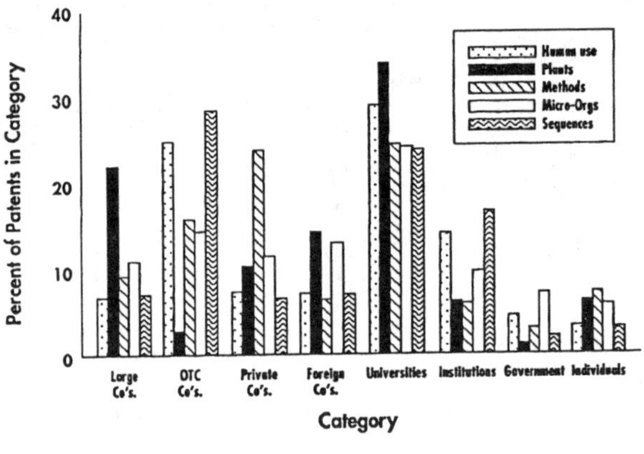

Figure 4. Organizational shares of patenting activity, by category.

first bar in each organizational cluster), which includes some patents double-counted under other categories of Figure 4, and indeed in all categories but one. In second place as originators of human-use patents, with 137 patents, were companies too small to be listed in the *Fortune* 1000 but well enough established to have their shares traded on the NASDAQ OTC market.

The OTC companies were the most prolific originators of patents that included human DNA sequence claims, with 108 such patents. The second rank among DNA sequence patents was held by universities (91 patents); third rank by hospitals and nonprofit research institutions (63 patents), fourth place by the U.S. affiliates of foreign corporations (29 patents), fifth place by large U.S. corporations (26 patents), and sixth place by private companies without publicly traded securities (24 patents).[40] The private-company group appears to have a noteworthy comparative advantage in generating DNA sequencing method, i.e., research tool, inventions.

The following list extracts from the organizational data in a different way, listing the leading eight corporate patent assignees in descending order of the total number of U.S.-origin sample patents received. The list presents both total patents and human-use patents.

Incyte Pharmaceuticals:	58 total, 56 human use
Human Genome Sciences:	35 total, 35 human use
Novartis (Switzerland):	17 total, 0 human use
Monsanto:	16 total, 0 human use
Novo Nordisk (Denmark):	15 total, 1 human use
Genencor:	15 total, 0 human use
Bayer (Germany):	13 total, 13 human use
Roche (Switzerland):	13 total, 3 human use

The largest number of patents went to Incyte Pharmaceuticals, traded over-the-counter, whose Worldwide Web home page claimed that the company is a leading provider of genomics information. Incyte was also said to be leveraging its database to make significant therapeutic discoveries. In second position was Human Genome Sciences, also traded over-the-counter, whose home page states that the company intends to commercialize novel compounds for treating human diseases based upon the identification and study of genes.[41] The posture of OTC-listed Genencor, with no human-use patents, seemed puzzling at first glance. However, its home page explains that the company's main products are industrial enzymes used, e.g., to reduce cellulose and grains, even though Genencor also works on human immunology problems. Although Hoffmann–LaRoche is best known as a leading pharmaceutical and vitamin supplier, its DNA patents (not including those assigned to its partly-owned Genentech subsidiary) were distributed across a broad array of functional categories.

The prospect of obtaining patent protection is probably most important to the financing and R&D investment incentives of small companies without publicly traded securities. Next most sensitive in this respect are likely to be the companies with securities traded on OTC markets. Here, however, a distinction should be made, which could not be made in the organization type codings, between fledgling companies with no commercial products providing internal cash flow and those (such as Chiron and Genzyme) with well-established product sales.[42] From the evidence at hand, it would appear that at most about 36% of the DNA sequence patents, or 132 patents in total, and 34% of the human-use patents defined more broadly, originated from these especially patent-sensitive organizations. Since some of the OTC companies have strong internal cash flows and interest in ensuring their survival through vigorous R&D programs, the realm of high patent dependence is probably smaller than these estimates.

To be sure, some research and hence patents from universities and not-for-profit institutes, which together dominate the surveyed patent categories, may depend upon R&D grants from industry, which in turn may be conditioned by the expectation of patent licenses, often exclusive. On this the direct evidence from our survey is sparse. Fifty-two of the patents in our U.S.-origin sample had multiple assignees, which means that the underlying research was jointly conducted or that a second assignee provided important financing. Among the 37 jointly-assigned patents in which a university was one assignee, the other assignees were private domestic companies in 14 cases, foreign companies in one case, other universities in eight cases, hospitals or free-standing research institutions in 14 cases, and government agencies in two cases. Among the 19 jointly-assigned patents with a hospital or research institute as one assignee, private companies were a second assignee in two cases. Thus, the cohort of not-for-profit institutional patents with direct private company co-assignment is small. It is certainly much smaller than the domain of all university and hospital R&D financed by industry, since split patent rights assignment is neither necessary for, nor customary in, the financing arrangements linking not-for-profit and private-sector industrial organizations.

One further element of perspective must be added. As we have seen, our survey covering 33 months of patent issues yielded 1,770 patents with the required combination of "DNA" or "nucleic acid" plus "sequence" terms in their abstracts. This, of course, is a subset of all patents resulting from biological research and development. It is difficult to tell how large the relevant patent universe is. Statistics from the U.S. Patent and Trademark Office reveal that in calendar years 1999 and 2000, 10,124 patents were issued in the four patent classes most closely corresponding to our sample:

class 435 (molecular biology and microbiology), 436 (immunological testing), 504 (plant protecting and regulating compositions), and 530 (natural resins, peptides, proteins, lignins, and reaction products thereof).[43] Excluded from this count are the large number of patents in traditional pharmaceutical classes, including class 424 (bio-affecting and body-treating compositions, with 5,844 patents). Within our sample, 1,185 patents were issued during the comparable years 1999 and 2000. Thus, our sample encompasses at most only one tenth of the biotech patent universe, and our sample of DNA sequence patents, one thirtieth of the relevant universe. Assuming no differences in the propensity to patent per million dollars of R&D, a point on which little information exists,[44] the research and development from which our sample of patents stemmed must also have been a relatively small fraction of all the R&D, academic and industrial, carried out in the biological disciplines. Clearly, as debate proceeds over how much patent protection human DNA sequences should receive, it must be recognized that it directly affects only a modest fraction of all research and development activity in biology.

INSIGHTS FROM ECONOMIC THEORY

During the past three decades numerous articles and books have been published on the economics of patent policy. Many ask how patent grants might be fine-tuned under specific conditions to secure the maximum surplus of economic benefits from innovation, after deducting production, R&D, and legal costs. In practice, there is far too much uncertainty, and the institutional apparatus is far too clumsy, to follow the fine-tuning guidance provided by this theoretical work. Nevertheless, some useful but crude insights can be extracted.

The pioneering contribution was by William Nordhaus, now at Yale University.[45] Nordhaus asked how, assuming that innovators maximize the discounted present value of their expected profits from process (i.e., cost-reducing) innovations, government might set a patent life that maximized the surplus of *social* value (i.e., the combined surpluses realized by consumers and producers) less R&D costs. He found that the optimal patent life varied widely from one to 34 years, depending upon how easily cost reductions flowed from R&D efforts and the price elasticity of demand for the product whose production process was being improved. The "easier" it was to achieve cost reductions and the more price-elastic product demand was, the shorter the optimal patent life.

Prospect versus Rent-seeking Models

Nordhaus' initial theoretical model has been extended to deal with questions of optimal patent breadth (or scope), compulsory licensing regimes, and much else. The most important extension for our present purposes was the questioning of Nordhaus' implicit assumption that inventors had what Edmund Kitch later called exclusive "prospect" rights to undertake their inventions.[46] That is, Nordhaus assumed that the firm conducting R&D on a particular problem was the only such firm working on the problem, and it could therefore choose the intensity of R&D effort that maximized the time-discounted surplus of what economists call quasi-rents (i.e., the surplus left over after production and marketing costs are subtracted from product sales) over alternative levels of R&D investment. This implied that the firm in question did not have to worry about other firms conducting parallel R&D and possibly preempting its invention.

An alternative view of the R&D world was first proposed by Yoram Barzel.[47] He argued that inventors had to compete with other inventors for priority and hence temporary monopoly profits from the relevant invention. Invoking an assumption from the theory of competition in price-setting, this inter-firm competition would drive R&D costs upward until the aggregate net profit from inventive activity, netting the probability-weighted costs of the losers against the gains of the winner, would on average be zero. This perspective falls under a larger set of contributions known as the theory of rent-seeking. However, the standard view of rent-seeking is unfavorable, e.g., visualizing lobbyists increasing their costly efforts to secure a particular legislative outcome until the net gains are zero. This view, I shall argue in a moment, is too pessimistic, and so one might characterize Barzel-type R&D escalation as virtuous rent-seeking.

Early applications of the Barzel view to the Nordhaus theory assumed to the contrary that rivalrous and pejoratively duplicative R&D activity was purely wasteful.[48] From this perspective, the profit (or more precisely, quasi-rent) gains from extending patent lives and spurring more expenditure on R&D were mostly dissipated as costs, and with smaller net gains, much shorter patent lives—e.g., in the range of a half year to one year—were warranted.

It is almost surely wrong, however, to view the competitive escalation of R&D as intrinsically wasteful. Especially at the early concept-forming stages of technological research, there is likely to be substantial uncertainty about the approach that will both yield technical success and best meet consumer demands.[49] The pursuit of multiple research paths increases the likelihood that good solutions will be found, and for both statistical and behavioral rivalry reasons, it accelerates the achievement of an acceptable solution. The greater the payoff from finding a solution, the larger is the optimal number of parallel paths, especially when ex ante uncertainties are substantial.[50] Thus, in the top-priority U.S. atomic bomb development program during World War II, five costly parallel paths to the separation of fissionable materials and two approaches to bomb design were pursued. Progress toward therapy for AIDS patients was undoubtedly accelerated by the large number of researchers attracted to the problem because of its enormous social importance. Also, since different researchers are likely to pursue qualitatively different approaches to a problem, competition among groups makes it likely that superior solutions will be found, or in the case where numerous solutions have merit, that the diversity of consumer wants will be more completely satisfied.[51] The large number of molecules with which anti-AIDS drugs can be formulated provides an example.

An Aggregate Perspective

Figure 5 provides a schematic illustration of the two polar theoretical models, their implications, and the implications of weaker or stronger intellectual property regimes. The heavy solid line labelled *C(RD) Function* is what William Nordhaus originally called an invention possibilities function. It shows how the number of new biological entities (NBEs) carried successfully through clinical testing depends upon the amount or cost of R&D done. With no warranty that the parameters chosen are correct, it is drawn to exhibit diminishing returns in the R&D–NBE relationship, at first weakly and then strongly. Diminishing returns set in because at any moment in time the progress of science offers a limited menu of promising therapeutic strategies, and as the level of R&D effort is increased, there will be more near-duplication of others' efforts and the pursuit of less promising technical possibilities, in both instances leading to a smaller incremental new product yield.

Three alternative quasi-rent regimes are illustrated, one (Q_1) assuming relatively weak intellectual property laws, another (Q_2) patents of longer duration and broader scope, and the third (Q_3) an extension of the intellectual property regime from the developed nations to the entire world, as under the TRIPS section of the Uruguay Round treaty. All three are also subject to diminishing marginal returns, since, despite huge uncertainties, profit-seeking enterprises will try to target first the biological entities with relatively large prospective payoffs and address lower-payoff projects only with relatively generous R&D budgets.

Under a Nordhaus or "prospect" view of the world, firms choose projects without competitive challenge so as to maximize their net profits, shown as the horizontal distance between the R&D cost function C(RD) and the relevant quasi-rent function (the dot-dot-dashed lines). With low-payoff quasi-rent function Q_1, the profit-maximizing equilibrium (at point N_1) entails R&D spending of roughly $1.5 billion, leading to seven new products per year and quasi-rents of $2.6 billion, with net discounted profits of $1.1 billion. As potential quasi-rents rise to Q_2 and then to Q_3,

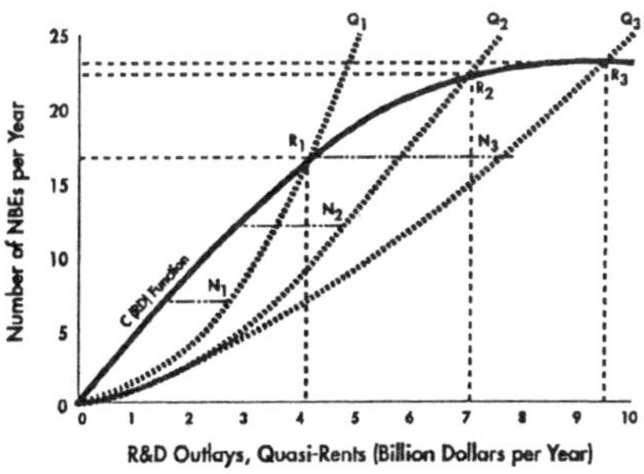

Figure 5. Illustration of how stronger patent rights affect NBEs (new biological entities) discovery. The horizontal axis measures two variables, both scaled in billions of dollars per year-the amount spent on biopharmaceutical research and development (R&D) and the discounted quasi-rents (i.e., therapeutic agent sales less costs of production and marketing) resulting from that R&D. The vertical axis measures the number of NBEs entering the market per year as a result of R&D.

R&D spending rises to $2.75 billion and then to $4.1 billion, leading to 13 and 17 new products per year at equilibrium points N_2 and N_3, respectively, with net profits of $2.0 billion and $3.4 billion.

R&D competitive in the Barzel sense leads to substantially more R&D, indeed up to the point where quasi-rent and R&D cost functions intersect so that their net value is zero, at equilibria R_1, R_2, and R_3. More new products will therefore be forthcoming on average, but with diminishing returns —e.g., 17 per year in the relatively weak intellectual property regime Q_1, 22.5 in regime Q_2, and one additional product in the regime Q_3 with the most wide-ranging scheme of intellectual property rights.[52]

Nearly-simultaneous Discovery

Between these two polar extremes, the rent-seeking model is almost surely the more realistic. An implication deserves repeated emphasis. Especially in those areas of potential therapy with high possible payoffs—i.e., where anticipated quasi-rents are well in excess of expected R&D costs—there is likely to be a good deal of what appears to be simultaneous or near-simultaneous discovery and a considerable number of NBEs meeting similar but not identical therapeutic needs. Indeed, the histories of science and technology are replete with additional chronicles of basic discoveries and technological inventions made more or less simultaneously—so much so that the distinguished sociologist Charles Ogburn suggested that some basic inventions might be "inevitable."[53] The work that led to the discovery of nuclear fission by

Hahn and Strassmann was being pursued along similar tracks in numerous research centers, leading physicist Robert Wilson to remark that "nuclear energy was a time bomb set for the human race, and eventually the human race would have had to reckon with it."[54] James Watson and Francis Crick correctly perceived themselves to be in a race to discover the structure of DNA with Maurice Wilkins and Rosalind Franklin at the University of London and Linus Pauling at California Institute of Technology.[55]

To conclude from this "softly deterministic"[56] view of the discovery process that *most* discoveries will be made fairly quickly by someone else if the credited discoverer were absent from the scene goes too far. Even in modern times, the genius of a Newton or an Einstein could make a difference of decades in at least some cases. But it is fair to say that the greater the salience attached to a problem by the scientific or technological community, determined in part by the importance of the problem and in part by the agenda-setting effect of prior knowledge advances, the more likely it is that a solution will emerge from one group or another within a few years. Exceptions include those challenges, such as counting the flux of neutrinos, exploring Mars, or harnessing the power of hydrogen fusion, requiring such enormous resources that at most one or a very few teams can work simultaneously, and then only if the climate of political and hence financial support is favorable. I consciously exclude the case of sequencing the human genome, on which the presence of high-stakes competition appears to have advanced the inevitable by at most a few years.

Sequential Invention and Patents

Mapping human genome sequences is important in the worlds of medicine and commerce because it helps point the way toward useful inventions—the synthesis of therapeutic molecules or the repair of existing molecules, probes to identify and perhaps correct genetically-based diseases, etc. The chain of discovery might also be reversed, e.g., from the identification of a genetically-based disease, one can pinpoint the genome sequences shared by persons with that disease and hence target the search for corrective therapies more narrowly. In either case, later discoveries and inventions, to paraphrase Newton, build upon the shoulders of prior discoveries. Scientific and technological advances are temporally linked.

The relationship between discoveries from basic research and the development of commercialized follow-on inventions is particularly close in biology. Surveying 108 business enterprises, Edwin Mansfield asked research laboratory heads what percentage of their new products introduced between 1986 and 1995 "could not have been developed (without substantial delay) in the absence of recent academic research."[57] For all surveyed industries, 15% of the new products were related in this strong way to prior academic research. In drugs and medical products, the proportion was 31%, the highest for any of seven reported industry groups. An additional 13% of commercialized drug and medical products were said to have received "substantial aid" from recent academic research.

A schematic view of these dynamic linkages is provided by Figure 6. Consider a key precipitating discovery, A, emerging from either academic or industrial basic research efforts. A itself undoubtedly builds upon the shoulders of predecessor discoveries, indicated by the dotted arrows to the left of A. But what is of most immediate interest is the impact A has on the subsequent discovery and/or development of B_i (and other products $B_1 \ldots B_N$, which are ignored here). How A affects B can vary over a two-dimensional spectrum: at the extreme, B may be impossible without A, or in less extreme variant cases, the existence of A can speed the development of B or lessen the cost of developing B, or both. B in turn sets the stage for further improvement developments, shown by the dotted arrow to the right of B_1 but these too are ignored here.[58]

The crucial question is, How do varying degrees of patent protection affect incentives for discovery and development at the sequential stages of this model?[59] The problem begins with recognition that A confers upon the developers and users of B (and subsequent inventions) what economists call "external benefits," that is, benefits that would not be realized in a regime of no intellectual property rights by an entity whose work gave rise to A but not subsequent innovations. In other words, the benefits are realized externally from whatever commercialization A achieves. If A is indispensable to the development of B, a correct assessment of the benefits from A includes not only any net benefits from the sale of products based directly upon A, but also the total surplus of consumer plus producer benefits less R&D and production costs from subsequent innovations such as B. If A merely accelerates the development of B or reduces the cost of developing B, the external benefits from A include the value to consumers of having B earlier and/or the (ap-

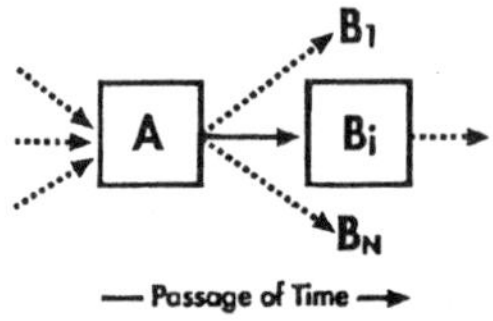

Figure 6. Time phasing of interdependent sequential interventions.

propriately discounted) reduction of B's underlying R&D cost.

A fundamental proposition of economics is that in the presence of such "externalities," markets may fail to send the signals needed for a proper allocation of resources. An extreme case arises when, as is not uncommon in scientific discovery for the reasons articulated by Thomas Jefferson, A has no commercial value in its own right, but is valuable only because of the inducement it provides to the development of B. Or in an only slightly milder case, the commercial exploitation of A alone yields profits (or more accurately, quasi-rents) too small to warrant the investment needed to discover A, but which, when added to the external benefits from B, more than justify that investment.

Instruments used by government to help solve this problem include research and development subsidies and the awarding of intellectual property rights. A principal rationale of the research grants system is that scientific discoveries confer substantial social benefits that cannot be "internalized" by the persons performing the desired research, which therefore has to be supported by an entity—the government or a philanthropic institution—that can take a broader society-wide view. As we have seen, roughly half of recent U.S. basic research in biology has been supported in this way. An alternative is to bestow upon the entity investing in the discovery of A intellectual property rights of sufficient breadth and duration to let the rights holder extract enough of the benefits from B's commercialization to repay or, to compensate for uncertainty, perhaps more than repay, its investment in A. These rights could be minimal, e.g., a natural head start toward developing B protected by temporary secrecy or non-transferable know-how from work on A that gives A's research team an advantage over others in the development of B.[60] Or they might be embodied in patents whose scope and duration cover improvement inventions such as B along with the immediate subject matter of A. It is here that the controversy over human gene sequence patents reaches its most acute focus.

One possible rights-based solution is to give the discoverer of A sufficiently broad and enduring rights totally to exclude others from commercializing derivative inventions such as B. Under the "prospect" theory of patent rights, the organization discovering A would have exclusive rights to "prospect" for B-type inventions, often with the further explicit or implicit assumption that by virtue of having discovered A, that entity is most competent to do the follow-on work. This view is almost surely wrong. For one, the kinds of competence needed for follow-on work may be quite different from what was needed to make the initial discovery. The different capabilities of university researchers as compared with industrial R&D teams are an obvious example. Biotech startup firms are relatively strong in basic research and discovery activities, but commonly lack clinical testing and (especially) marketing skills. This imbalance leads them to license their patents to, cooperate with, or sell out to well-established pharmaceutical enterprises. Also, the directions in which a discovery may be applied are often myriad, and a single entity is not likely to perceive and back financially all the various derivative development possibilities.[61] It is more the norm than the exception in the history of technology for the firms introducing significant derivatives of and improvements upon a basic discovery to be other than the original discoverer.[62] To ensure that all valuable commercial derivatives of a discovery are exploited, there is no better maxim than that of Mao Tse-tung, "Let one hundred flowers bloom."

A solution to this problem is to confer patent rights that cover follow-on B-type developments, but to allow or even (through governmental policy) encourage the licensing of rights to utilize the science or technology underlying A to other entities wishing to build upon A's shoulders.[63] Licenses might be non-exclusive, as in the licensing of the basic Cohen–Boyer gene splicing patents by Stanford University to several hundred users,[64] or exclusive, as might be necessary when development of a therapeutic molecule—requiring clinical testing outlays measured in the tens of millions of dollars—follows directly from a scientific discovery. Here too, however, bad things can happen. Determining the division of rents between the original discoverer and follow-on developers requires bargaining, and solutions may materialize that either stalemate further progress or undermine incentives for additional private investment in basic discovery.[65] Bargaining stalemates are especially likely when the discoverer of A has broad rights covering follow-on developments, but when A, like many basic scientific discoveries, has little or no commercial value by itself.[66] Then the follow-on developer, which may have to invest substantial sums in development and testing, may insist upon the lion's share of the economic rents attributable to its undertaking, while the discover of A, with the power to block development completely, demands an incompatibly large share. Merges and Nelson document numerous important cases in which such patent stalemates significantly delayed, or without permissive regulatory interventions could have blocked, desirable technological advances.[67] If the balance of power lies too much on the side of the stage-A discoverer and too little on stage-B developers, there could be too little investment in practical implementations $B_1 \ldots B_N$ and too much (at least in relative terms) in early-stage research. We address in a subsequent section the kinds of government policy interventions that might minimize the resource allocation distortions resulting from such problems.

The A → B schema, however, may present too simplified a picture of the relationships between early-stage discovery and commercialization of the derivative inventions. A National Science Foundation–backed study of the scientific and technological "events" that led to five new technologies, including the first oral contraceptive pill, demonstrated the large number of research streams that had to converge to yield ultimate practical embodiments.[68] For "the pill," approved in 1960 by the Food and Drug Administration, many of the antecedent events occurred between 1940 and 1960, and if they had received strong patent protection, the patent assignees might have been in a position to block or levy a heavy toll on the final development effort. The compression of time spans between scientific advance and commercialization since then means that even more potentially blocking events might occur, each with 20-year patent rights extending into the period of feasible commercial innovation for biotechnological advances.

When numerous patented technologies are precursors to an innovation, each seeking to extract substantial royalties from derivative efforts, the combined demands could accumulate to a level at which investments in commercialization or next-step research are seriously impeded.[69] The problem is analogous to conditions on the Rhine River during the 18th Century. Over the 85-kilometer stretch between Mainz and Koblenz in 1780, there were nine toll stations, and from there to the Dutch border, there were 16 more.[70] Each acted as a partial monopolist over the rights of traffic to travel the Rhine by its fortified *Raubritter* (robber baron) castle, seeking to extract the profit-maximizing toll from passers-by. The multiplication of tolls suppressed most of the traffic that otherwise would have traversed that artery—absent tolls, the least costly means of travel in an era of hopelessly bad roads—and thereby impeded German economic development. Not until 1831 were treaties concluded that allowed essentially free navigation and hence full development of the waterway's potential.

One way of breaking such impasses is for the participants in an industry to enter into cross-licensing accords under which each patent holder contributes its patents to a generally available pool, with or without modest compensation, on the understanding that all others will behave similarly. This was done at government insistence in the early years of the aircraft and semiconductor industries.[71] More recently in semiconductors, cross-licenses have again been the solution to a logjam comprising thousands of patents, any of which might emerge to challenge or even enjoin a firm's right to produce a complex microprocessor or application-specific integrated circuit.[72] In biotechnology, the asymmetry of relevant actors' positions—ranging from university scientists through genome-researching firms, vector providers, and instrumentation makers to specific biopharmaceutical developers—is likely to make it more difficult to find a sufficient community of interest to organize comprehensive low-royalty cross-licensing.[73] Even in semiconductors, market power and utilization asymmetries between general-purpose semiconductor producers, microprocessor specialists, and computer makers led to bitter and expensive litigation, resolved in part through Federal Trade Commission intervention.[74] Thus, there is a genuine danger that strong and extensively overlapping patent positions could impede the progress of therapies based upon molecular biology.

Policy Options

Various alternative policies might be pursued toward the patenting of human gene sequences. Thus far, it would appear, the U.S. Patent and Trademark Office has not been issuing patents for human DNA sequences without additional claims showing their utility, e.g., through identification of the proteins they express or the medical treatments they enable. It is doubtful that allowing the patenting of DNA sequences without additional utility proofs would be beneficial to anyone other than persons who have such applications pending. Preliminary sequencing of the entire human genome has been completed, partly through a government-sponsored effort and partially at private initiative. Offering patent rights cannot motivate what has already been done; it can only confer windfalls for past investments made without the clear expectation that patent rights could be secured. Unless errors are found in the two existing sequence sets, any non-pathological sequences discovered in the future are likely to be barred from patentability by the existence of prior art. Therefore, changes toward more permissive patent standards are unlikely to induce positive effects, but could cause damage by impeding further research on links between gene sequences and medical treatment modalities.

Under the policies that have been adopted by the Patent Office, thousands of patents have been issued on human genome sequences and their claimed links to proteins, diseases based upon genetic defects, diagnostic probes, and the like. It is unknown how well founded the claims of utility are. Not allowing such patents in the future would discourage some research supported by private-sector investment. How much discouragement there would be is highly uncertain. As we have seen, the kinds of patents that did satisfy Patent Office criteria between 1998 and 2001 were only a small subset of all patents generated by industry and academic research in the biological fields—almost surely, after pharmaceutical patents are counted, less than 5–10% of the universe. At most half of the underlying research within that subset was by institutions whose efforts depend significantly upon patent protection; the rest was financed by government grants and/or motivated by well-financed companies' need to remain competitive by staying abreast of the relevant science and introducing new products. And if some research were discouraged by policies that lean on the side of withholding basic genome sequence rights, it would not necessarily mean the perpetual loss of life-saving therapies. It is more likely that the seminal inventions would be delayed, not lost forever. To illustrate hypothetically, assuming a discount rate of 7%, a five-year delay in the availability of a key invention implies a social loss on the order of 30% relative to the discounted benefits that would be realized if the invention were available immediately.[75]

To ensure that promising future lines of research are not impeded, existing genome patent claims and any claims allowed in the future ought to be interpreted narrowly. In ruling whether basic patents have been infringed, the courts sometimes apply a doctrine of equivalents that permits a broad reading of the scope of the subject patents with respect to follow-on inventions,[76] and thus allows the original patent to extract royalties from, or even to enjoin, covered follow-on inventions. Alternatively, when a follow-on invention entails a substantial advance in the state of the art, the courts have the option of applying a reverse equivalents doctrine, construing earlier patents narrowly and hence finding no infringement. The technological achievements that will be built upon knowledge of the human DNA structure are likely to be so important to human welfare that it would appear preferable to interpret existing and future utility-based claims narrowly. For instance, there is reason to believe that a single DNA sequence, perhaps in combination with other sequences, will define more than one protein. If a patent claims a DNA sequence and some protein with proven or suspected functions in the human body plus "all analogous proteins that might be expressed," the benefits from ensuring that all new therapeutic opportunities are thoroughly explored would argue for a narrow interpretation of claims and the exclusion of speculative "analogous" claims. Construing claims broadly could block important avenues of research or appreciably increase their cost and risk.

Because it is vital to keep future research possibilities open to the maximum possible degree, legal precedents that exempt from injunction and the payment of royalties the use of patented technologies solely for research purposes are highly desirable. Their continuation apparently stands in jeopardy as a result of new appellate court interpretations.[77] The research exemption is an imperfect policy because of its asymmetry: the use of patented concepts or methods in research may carry no royalty burden, but patented research instruments, vectors, and the like are purchased at prices elevated by whatever monopoly power patent grants confer. And as our survey has revealed, a considerable fraction of the patents relevant to genome sequencing cover research tools. However, as has long been recognized, perfect intellectual property policies are unattainable. When research that has been exempted from patent claims leads to commercialized products whose production requires the use of others' patented technology, those patent holders will presumably be able to reach through and collect their toll on product sales.

Even when patent claims are interpreted to bind upon new and improved products developed by entities other than the patent owner, it is important to ensure that new product development is not blocked or severely retarded by a multiplicity of prior patent claims. Developing new products that enhance human health and welfare should not be like walking through a mine field, with risk of severe consequences should a loosely-related patent claim be infringed, or like cruising along the Rhine River in the 1770s, with toll collectors every few kilometers. Means of breaking possible bargaining stalemates should be in place. One possible policy, consistent with the Federal Trade Commission's settlement of the *Intel* case, would be to prohibit outright injunctions against drugs that have been accepted by the Food and Drug Administration for clinical testing and/or approved for marketing to humans. Rather, all disputes would be settled with a determination of damages, if infringement were found to have occurred. A stronger and preferable policy would be to require that disputes threatening the availability of new therapeutic modalities be settled through mandatory arbitration, with the standard of damages for infringement being "reasonable royalties" rather than the "lost profits" standard applied in infringement cases.[78] Therapeutic innovators found to be infringing background technology patents could be required to provide a cross-license to their own product improvement patents in exchange for a royalty-bearing license to the earlier background technology. This approach would strengthen the rewards to original patent holders while facilitating the maximum diffusion of biological technology for medical purposes.

This work was supported by a grant from the Alfred P. Sloan Foundation.

ENDNOTES

[1]"Profiting from the Human Genome," *Harvard Magazine*, July–August 2001, p. 70.

[2]In the limiting case of purely competitive pricing, with price equal to the $4 per-unit cost, consumers' surplus would be approximately $2.4 million per year, compared with $640,000 under monopoly pricing.

[3]See Edwin Mansfield et al., "Imitation Costs and Patents: An Empirical Study," *Economic Journal*, vol. 91 (December 1981), pp. 907–918.

[4]See F. M. Scherer, *Industry Structure, Strategy, and Public Policy* (New York: HarperCollins, 1996), p. 211.

[5]The original research pinpointing this phenomenon was Ronald Bond and David Lean, *Sales, Promotion, and Product Differentiation in Two Prescription Drug Markets*, Federal Trade Commission staff report (Washington: February 1977); and Robert D. Buzzell and Paul W. Farris, "Marketing Costs in Consumer Goods Industries," in Hans Thorelli, ed., *Strategy + Structure = Performance* (Indiana University Press, 1977), pp. 122–124.

For some underlying theory, see Richard Schmalensee, "Product Differentiation Advantages of Pioneering Brands," *American Economic Review*, vol. 72 (June 1982), pp. 349–365.

[6]Richard Levin, Alvin Klevorick, Richard R. Nelson, and Sidney Winter, "Appropriating the Returns from Industrial Research and Development," *Brookings Papers on Economic Activity: Microeconomics* (Washington: 1987), pp. 783–820.

[7]Wesley J. Cohen, Richard R. Nelson, and John P. Walsh, "Protecting Their Intellectual Assets: Appropriability Conditions and Why U.S. Manufacturing Firms Patent (or Not)," working paper, Carnegie–Mellon University, January 2000.

[8]Edwin Mansfield, "Patents and Innovation: An Empirical Study," *Management Science*, vol. 32 (February 1986), p. 175.

[9]Joseph A. DiMasi, Ronald W. Hansen, Henry G. Grabowski, and Louis Lasagna, "Research and Development Costs for New Drugs by Therapeutic Category," *PharmacoEconomics*, vol. 7 (February 1995), p. 165. The estimates assume 1993 general price levels.

[10]Since these outlays precede clinical testing outlays, and hence occur from 8 to 15 years on average before the ultimate products are marketed, their magnitude is raised much more than clinical testing costs by capitalization to the time of marketing.

[11]Tufts University Center for the Study of Drug Development, News Release, November 30, 2001, summarizing results presented at a November 30 conference in Philadelphia attended by the author. Although the survey included seven biological entities (out of a total of 68 entities), all were originated by traditional pharmaceutical enterprises. Because therapeutic substances originated during the 1980s and early 1990s by biotechnology firms required on average fewer human test subjects and shorter testing periods than traditional "small molecule" entities, the average capitalized cost of approved biotech-based pharmaceuticals would undoubtedly be found to be lower than $800 million. See Tufts University Center for the Study of Drug Development, News Release, July 16, 2001.

[12]E.g., the seven-year period of marketing exclusivity accorded orphan drugs in their orphan indication(s).

[13]This account is based upon the Harbridge House report, "Effects of Government Patent Policy on Drug Research and New Product Development," prepared under a contract with the Committee on Government Patent Policy, Federal Council for Science and Technology (Boston: May 1967), especially Sections I and IV.

[14]*Ibid.*, p. I-2. The nearly unanimous rejection is puzzling, given evidence that average clinical testing costs were much lower then—roughly $1 million per successful drug—than after testing requirements escalated following passage of the Kefauver–Harris Act in 1962. See Edwin Mansfield, "Comment," in Joseph D. Cooper, ed., *The Economics of Drug Innovation* (Washington: American University, 1970), p. 151.

[15]See Rebecca S. Eisenberg, "Public Research and Private Development: Patents and Technology Transfer in Government-Sponsored Research, *Virginia Law Review*, vol. 82 (November 1996), pp. 1682–1683.

[16]The summary here is drawn from my monograph, *The Economic Effects of Compulsory Patent Licensing*, New York University, Monograph Series in Finance and Economics, Monograph 1977-2 (New York: 1977), pp. 79–82. See also Harbridge House, Inc., *Government Patent Policy Study*, volumes I–IV (Boston, 1968).

[17]On the complex history of the acts, see Eisenberg, "Public Research and Private Development," pp. 1662–1727.

[18]*Venture Capital Yearbook: 1998* (Wellesley Hills, MA: Venture Economics Inc., 1999), p. 58.

[19]"Disbursements per Company by Industry: 1999," *Venture Capital Journal*, April 2000, p. 45.

[20]"Move Over, Dot-Coms, Biotech Is Back," *Business Week*, March 6, 2000, p. 120, citing data collected by Ernst & Young.

[21]*Ibid.*, p. 118.

[22]David H. Hsu and Tim Bernstein, "Managing the University Technology Licensing Process: Findings from Case Studies," *Journal of the Association of University Technology Managers*, vol. 9 (1997), pp. 1–33.

[23]Patricia M. Danzon, Jeffrey McCullough, and Sean Nicholson, "Efficiency in the Market for Biotech-Pharmaceutical Alliances," working paper, University of Pennsylvania, June 2001.

[24]Joshua S. Gans, David H. Hsu, and Scott Stern, "When Does Start-up Innovation Spur the Gale of Creative Destruction?" National Bureau of Economic Research working paper 7851 (August 2000).

[25]Cohen et al., "Protecting Their Intellectual Assets."

[26]I am reminded of how General Georges Doriot, president of the first modern high-technology venture capital fund, responded when I told him that our research group at Harvard Business School would study the impact of patents on corporate R&D decisions: "Patents are just a weapon with which big firms bludgeon my startup companies," he said. On the other hand, in an analysis of the first 148 patent infringement cases decided by the Appellate Court for the Federal Circuit, created in 1983, I found that firms with sales of less than $25 million did not experience disproportionately unfavorable outcomes. "Changing Perspectives on the Firm Size Problem," in Zoltan J. Acs and David B. Audretsch, eds. *Innovation and Technological Change* (Harvester Wheatsheaf: 1991), pp. 34–35.

[27]See, e.g., F. M. Scherer, "The Link from Gross Profitability to Pharmaceutical R&D Spending," *Health Affairs*, September/October 2001.

[28]From an 1813 letter to Isaac McPherson reproduced in John P. Foley, ed., *The Jeffersonian Cyclopedia* (New York: Russell and Russell, 1967), p. 433. Jefferson goes on to recognize the ambiguous case for providing patent protection on the ideas embodied in inventions.

[29]National Science Board, *Science and Engineering Indicators: 2000*, pp. A-16 to A-25.

[30]See especially *Business Week*, March 6, 2000, p. 120.

[31]National Science Board, *Research and Engineering Indicators: 2000*, pp. A-60 and A-72; and U.S. National Research Council, Board on Science, Technology, and Economic Policy, *Trends in Federal Support of Research and Graduate Education* (Washington: pre-publication version, 2001), p. 122. The figures from the first source have been adjusted to reflect the more recent benchmarks reported by the second source.

[32]National Research Council, *Trends in Federal Support*, p. 139.

[33]National Research Council, *Trends in Federal Support*, p. 128. This estimate undoubtedly contains a substantial but unknown fraction of the research done in university-affiliated hospitals.

[34]National Research Council, *Trends in Federal Support*, pp. 144–145.

[35]National Science Board, *Science & Engineering Indicators: 2000*, p. A-319.

[36]U.S. National Science Board, *Science and Engineering Indicators: 1996*, p. 249; and *Science and Engineering Indicators: 2000*, p. A-476. A search by Lucent Technologies revealed 46 cases in which universities asserted their patents through law suits against private firms. Testimony of Jim Finnegan at the National Academies Board on Science, Technology, and Economic Policy workshop on Academic IP, April 17, 2001, p. 40 of the proceedings transcript.

[37]Jerry G. Thursby and Marie C. Thursby, "Who Is Selling the Ivory Tower? Sources of Growth in University Licensing," National Bureau of Economic Research working paper 7718 (May 2000), pp. 15–17. On the response of Stanford and the University of California, see David C. Mowery and Arvids A. Ziedonis, "The Effects of the Bayh-Dole Act on U.S. University Technology Research and Technology Transfer," working paper, July 1999.

[38]The author is indebted to Andrew Mulcahy, a biology major at Princeton University, who carried out the search.

A parallel search of pending patent *applications* published between March 15, 2001, when the Patent Office began publishing applications, and August 10, 2001, retrieved only 29 applications. A curiosity of this meager trove was that few disclosures revealed the ultimate patent assignee, which will make it difficult to obtain early insight into individual companies' inventive activity from published patent applications.

[39] Statistical analyses revealed that European firms were more likely to use "DNA" than "nucleic acid" in their patent abstracts. Patents received by hospitals and nonprofit research institutions, patents with direct implications for human health, and methods patents were more likely to use "nucleic acid" than "DNA," all else equal.

[40] This category is subject to some imprecision because information about such firms' structures is sparse. It is possible that some small companies with foreign home bases were classified as private U.S. companies.

[41] The Web page for Incyte Genomics, Inc., corporation is ⟨http://www.incyte.com⟩, and that for Human Genome Sciences, Inc., is ⟨http://www.hgsi.com⟩ (both accessed August 29, 2002).

[42] Biotech specialists Amgen, Calgene (controlled by Monsanto), Genentech (partly owned by Hoffmann-LaRoche), and Genetics Institute (controlled by American Home Products) were counted in the large-company group.

[43] U.S. Patent and Trademark Office, Information Products Division/TAF Branch, Web site report, "Patent Counts by Class by Year: January 1977–December 2000," March 2001. The largest number of patents, 7,530, were issued in class 435.

[44] See F. M. Scherer, "The Propensity to Patent," *International Journal of Industrial Organization*, vol. 1 (March 1983), p. 110, showing that the pharmaceutical industry received 1.93 1976–77 patents per million dollars of 1974 R&D, compared with 1.70 patents for all U.S. industrial firms combined.

[45] William D. Nordhaus, *Invention, Growth, and Welfare: A Theoretical Treatment of Technological Change* (Cambridge, MA: MIT Press, 1969), Chapter 5. For an early elaboration, see F. M. Scherer, "Nordhaus' Theory of Optimal Patent Life: A Geometric Reinterpretation," *American Economic Review*, vol. 62 (June 1972), pp. 422–427.

[46] Edmund W. Kitch, "The Nature and Function of the Patent System," *Journal of Law & Economics*, vol. 20 (October 1977), pp. 265–290.

[47] "Optimal Timing of Innovations," *Review of Economics and Statistics*, vol. 50 (August 1968), pp. 348–355.

[48] See Michael Berkowitz and Y. Kotowitz, "Patent Policy in an Open Economy," *Canadian Journal of Economics*, vol. 15 (February 1982), pp. 1–17; and the debate between Donald McFetridge and M. Rafiquzzaman on one hand and Roger Beck in John Palmer, ed., *Research in Law and Economics*, vol. 8 (1986) (Greenwich, CT: JAI Press), pp. 91–129.

[49] On the financial outcomes of innovation projects, see F. M. Scherer, Dietmar Harhoff, and Jörg Kukies, "Uncertainty and the Size Distribution of Rewards from Innovation," *Journal of Evolutionary Economics*, vol. 10 (2000), pp. 175–200.

[50] The earliest modern theoretical demonstration of that point was F. M. Scherer, "Time–Cost Tradeoffs in Uncertain Empirical Research Projects," *Naval Research Logistics Quarterly*, vol. 13 (March 1966), pp. 71–82. There is reason to believe that the same insight was presented a century ago by philosopher Charles Sanders Peirce, but no citation is available.

[51] See, e.g., Burton Klein, *Dynamic Economics* (Harvard University Press, 1977); Michael Spence, "Product Differentiation and Welfare," *American Economic Review*, vol. 66 (May 1976), pp. 407–414; and F. M. Scherer, *International High-Technology Competition* (Harvard University Press, 1992), pp. 35–40 and 178–181.

[52] The small incremental gain from R_2 to R_3 comes from the severely diminishing returns assumed in the *C(RD)* function's final segment. They may be exaggerated. This is a point on which we have sparse empirical evidence.

[53] William F. Ogburn and D. S. Thomas, "Are Inventions Inevitable?" *Political Science Quarterly*, vol. 37 (1922), pp. 83–98. See also Robert K. Merton, "Singletons and Multiples in Scientific Discovery," *Proceedings of the American Philosophical Society*, vol. 105 (1961), pp. 470–486; Merton, "Priorities in Scientific Discovery: A Chapter in the Sociology of Science," in Bernard Barber and Walter Hirsch, eds., *The Sociology of Science* (Glencoe, IL: The Free Press, 1962), pp. 447–485; Thomas S. Kuhn, *The Structure of Scientific Revolutions* (University of Chicago Press, 1962), Chapters VII and VIII; and Abbott P. Usher, *A History of Mechanical Inventions*, rev. ed. (Harvard University Press, 1954), Chapter IV.

[54] From a book review in the *Scientific American*, December 1958, p. 146.

[55] James D. Watson, *The Double Helix* (New York: Atheneum, 1968).

[56] The term is Usher's, following William James.

[57] Edwin Mansfield, "Industry–University R and D Linkages and Technological Innovation," paper presented at the annual meetings of the American Economic Association, January 1996, updating "Academic Research Underlying Industrial Innovations: Sources, Characteristics, and Financing," *Review of Economics and Statistics*, vol. 77 (February 1995), pp. 55–65. "Recent academic research" was defined as research within 15 years of the relevant innovation's commercialization.

[58] For a pioneering analysis of the multi-stage invention problem that has been almost totally ignored in more recent scholarship, see C. C. von Weizsäcker, *Barriers to Entry: A Theoretical Treatment* (Berlin: Springer, 1980), Chapters 8 and 9. Von Weizsäcker's principal conclusion is that in an environment of rapid technological change and obsolescence, the series-relatedness of inventions calls for stronger intellectual property protection of inventions at any given stage. He recognizes that variations upon his stylized model can lead to different inferences, as we shall see here.

[59] For seminal treatments of this problem, see Suzanne Scotchmer, "Standing on the Shoulders of Giants: Cumulative Research and the Patent Law," *Journal of Economic Perspectives*, vol. 5 (Winter 1991), pp. 29–41; and Jerry Green and Suzanne Scotchmer, "On the Division of Profit in Sequential Innovation," *RAND Journal of Economics*, vol. 26 (Spring 1995), pp. 20–33.

[60] Secrecy with respect to fundamental concepts as a more lasting source of advantage to the discovers of A seems a less practical alternative. Universities such as Harvard permit their staffs to delay publication of research results, e.g., while patent protection is being sought, by at most a few months. See the testimony of Wesley Cohen at the National Academies workshop on Acedemic IP, transcript p. 158, which estimates the average publication delay at 4.7 months. Business firms in fields such as biotechnology would find it difficult to recruit able staff if they sought to delay indefinitely the publication of scientific discoveries.

[61] As Jack L. Tribble, patent counsel to Merck & Co., observed in the National Academies Academic IP workshop, "99 percent of everything exciting that happens will happen outside of your own research labs." Workshop transcript p. 113.

[62] See, e.g., John Jewkes, David Sawers, and Richard Stillerman, *The Sources of Invention*, 2nd ed. (New York: Norton, 1969); Willard F. Mueller, "The Origins of the Basic Inventions Underlying DuPont's Major Product and Process Innovations, 1920 to 1950," in the National Bureau of Economic Research conference volume, *The Rate and Direction of Inventive Activity* (Princeton University Press, 1962), pp. 323–358; Clayton M. Christensen, *The Innovator's Dilemma: When New Technologies Cause Great Firms to Fail* (Harvard Business School Press, 1997); and Klein, *Dynamic Economics*, Chapter 5.

[63] Another alternative is an R&D joint venture between the discover of A and a follow-on developer. For its advantages, see Scotchmer, "Standing on the Shoulders of Giants." One of many possible problems is that when a would-be co-developer approaches the original patent holder with its commercialization idea, the basic patent holder may choose not to grant the necessary rights but to start its own solo development effort.

[64] See Susan Aldridge, *The Thread of Life: The Story of Genes and Genetic Engineering* (Cambridge University Press, 1996), pp. 104–108.

[65] This point is powerfully argued by Robert P. Merges and Richard R. Nel-

son. "On the Complex Economics of Patent Scope," *Columbia Law Review*, vol. 90 (May 1990), pp. 839–916.

[66]Merges and Nelson, p. 866.

[67]*Ibid.*, pp. 884–908.

[68]Illinois Institute of Technology Research Institute, *Technology in Retrospect and Critical Events in Science* (the TRACES study), vol. 1 (Chicago: December 1968), Figure 8.

[69]That this problem exists at a serious level in biotechnology is argued by Michael A. Heller and Rebecca Eisenberg, "Can Patents Deter Innovation? The Anticommons in Biomedical Research," *Science*, vol. 280 (May 1, 1998), pp. 698–701.

[70]Elliot Forbes, ed., *Thayer's Life of Beethoven* (Princeton University Press, 1967), p. 38.

[71]See Merges and Nelson, "On the Complex Economics of Patent Scope," pp. 890–896.

[72]See Rosmarie Ham Ziedonis and Bronwyn Hall, "The Effects of Strengthening Patent Rights in Firms Engaged in Cumulative Innovation: Insights from the Semiconductor Industry," University of Pennsylvania and University of California at Berkeley, May 2001.

[73]On the differences between academic and industrial laboratories, see the testimony of Jim Finnegan at the National Academies Academic IP workshop, transcript pp. 40–41.

[74]Federal Trade Commission, *In the Matter of Intel Corporation*, Agreement Containing Consent Order, Docket No. 9288 (March 1999). The author was economic expert for the FTC in the proceedings that preceded consent settlement.

[75]It is assumed that once the key invention is made, the benefits do not reach their full potential immediately, but achieve $(1 - e^{0.2T})$ of their potential in early years, where T is the number of years from the initial availability of the invention. Thus, by year 10, they have reached 86% of their potential.

[76]See Merges and Nelson, "On the Complex Economics of Patent Scope;" and Merges, "Intellectual Property Rights and Bargaining Breakdown: The Case of Blocking Patents," *Tennessee Law Review*, vol. 62 (Fall 1994), pp. 75–106.

[77]*Embrex Inc. v. Service Engineering Corp.*, 216 F. 3d 1343 (2000) (Court of Appeals for the Federal Circuit).

[78]On the differences between royalty standards, see F. M. Scherer and Jayashree Watal, "Post-TRIPS Options for Access to Patented Medicines in Developing Nations," J International Law. 2002.

[5]

The Pharmaceutical Industry and World Intellectual Property Standards

F. M. Scherer[*]

When I was a high school student during the late 1940s, the first so-called "wonder drugs"—initially penicillin and then the broad-spectrum antibiotics such as tetracycline—were entering the U.S. market. From their profitable experience developing the broad-spectrum antibiotics, the leading pharmaceutical companies of America and Europe acquired a strong research orientation that led to a cascade of new therapeutic entities, including additional anti-infectives, vaccines, diuretics, and then other agents to reduce heart attack risks, tranquilizers, antidepressants, birth control pills, anti-fungal agents, immunosuppressants, cortico-steroids, AIDS inhibitors, powerful pain relief agents, and many other agents effective against specific diseases. Thanks to this pharmaceutical revolution, life spans have been prolonged, the incidence and duration of hospital stays have been reduced, and the quality of countless citizens' lives has been enhanced.[1]

The benefits of modern pharmaceutical therapy have accrued mainly to the citizens of the world's more prosperous nations. United Nations staff have estimated that average purchases per capita of modern pharmaceutical products (excluding traditional

---

* Aetna Professor of Public Policy Emeritus, Harvard University; Visiting Professor, Princeton University.

1. *See* FRANK R. LICHTENBERG, THE EFFECT OF PHARMACEUTICAL UTILIZATION AND INNOVATION ON HOSPITALIZATION AND MORTALITY 23, (National Bureau of Economic Research Working Paper No. w5418, 1996) (estimating that a $1 increase in pharmaceutical purchases was associated on average with a $2.65 reduction in hospitalization expenditures), *available at* http://papers.nber.org/papers/w5418.

medicines) in 1990 (calculated at prevailing exchange rates) in diverse parts of the world were as follows:[2]

North America	$123.90
European Community	102.90
Other Western Europe	85.70
Japan	276.60
South and East Asia	5.00
China	4.80
Latin America	20.30
Sub-Saharan Africa	3.30

A rough extrapolation of these figures reveals that the 73 percent of the world's 1990 population located in south and east Asia, including China, Sub-Saharan Africa, and Latin America, consumes only 16.2 percent of modern pharmaceutical output by dollar volume. One consequence of the inadequate purchasing power that limits such nations' ability to consume pharmaceuticals is a higher rate of morbidity and debility, which in turn impairs the growth of income so that pharmaceuticals can be afforded—a vicious cycle.

Nearly all of the research-oriented pharmaceutical companies responsible for innovations in drug therapy have their home bases in the United States, the European Community nations, or Japan, where demand is most intense and highly able scientists interacting with first-rate universities are at hand. Excepting those of Japan, the research-oriented pharmaceutical companies are among the most multinationally oriented enterprises in the world.

Discovering a new drug and carrying it through the tests required to obtain marketing approval from regulatory agencies in the United States and Europe costs upwards of $100 million per successful new chemical entity. Once such a large investment has been made, there are powerful incentives to obtain requisite regulatory approvals in other nations and sell the product as widely as possible. Foreign markets are served both by exporting, often from a tax haven such as Puerto Rico, Ireland, or Singapore, and through direct plant investment in consuming nations. According to United Nations estimates, pharmaceutical imports averaged 8.2 percent of domestic consumption during 1989 in developed nations and 19.8

---

2. *See* ROBERT BALLANCE, THE WORLD'S PHARMACEUTICAL INDUSTRIES: AN INTERNATIONAL PERSPECTIVE ON INNOVATION, COMPETITION AND POLICY 30-31 (1992).

percent in less-developed nations.[3] In 1980, approximately 27 percent of the world's demand was satisfied through local production by foreign-owned companies.[4] Since then, the extent of multinational operation has increased, in part due to numerous cross-border mergers. In 1995, members of the Pharmaceutical Research and Manufacturers of America trade association recorded prescription drug sales of $65 billion within the United States and $37 billion outside the United States.[5]

Most of the R&D outlays incurred by pharmaceutical companies are made to discover therapeutically interesting molecules and prove their efficacy and safety through extensive human trials—i.e., to create knowledge that approximates what economists call a pure public good. Absent legal barriers to copying, once a drug has been found to be safe and effective, another firm might come up with a generic equivalent by spending roughly a million dollars on production process methods and formulation and begin to compete with the pioneering firm. If such generic imitation were widespread and rapid, surplus revenues that repay pioneers' initial R&D outlays and make them worthwhile would be severely eroded, undermining incentives to invest in research and product testing. Because of the huge disparity between drug finding and imitation costs, multi-industry surveys show, pharmaceutical manufacturers attach unusually high importance to the patent system, which in effect grants them 20 years of exclusive rights to their invention from the time a patent application is filed, as a means of recouping their R&D expenditures.[6]

The combination of multinationality and heavy stress on patent protection set the stage for a conflict between the pharmaceutical manufacturers and the world's developing nations. Under the Paris Convention to which most of the nations with patent systems adhered, nations were free to structure their patent laws however they desired, as long as they did not discriminate between local and foreign inventors. Many nations excluded drug products from patentability because they considered drugs (and for analogous reasons, food products) to be of such great importance to the national welfare. Even Switzerland, home to three of the world's leading pharmaceutical companies, abstained until 1977 from

3. *See id.* at 52.

4. *See id.* at 68-69.

5. *See* PHARMACEUTICAL RESEARCH AND MANUFACTURERS OF AMERICA, 1998 INDUSTRY PROFILE 100 (1998).

6. *See* Richard C. Levin et al., *Appropriating the Returns from Industrial Research and Development,* 3 BROOKINGS PAPERS ON ECONOMIC ACTIVITY 783, 807-12 (1987).

granting drug product patents. Most less-developed countries ("LDCs") followed that pattern and tended more generally to provide weaker patent protection than the most industrialized nations—actions perfectly compatible with the Paris Convention.

For the multinational pharmaceutical companies, the inability to obtain universal patent protection for their new products was seen as a constraint on their world wide sales and profits, limited though the potential in LDCs might be due to consumers' meager purchasing power. An attempt to alter the situation at a World Intellectual Property Organization conference in Kenya during the 1970s proved unsuccessful. The drug makers then sought other fora to pursue their quest. Led by Edward Pratt, CEO of Pfizer Inc. and a member of President Reagan's business advisory committee on international trade, the U.S. pharmaceutical companies commenced a superbly orchestrated campaign to bring the U.S. government into the dispute.[7] They fanned out to enlist the support of motion picture, music recording, and software industry leaders who were equally unhappy about the copyright protection they received abroad. The U.S. pharmaceutical makers in turn enlisted the lobbying support of their European counterparts through the so-called Dolder Group (named after a Swiss hotel at which pharmaceutical executives met regularly) and also Japanese companies. An astute public relations campaign succeeded in making "piracy" the accepted term to describe the imitation of drug or software products in nations with permissive intellectual property laws, even though "knock-off" production was quite legal under accepted international conventions. Lobbying efforts at the U.S. Congress led to amendments in Section 301 of the U.S. international trade code, defining other nations' failure to grant full patent or copyright protection for U.S. products as an unfair trade practice and charging the President's Trade Representative Office with monitoring foreign intellectual property practices and identifying for unilaterally imposed trade sanctions the most egregious perceived offenders.[8] Faced with increased tariffs, loss of most-favored-nation status, or the threat of such sanctions on the products they sought to export to the United

7. *See* M.P. RYAN, KNOWLEDGE DIPLOMACY: GLOBAL COMPETITION AND THE POLITICS OF INTELLECTUAL PROPERTY 67-69 (1998); LYNN SHARP PAINE & MICHAEL SANTORO, PFIZER: PROTECTING INTELLECTUAL PROPERTY IN A GLOBAL MARKETPLACE (Harvard Business School case study N9-392-072 1995), *available at* http://www.hbsp.harvard.edu.

8. *See generally* AGGRESSIVE UNILATERALISM: AMERICA'S 301 TRADE POLICY AND THE WORLD TRADING SYSTEM (Jagdish Bhagwati & Hugh Patrick eds., 1990) (detailing and evaluating section 301).

States, some nations such as Korea, Thailand, and Brazil strengthened their patent and copyright laws and began enforcing them assiduously.

Prodded by industrial lobbies, the governments of the United States, European Union member nations, and Japan insisted that harmonization of national intellectual property laws be made a key agenda item in the Uruguay Round of international trade negotiations, initiated in 1986. That effort pitted the leading industrialized nations working en bloc against the less-developed countries.

LDCs were opposed to granting stronger intellectual property rights because such grants seemed so clearly against their national interests. If they allowed patents on drug products, the multi-national enterprises ("MNE") would have stronger monopoly positions in selling their products to the limited array of domestic consumers able and willing to pay, so prices would be elevated relative to those charged by local drug makers able to "knock off" the most advanced products invented elsewhere and sell them at more competitive prices. Higher prices mean smaller quantities demanded and hence lower health benefits from the newest therapies. If imports were the vehicle through which the MNEs sold their newest drugs, and probably also if the drugs were produced by local branch plants, the profits realized would be repatriated to the MNEs' home countries, imposing a drain on scarce hard-currency reserves. To be sure, higher sales and profits for the multinationals would mean stronger incentives to invest in R&D (conducted almost entirely in highly industrialized nations) and hence the emergence of more new products. A theoretical analysis of less-developed nations' incentives reveals that to compensate for the loss of domestic surplus (assuming linear demand functions) as a result of high drug prices for existing patented products, a threefold increase in the number of equivalent new products reaching the LDCs would be required.[9] Such a large increase in new products driven solely by the modest sales increase attributable to newly-authorized patenting in low-income nations seemed highly improbable.

The tradeoff might become more favorable for less-developed countries if granting patent protection on new drug products strengthened the incentives of indigenous pharmaceutical compa-

9. *See* F. M. SCHERER, INDUSTRY STRUCTURE, STRATEGY, AND PUBLIC POLICY 363-65 (1996). Viewing the situation from the perspective of the world as a whole rather than that of a single LDC, an increase of less than one-third in the number of equivalent new products will compensate for the loss of consumers' surplus as a result of monopolized, as compared to competitive, pricing of the new products that would reach the market without patent protection.

nies so much that those firms began developing innovative drugs, selling them at home and perhaps exporting them too. The benefits from granting product patents would also be enhanced if multinational companies were spurred to develop new drugs explicitly targeted toward diseases prevalent mainly in low-income nations.

Evidence on the first of these possibilities is provided by the experience of Italy. During the 1950s and 1960s, Italy did not grant drug product patents. A thriving "knock-off" drug industry emerged, selling drugs at bargain prices in the home market and becoming the world's leading exporter of new drugs to other nations that also denied patent protection to drug products.[10] During the 1970s, however, multinational enterprises challenged the Italian law, and in 1978, the Italian Supreme Court ruled that the law denying drug product patents was unconstitutional. It ordered that the law be amended and that the Italian authorities begin accepting drug patent applications immediately. Analyses of events during the decade that followed yield three principal conclusions: (1) no significant increase in Italian drug R&D expenditures relative to world trends; (2) no significant increase in the number of new drug entities introduced by Italian firms; and (3) a sharp deterioration of the Italian trade balance in drugs into the negative realm as export sales faltered and multinational firms imported many of their products into Italy from elsewhere in Europe. In addition, numerous Italian drug manufacturers were acquired by multinational firms seeking to strengthen their foothold in the Italian market. It is unclear why Italy failed to make the transition from drug imitator to drug innovator. It may be that a decade is too short a time to do so, or price controls may have impaired domestic market incentives for the development of pioneering drugs. Also, Italy lacked the university research infrastructure and cooperative university-industry relationships needed to nourish drug innovation. What is clear is that leadership in the production and export of knock-off drugs to nations lacking product patent protection shifted to India.

There is little evidence on the second possibility—an increase in the targeting of drug development efforts toward low-income nations' health problems by multinational drug companies. An early investigation of this possibility and also the hypothesis that the anticipation of domestic future product patent protection

10. Indeed, in the late 1950s, frustrated by the high prices charged for tetracycline by licensees to the patent held by Pfizer, the U.S. armed forces invoked their statutory right to infringe domestic patents and purchased tetracycline from Italian firms.

stimulated Indian manufacturers' new drug discovery efforts up to 1997 yielded equivocal findings.[11]

What some multinational drug companies have done is to *donate* already developed drugs such as Zithromax (effective against trachoma blindness), Ivermectin (effective against river blindness), and Albendazole (effective against lymphatic filariasis parasites) to especially poor nations with a high incidence of such diseases.[12] Four pharmaceutical companies led the list of U.S. corporations, ranked by the total amount of money devoted to philanthropic giving in 1998.[13] Presumably, the market value of the drugs donated in this way could be claimed as tax deductions by the donor corporations. Since the marginal cost of a typical still-patented drug tends to be on the order of 20 percent of the U.S. wholesale price, and since for most corporations (but perhaps, because of tax credits on Puerto Rican production, not the drug makers) the marginal corporate income tax rate is 34 percent, it is conceivable that the drug companies' charity actually increased the donors' after-tax profits. If this conjecture is correct, the donation programs may in fact be a subtle form of foreign aid subsidized by the U.S. Treasury.

As the Uruguay Round negotiations culminated in 1993, the industrialized nation coalition demanding patent law unification held firm, and the final draft included a strong Trade-Related Intellectual Property Rights ("TRIPS") chapter. Less-developed nations were presented with hard choices. On the one hand, they could agree to strengthen their intellectual property laws and accept the consequences outlined above. On the other hand, they could opt out and lose the benefits of World Trade Organization ("WTO") membership, including gradual relaxation of industrialized nation import restraints against agricultural products and phase-out of the Multi-Fibre Agreement limiting the export of textiles and apparel—fields in which many LDCs had found comparative advantage. The best the LDCs could do was to negotiate compromise clauses in the final draft—notably, allowing LDCs ten years from January 1995 to implement full drug patent rights and providing exceptions under which national governments could abridge the rights granted. Exceptions can occur when monopolistic abuses of patent rights are proven, or non-exclusive compulsory licensing of

11. *See* JEAN O. LANJOUW & IAIN COCKBURN, *Do Patents Matter?: Empirical Evidence After GATT* 1-4, (National Bureau of Econ. Research Working Paper No. 7495, 2000), *available at* http://www.nber.org/papers/w7495.pdf.

12. *See id.* at 22-23.

13. *The List: Corporate Giving*, BUS. WK., Jan. 24, 1990, at 8.

patents with "adequate remuneration" is permitted when a proposed patent licensor has been unable within a reasonable period of time to "obtain authorization [for domestic use] from the right holder on reasonable commercial terms and conditions." Exactly what these loopholes mean is at best unclear; the provisions are likely to be contested in future proceedings before the WTO.[14] Meanwhile, the Uruguay Round treaty requires that less-developed nation signatories accept interim patent applications in a so-called letter box, granting to the patent applicant exclusive marketing rights for five years beginning not later than the year 2000 for new drugs covered by pending applications.

Some nations, such as Brazil, have complied fully with the TRIPS mandates. In other nations compliance remains bitterly disputed and the subject of U.S. complaints to the WTO. In Argentina, for example, implementing legislation proposed by President Menem was at first rejected by the Parliament and then passed in a weakened form that, according to U.S. trade officials, fails to comply with the Uruguay Round treaty language. As of 1999, India had not passed implementing legislation or delineated procedures for accepting letter box applications. In April 1999, the United States government launched a special review threatening trade sanctions against South Africa for its efforts to import drugs at bargain prices from nations with no drug product patent protection and to issue compulsory licenses permitting domestic manufacture of Taxol (effective against certain forms of cancer) and drugs that could help combat an AIDS epidemic. A contemplated complaint to the World Trade Organization concerning South Africa's policies was apparently headed off when AIDS activists threatened to make the U.S. policies a contested issue in Al Gore's campaign to win the U.S. presidency.[15] Had the Seattle round of international trade negotiations not collapsed in debacle during December 1999, it was likely that less-developed nations would have sought reconsideration and weakening of the Uruguay Round TRIPS agreements.

The debate over drug product patent extension into the Third World is a classic problem of federalism, questioning whether

---

14. Exploitation of the loopholes is encouraged in a World Health Organization staff report, to which the U.S. pharmaceutical company trade association and the U.S. Food and Drug Administration have taken pointed exception. *See* World Health Organization Staff Report, *Globalization and Access to Drugs: Implications of the WTO/TRIPS Agreement* 32-36 (1998), *available at* http://whqlibdoc.who.int/hq/1998/WHO_DAP_98.9_revised.pdf.

15. *See* Kathy Chenault, *Will the AIDS Plague Change U.S. Trade Policy?* BUS. WK., Sept. 13, 1999, at 58.

economic policy choices should be made locally or by a supervening governmental entity. Usually decentralization is favored on allocative efficiency grounds while the possibilities of redistributive equity pull toward centralization. However, in the TRIPS debate, the opposite is true. Decentralization of patent policies allows poor nations to address their health problems despite limited resources by free-riding on wealthy nations' inventions, while the suppression of free-riding under a centralized policy improves the allocation of world R&D resources. Curiously, while the TRIPS debate was proceeding, the U.S. Supreme Court handed down a patent law decision logically inconsistent with the U.S. position on drug product patents in international fora.

The U.S. Patent Code includes provisions allowing the federal government to infringe valid patents when infringement serves the national interest and reasonable compensation is paid.[16] The requirements for reasonable compensation have been interpreted historically to imply much lower payments than the foregone monopoly profits standard imposed in private patent infringement damages cases. In *Florida Prepaid Postsecondary Education Expense Board v. College Savings Bank*, a majority of the Supreme Court ruled (over the strong dissent of a four-member minority) that Congress had exceeded its constitutional powers in nullifying individual states' sovereign immunity from patent infringement suits in federal courts, provided that the states maintain their own in-state legal procedures (either legislative or judicial) to ensure that the patent holders' property is not taken without due process of law.[17] Armed with that precedent, the states might in the future seek to alleviate their escalating drug cost burdens under Medicaid (and possibly in the future Medicare) by setting up patent-infringing generic drug production operations or importing drugs from no-patent jurisdictions and offering to pay "reasonable compensation" for the use of the infringed patents. Compensation substantially less than the monopoly profits foregone by the patent holders might arguably satisfy the due process requirement.[18]

---

16. *See* 28 U.S.C. § 1498 (1994). For an illustration of an actual case, see *supra* note 10.

17. Florida Prepaid Postsecondary Educ. Expense Bd. V. College Sav. Bank, 527 U.S. 627 (1999).

18. An example is the position taken by the courts of Canada in implementing the compulsory licensing of drugs up to 1987. When Hoffmann-LaRoche requested a flat fee royalty that would have amounted to 30 percent of the price of Valium *before* the emergence of generic competition, the Canadian Exchequer Court set an *ad valorem* royalty of four percent on sales *after* the emergence of generic competition. The Court provided that in setting the terms of the license and fixing the amount of royalty or other consideration payable, "the Commissioner shall have regard to the desirability of making the medicine available to the public at the lowest possible

One might argue that an analogous bargain must govern the division of powers between a supranational federalist organization such as the WTO and its member nations, especially when issues that are literally a matter of life or death are joined. Indeed, from the U.S. Senate debate that preceded U.S. ratification of the Uruguay Round treaty, it was clear that U.S. legislators were reluctant to cede such life-or-death powers to a supranational federal body. For the United States to insist that other member nations should not infringe patents issued within their home jurisdictions, paying compensation deemed reasonable under their own judicial or administrative processes, would be hypocritical, given the Supreme Court's strong states' rights position in the *Florida Prepaid Postsecondary Education* case.

What seems clear is that controversy over the Uruguay Round TRIPS mandates will continue and indeed intensify.

---

price consistent with giving to the patentee due reward for the research leading to the invention." *See* Hoffmann-LaRoche Ltd. v. Frank W. Horner Ltd., (1970) C.P.R. 107-108. Four percent royalty rates became standard in subsequent drug patent licenses. *See* F. M. Scherer, *The Economic Effects of Compulsory Patent Licensing, in* MONOGRAPH SERIES IN FINANCE AND ECONOMICS 1977, at 47 (N.Y.U. Graduate Sch. Bus. Admin. Center for Study Fin. Inst. Monograph 1977-2, 1977).

F. M. Scherer and Sandy Weisburst***

Economic Effects of Strengthening Pharmaceutical Patent Protection in Italy***

I. Introduction

In March 1978, the Italian Supreme Court declared unconstitutional a 1939 law that had excluded pharmaceutical products from eligibility for patent protection within Italy.[1] In 1982 the Italian Parliament passed a new law implementing the Court's decision. Because the Supreme Court ruling permitted immediate filing of drug patent applications, pharmaceutical inventors enjoyed a new regime of patent protection on eligible inventions beginning in 1978, even though full implementation occurrred four years later.

The Italian experience is particularly interesting because it presages legal changes on a much broader international front. The Uruguay Round of international trade negotiations, culminating in the Treaty signed at Marrakech on April 15, 1994, requires nations to offer intellectual property protection satisfying agreed-upon minimum standards as a condition for membership in GATT and the newly-formed World Trade Organization. Up to that time, dozens of nations, and especially less-developed countries, excluded pharmaceutical products from receiving patent protection, as Italy had until 1978. The Marrakech Treaty lays out a complex time framework within which national patent systems must be harmonized. For nations that previously granted no pharmaceutical product patents, such patents must be available within five years, although for the least-developed nations, implementation can be delayed for a maximum of ten years from 1994. Thus, India, Thailand, Egypt, Brazil, and Argentina will be among the many nations making a transition to a regime of pharmaceutical patent protection like that experienced by Italy in the early 1980s.

In the decades prior to the Italian Supreme Court's 1978 decision, Italy harbored a thriving pharmaceutical industry, including both national producers and the subsidiaries of multinational enterprises. Domestic producers took advantage of the lax intellectual property environment

* Professor of Business and Government, Harvard University.
** University of Chicago Law School.
*** This paper is drawn from a 1995 senior honors thesis at Harvard University, "Strengthening Patent Protection in Italy," written by Sandy Weisburst and advised by F. M. Scherer.
1 Constitutional Court, March 9, 1978, n. 20, published in Gazetta Ufficiale.

 0018-9855/95/0612-1009$5.00+.25/0

by imitating drugs that had been invented and patented outside Italy, selling those drugs at relatively low prices within Italy and aggressively exporting them to other nations that allowed no pharmaceutical product patents.[2]

Italian producers were much less active in developing important new drugs. They originated only 5 of the 900 new chemical entities introduced into the United States market between 1940 and 1975, generating 0.14 new drugs per billion dollars of annual gross national product (averaged over the years 1953, 1958, and 1962).[3] For the 17 nations to which those 900 new drugs were traced, the average number of new drugs per billion dollars of GNP was 1.42. Only India, Canada, and Argentina had lower non-zero new drug/GNP ratios than Italy. More recently, India has displaced Italy as the leading exporter of imitative drugs.[4]

For insights into how drug development efforts might change in nations required by the Marrakech Treaty to offer drug product patent protection, we investigate here how Italian producers adapted to the intellectual property regime changes of 1978. In particular, we undertake a detailed statistical analysis of changes in drug research and development expenditures and patenting, supplementing that inquiry with information on new drug product introductions, foreign direct investment by multinational enterprises, and import-export balances.

II. Theory and Hypotheses

Prior research has shown the expectation of patent protection to be important to the development of new pharmaceutical products. Surveying 650 research and development executives from 130 U.S. industries, Richard Levin and associates found that only in the pharmaceuticals industry were product patents regarded by a majority of the respondents to be strictly more effective in protecting the competitive advantages from innovation than such barriers to imitation as high duplication cost, head start advantages, and secrecy.[5]

The importance of patent protection in new drug development is attributable to four main phenomena. First, the cost of developing and testing new chemical entities to satisfy regulatory agencies' safety and

2 Vittorio Sironi, "Le Officine delle Salute: Storia del Farmaco e della Sua Industria in Italia dall 'Unita al Mercato Unico Europeo (1861–1992)" (Bari, Editori Laterza, 1992).

3 F. M. Scherer, "The Economic Effects of Compulsory Patent Licensing" 38 (New York University Graduate School of Business Administration, 1977).

4 Robert Ballance, et al., "The World's Pharmaceutical Industries" 61–68 (Hants, UK, Edward Elgar, 1992).

5 Levin, et al., "Appropriating the Returns from Industrial Research and Development," 1977 Brookings Papers on Economic Activity, Microeconomics 796.

efficacy standards is high, averaging nearly $ 100 million out-of-pocket (including the cost of unsuccessful projects) for products introduced into the U.S. market during the 1970s and early 1980s.[6] Second, once a new chemical entity has been shown to be effective and safe, the cost of imitation is much lower – as little as $ 1 million for the chemical engineering required to devise a me-too production process. Absent patent protection, competitive imitation could be swift. Third, it is difficult to invent around drug product patents, since the patents cover a specific and well-defined molecular structure. If an alternative molecule is devised, tests of safety and efficacy must be repeated at high cost. Fourth, because of the high perceived value of many drug therapies, because drug purchase outlays are often reimbursed by insurance, and because of information imperfections, the demand for drugs is commonly price-inelastic over a broad range of prices; that is, consumption is reduced little even when substantial price increases are effected. As a result, the company marketing a novel drug covered by one or more product patents typically enjoys a substantial degree of monopoly power.

The fourth of these phenomena has made the harmonization of patent standards under GATT bitterly controversial.[7] For a representative less-developed nation with a vigorously imitative domestic drug industry, the consequences of the Uruguay Round can be portrayed schematically by Figure 1.[8] D represents the demand curve for a drug chemical entity; MC the cost of producing and distributing the drug (excluding research and testing outlays). The demand and cost functions are assumed for simplicity to be linear. If product patents are available, a multinational enterprise (MNE) developing the drug will be able to charge the monopoly rent-maximizing price P_M, restricting the quantity of the drug sold to Q_M. Citizens of the subject LDC will realize from their purchases of the drug a consumers' surplus measured by diagonally shaded area A. The patent-holding MNE will gain a producer's surplus, or contribution to its worldwide profits, measured by horizontally cross-hatched area B. If there is no patent protection, competition in the production and sale of knock-off imitations will drive the price down to approximately the

6 Joseph A. DiMasi, et al., "Cost of Innovation in the Pharmaceutical Industry," 10 Journal of Health Economics (July 1991) 107–142.

7 On the pressures that led to according intellectual property questions high priority in the Uruguay Round negotiations, *see* Michael Santoro, "Pfizer: Protecting Intellectual Property in a Global Marketplace," Harvard Business School Case Study N9-392-071 (February 1992). When Peter Sutherland, the director-general of GATT responsible for negotiating the final Uruguay Round compromise in December 1993, visited India in early 1994, militia forces had to be summoned to protect him from protesting citizens.

8 The analysis here follows Alan Deardorff, "Welfare Economics of Patent Protection," 59 Economica (February 1992) 35–51.

Figure 1
Pricing and Surplus Distribution Under Alternative Patent Regimes

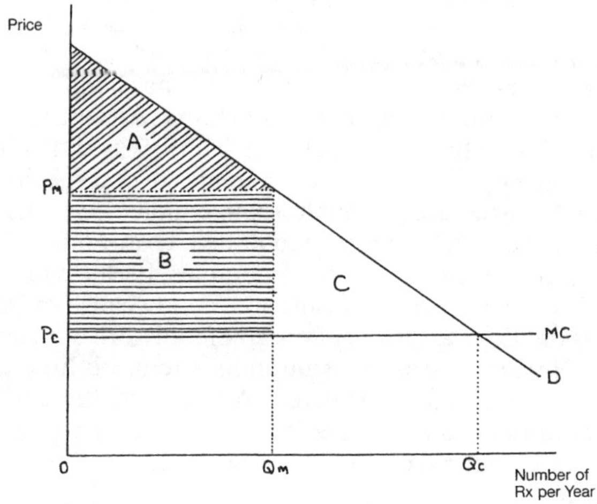

level of production cost P_C. In this competitive case, consumers' surplus amounts to the sum of areas A (the surplus that would be realized by consumers under monopoly pricing), B (redistributed from monopoly rent), and C (realized as the price reduction leads to additional consumption and hence the elimination of what would otherwise be a deadweight loss). With linear demand and cost functions, the total surplus accruing to LDC consumers under competitive imitation is four times the surplus they realize under monopoly pricing. Not surprisingly, the citizens of less-developed nations have expressed discontent over their governments' acceptance of Uruguay Round patent system harmonization provisions.[9]

To this stark conclusion several possible qualifications must be recognized. First, citizens of the LDC realize surplus from the drug in question only if it is invented, developed, and tested. Lacking the patent protection that permits them to appropriate surplus B, multinational enterprises might conceivably reduce their research and development expenditures so that the drug in question is not developed and made available at all to the world, with the consequent loss of surplus A to the LDC's citizens. However, the legislators of many small and/or poor nations have viewed the addition to MNE profits resulting from their granting, as opposed to not offering, drug product patent protection within their borders as so small relative to the profits realized by MNEs in other parts of the world that their marginal impact on MNE research

9 On the situation in Argentina, *see* "Argentine President Vetoes Patent Measure," New York Times, April 19, 1995, p. D5.

and testing decisions is negligibly small. In other words, they see themselves as free riders on foreign firms' research and development efforts that would be undertaken, whether or not they offer patent protection. Given this perspective, to leave the LDC's citizens as well off under a regime of patent monopolies as under a regime of imitative competition, the number of equivalent new drugs developed and sold in the LDC under monopoly (each yielding consumers' surplus A) would have to be at least four times the number available to LDC inhabitants through free riding under competition (yielding consumers' surplus A + B + C). Since there are diminishing returns in the development of new pharmaceutical entities, Alan Deardorff has argued, it is highly unlikely that the extension of drug product patent protection to all LDCs would increase MNE drug developers' profits sufficiently to induce the required three-fold increase in new drug products relative to the number available without such protection. Therefore, as a first approximation, LDC citizens are worse off granting drug product patents than they would be denying patent protection.

The balance changes if the granting of drug product patents leads pharmaceutical companies within the LDC to begin developing significant new drug entities. For a new drug developed by an indigenous firm and patented in the LDC, the gain to inhabitants of the LDC is consumers' surplus triangle A plus rectangle B, which now accrues to the (presumably local) owners of the local firm rather than to the remote shareholders of a foreign-based LDC. If product patent protection induced the development of as many equivalent new products by domestic firms as the number of new products that would be patented and supplied monopolistically by foreign MNEs, granting product patents would be as beneficial to the LDC's citizens as pursuing a strict no-patent free-riding policy.[10]

A key question therefore is how conferring product patents affects the new drug development efforts of domestic firms. The increase in domestically generated new products might be small if the domestic industry lacked the skills or scientific infrastructure needed to make the leap from imitation to significant new drug development. It might also be small if, before the change in patent regimes, domestic firms perceived their market prospects as worldwide, so that the additional prof-

10 Following the logic of Figure 1, let the surplus captured domestically be 1 for products sold monopolistically by foreign firms, 4 for foreign firm products sold competitively by domestic knock-off firms, and 3 for products sold monopolistically by domestic firms. Let a be the number of new drugs developed by domestic firms and n the number of new drugs developed by foreign MNEs (assumed invariant with the host nation's patent policy). For what value of a is $1n + 3a = 4n$, *i.e.*, with the benefits of alternative policies equalized? Clearly, when $a = n$; that is, when domestic firms develop as many new products as foreign firms.

its attainable locally as a result of new patent protection were small relative to the profits realizable through sales in product-patent-granting jurisdictions abroad. MNEs at home in Switzerland, for example, were among the leading originators of significant new drugs even before Switzerland began granting drug product patents in 1976.

In the empirical analysis that follows, we focus on ascertaining how Italian pharmaceutical firms' R&D, patenting, and new product introduction propensities changed with the shift in patent regimes. Setting up the key hypotheses in their customary null form, we hypothesize that: (1) there was no significant trend increase in Italian drug R&D expenditures following the change in patent regimes; (2) there was no significant trend change in Italian drug firms' patenting of new drug products in jurisdictions where patents were available both before and after the Italian legal transition; and (3) there was no significant trend increase in the Italian firms' introduction of new drug chemical entities following the change in Italian law.

The conclusions drawn from our Figure 1 analysis are altered in complex ways when the possible impacts of patent law changes on foreign direct investment by MNEs are taken into account. For instance, there could be adjustments in the location of production and perhaps also the location of research activities. The relocation of MNE research laboratories to patent-granting nations could provide favorable spillover benefits to the new host nations, disseminating knowledge and imparting skills that can be absorbed or hired by domestic pharmaceutical companies. Economic theory provides ambiguous guidance on the investment locational consequences of patent protection.[11] On one hand, the availability of strong patents may lead foreign multinational enterprise heads to view a nation as a more secure and hence more desirable location for investment. On the other hand, patent protection may shield product imports from imitative local competition, permitting production to be centralized in some other nation where economies of scale are exploited more fully or wage conditions are more favorable. Lacking such patent protection, MNEs may feel compelled to commence on-shore production in order to neutralize "made at home" advantages enjoyed by indigenous rivals. The Marrakech Treaty will strengthen the propensity to centralize production, since nations may no longer threaten compulsory

11 For conflicting evidence from broader statistical analyses, *cf.* EDSON KONDO, "Patent Laws and Foreign Direct Investment: An Empirical Investigation" (Diss., Harvard University, 1994), finding at most insignificant FDI increases in strong-patent jurisdictions; and EDWIN MANSFIELD, "Intellectual Property Protection and U.S. Foreign Direct Investment," 77 Review of Economics and Statistics (1995), showing increases in FDI with stronger patent protection.

patent licensing unless the patent holder manufactures the patented product domestically. Because the balance of MNE incentives could tip in either direction, we must look to the evidence on what happened in Italy after the change in patent regimes.

Dow Chemical Company's acquisition of Gruppo Lepetit S.p.a. in 1963 marked the first major entry of a foreign pharmaceutical company into Italy. Multinational entry continued throughout the 1960s and appears to have accelerated after the change in patent law. In 1993, foreign owners controlled a reported 64 % share of the Italian-based industry, compared to 56 % in France, 50 % in Germany, and 23 % in the United States.[12] How the pattern of foreign ownership altered with the changing patent regime will be investigated more fully as a by-product of our main data analysis.

III. Data and Hypothesis Tests

We seek to determine how trends in research and development spending, the introduction of new products, and the patenting of pharmaceutical products changed following the Italian Supreme Court's decision in 1978. Data on pharmaceutical research and development expenditures within Italy were drawn from an annual publication of Farmindustria, the Italian pharmaceutical industry association.[13] The R&D expenditures were deflated, using the Italian gross domestic product deflator, to constant 1990 Lira. The Lira values were then converted to dollars using the mean market exchange rate for 1990.[14] Italian R&D outlays might have increased even if there had been no change in patent regimes. For a comparative benchmark, we use R&D expenditures (measured in constant 1990 dollars) in seven leading pharmaceutical-producing nations.[15]

Patent statistics have been widely employed by scholars as a measure of inventive activity. For Italy, of course, no pharmaceutical product patents were issued before 1982, and so no reliable pre-and-post data series exists. However, comparative studies of inventive activity across nations have typically focused on patenting in the United States, the world's largest market covered by a uniform patent law.[16] Inventions of more than

12 FARMINDUSTRIA, "Indicatori Farmaceutici Giugno 1994" 52 (Rome 1994). For a lower (32 %) reported MNE share of Italian sales for 1987–88, *see* BALLANCE, ET AL., *supra* note 4, at 71.

13 FARMINDUSTRIA, "Indicatori Farmaceutici Giugno 1994" (Rome 1994).

14 The GDP deflator and exchange rates were drawn from INTERNATIONAL MONETARY FUND, "International Financial Statistics Yearbook: 1994" (Washington 1994).

15 The latter are derived from BALLANCE, ET AL., *supra* note 4, at 89.

16 *See* DANIELE ARCHIBUGI, "Patenting as an Indicator of Technological Innovation," Science and Public Policy, December 1992, at 357–368.

local significance tend to be patented in the United States if they are patented at all, while for non-U.S. residents, the cost of pursuing U.S. applications is sufficiently high so that unimportant inventions are filtered out. Thus, patenting in the United States provides a tolerably consistent index of significant inventive activity both inside and outside the United States.

The U.S. Patent Office's Automated Patent System was used to identify patents issued within U.S. patent class 514 (Drugs and Bio-affecting Treatments) between 1972 and 1992 and originating from inventors residing in Italy. Individual patent specifications were consulted to eliminate from the time series patents covering pharmaceutical processes and compounds not tested for human use. If the assignee company was Italian-owned, the patent was considered to be an "Italian-owned" (ITPAT) patent. If the assignee company was not Italian-owned but had a majority of inventors listed as Italian residents, the patent was considered to be a "multinational-owned" patent (MULTPAT) derived from research presumably conducted in Italy. The data were aggregated into an annual time series covering all companies operating in Italy. They were also disaggregated into individual company time series.

The number of U.S. patents issued to Italian inventors could vary from year to year not only because of changes in the number of drug product inventions, but also because companies became more or less sensitive to obtaining patent protection (*i.e.*, they experienced changes in "the propensity to patent"), because of fluctuations in the U.S. Patent Office's processing backlog, or for other reasons. To control for influences affecting all drug patent issues, the variable UNIVPAT (universe patents) was defined. It counts all patents issued by year in Class 514, including process inventions and patents issued to Italian inventors. The inclusion of Italian inventors imparts no appreciable bias to this benchmark series, since patents of Italian origin consistently amounted to 2 % or less of the UNIVPAT total. To investigate changes in the propensity to patent, we divide the number of patents issued by appropriately lagged pharmaceutical research and development outlays – for Italian patents, by Italian R&D; and for the UNIVPAT series, by total pharmaceutical R&D outlays in seven leading nations.[17] According to studies by DiMasi et al.,[18] approximately half of all pharmaceutical R&D outlays occur during pre-clinical stages and half in the clinical testing stages. Patents are usually applied for near the end of pre-clinical work and issue early in the clinical testing stage. Therefore, we assume a lag of zero in computing the propensity to patent variables.

17 *See* F. M. SCHERER, "The Propensity To Patent," 1 International Journal of Industrial Organization 107–128 (1983).

18 *Supra* note 6.

Summary of Variables

Summarizing, the principal time series variables analyzed are as follows:

ITRD – Aggregate pharmaceutical R&D expenditures by all companies in Italy.

UNIVRD – Aggregate pharmaceutical R&D expenditures in seven leading nations, including Italy.

PAT – Class 514 drug product patents with at least one Italian inventor and assigned to a company.

ITPAT – Class 514 drug product patents assigned to an Italian-owned company.

MULTPAT – PAT minus ITPAT, *i.e.*, drug patents of Italian origin assigned to a multinational enterprise.

UNIVPAT – All Class 514 U.S. patents.

ITPROP – PAT divided by ITRD, or the number of Italian origin patents per billion dollars of R&D.

UNIVPROP – UNIVPAT divided by UNIVRD, or the number of Class 514 patents per billion dollars of seven-nation pharmaceutical R&D.

Statistical Results

Figures 2, 3, 4 and 5 provide a broad overview of trends in the principal time series. Differing scales have been used for the Italian and universe data so that similarities and differences in the trends can be seen clearly.

Figure 2
Trends in Italian and Universe Patents

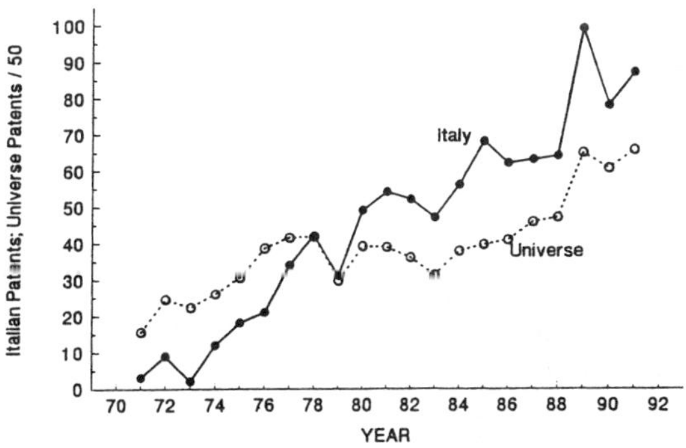

In Figure 2, Italian drug product patenting exhibits a strong upward trend well before the change in patent regimes and indeed before the challenge to earlier Italian law was initiated (in 1975). The (differently-scaled) Italian time series moves from below the universe trend line to

Figure 3
Trends in Italian and Universe R&D

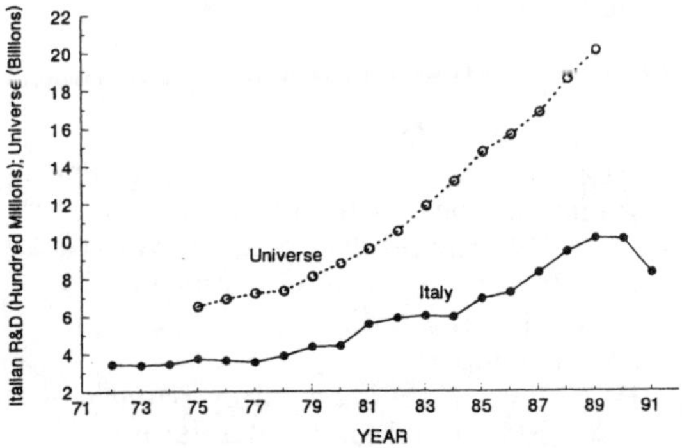

Figure 4
Trends in the Propensity To Patent

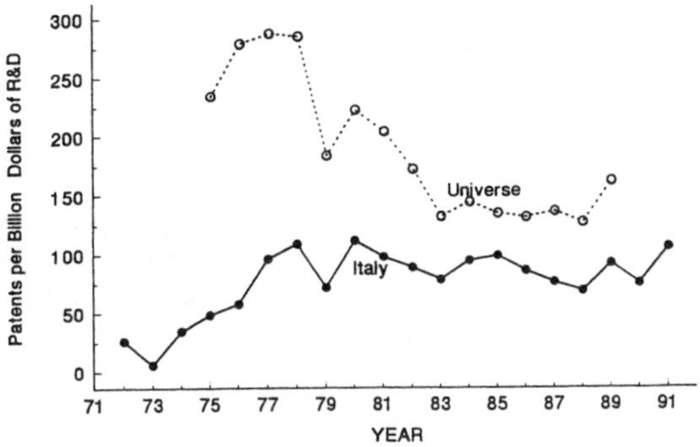

above it in the early 1980s, shortly after the change in patent law regimes. For Italy (except in 1991) and the universe of seven nations (Figure 3), the trend in inflation-adjusted R&D spending was upward. Within the years 1975–89 for which comparable data were available, the trend rate of growth (in constant dollars) was 7.84 % for Italy and 8.52 % for the seven nations. A more fine-grained analysis is needed to discern whether the change in patent regimes had any impact. Figure 4 reveals a marked decline in the propensity to patent, measured by the number of patents per billion dollars of R&D, for the seven-nation universe, but relative stability for Italy after 1980. In Figure 5, the Italian patent series is divided into two subgroups – those issued to domestic companies and those issued to multinationals. For the multinationals, there is a marked increase after 1982 which merits more precise analysis.

To tease further implications from the data, time series regression analyses were estimated by ordinary least squares. The effect of the change

Figure 5
Domestic and Multinational Drug Patents Originating in Italy

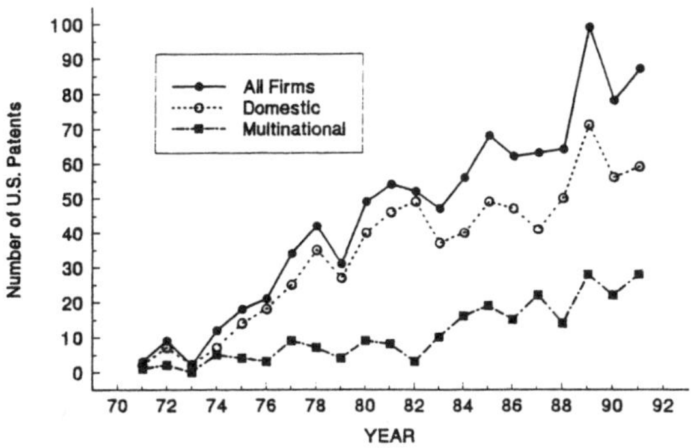

in legal regimes is tested by introducing a dummy variable COURT, with a value of zero before 1978 and unity thereafter.[19] For observations analyzed without transformation, a significant positive coefficient for COURT implies an abrupt one-time upward shift in the relevant time series. For R&D observations transformed logarithmically, a significantly positive COURT coefficient implies a change in the *slope* of the time series, with the absolute difference in R&D outlays attributable to the change of regimes increasing over time. Slope versions of the COURT dummy variable were also tested with untransformed patent counts. They had generally lower explanatory power than the intercept version and, when combined with intercept dummy variables, reduced unexplained variance of the time series by a statistically insignificant amount (measured in F-ratio tests).

To test whether the number of U.S. pharmaceutical patents issued to Italian firms increased with the change in patent regimes, we regress the count of Italian firms' patents on the Class 514 universe count plus the COURT dummy variable:

$$(1)\quad \text{PAT} = \underset{(5.0)}{-24.5} + \underset{(9.29)}{0.027}\ \text{UNIVPAT} + \underset{(6.43)}{25.3}\ \text{COURT};\ N = 21,$$

$$R^2 = 0.94,\ \text{D.W.} = 1.36;$$

with t-ratios reported in subscripted parentheses. Italian patenting rose as expected with the general growth in the number of all Class 514 drug

19 For the patent series, one would expect the change in legal regimes to materialize with a lag of roughly three years – the average time required to bring a patent application at the U.S. Patent Office to issue. Thus, calibrating COURT at 1978 implies a change in patent application behavior around 1975, when the suit challenging the Italian law was filed. An alternative dummy variable LCOURT with zero values before 1982 was tested with results (not reported here) that were only slightly inferior.

patents issued. Controlling for that effect, we observe an increase in the number of Italian company patents by approximately 25 per year. The dummy variable coefficient revealing this change is significant at better than the 99 % confidence level.[20]

We consider next changes in Italian pharmaceutical R&D as a function of R&D outlays in seven leading nations. Because of irregularities in the distribution of residuals from a regression equation in which untransformed observations were used, we take logarithms (to the base) of the observations:

$$\text{(2) Ln ITRD} = \underset{(20.7)}{-2.75} + \underset{(13.78)}{0.90}\ \text{Ln UNIVRD} + \underset{(0.33)}{0.020}\ \text{COURT; N} = 15,$$

$$R^2 = 0.96,\ \text{D.W.} = 1.15.$$

In this case, we find no significant break from the upward trend in Italian drug R&D outlays, which is highly correlated with world wide R&D trends. Italy's patent regime change apparently had little or no impact on the trend of inflation-adjusted R&D expenditures.

If Italian firms' drug patenting rose following the shift in patent regimes but R&D trend growth did not, the implication is that a change occurred in the Italian firms' propensity to patent. This is verified by regressing the ratios of Italian patents per billion dollars of R&D to comparable universe ratios plus a shift dummy variable:

$$\text{(3) ITPROP} = \underset{(0.37)}{8.4} + \underset{(2.75)}{0.22}\ \text{UNIVPROP} + \underset{(3.75)}{43.4}\ \text{COURT; N} = 15,$$

$$R^2 = 0.54,\ \text{D.W.} = 2.06.$$

After 1977 there was a sharp and statistically significant upward jump in the number of Italian patents received per billion dollars of R&D outlays. For this apparent behavioral change we suggest the following explanation. Before drug product patents could be obtained under Italian law, Italian firms had little reason to hire patent attorneys and seek protection for new chemical entities they developed at home. Since they had at best modest sales in nations awarding strong drug patents, applying for patents abroad, *e.g.*, in the United States, seldom promised benefits outweighing incremental costs. However, once patents could be sought in Italy, it became worthwhile to file for domestic patent protection. And once the fixed cost of prior art searches and filing in Italy was incurred, the marginal cost of obtaining U.S. patents was not so high as to discourage overseas applications.

20 Similar results were obtained when disaggregated individual company patent counts were regressed on UNIVPAT, COURT, and 25 company slope and intercept dummy variables. The results remained robust when the largest Italian patent recipient, Farmitalia Carlo Erba, was excluded from both the aggregated and disaggregated samples.

Figure 5 divides the patent data into two categories – patents obtained by firms whose home base was Italy, and patents with Italian inventors but assigned to a multinational enterprise. Equation (4) regresses the count of Italian firm patents on the Class 514 universe totals and COURT dummy variable:

$$(4)\ \text{ITPAT} = \underset{(3.5)}{-12.9} + \underset{(7.57)}{0.017}\ \text{UNIVPAT} + \underset{(7.55)}{22.4}\ \text{COURT};\ N = 21,$$
$$R^2 = 0.94,\ \text{D.W.} = 1.62.$$

The results are quite similar to those for all patents from Italian corporate inventors (equation (1)). For the patents assigned to multinational corporations, a superior "fit" was obtained using a slope dummy variable SLOPE81 rather than a simple intercept shift variable. Also, the fit was better when the slope variable had non-zero values only after 1980, rather than after 1977 as before:

$$(5)\ \text{MULTPAT} = \underset{(2.1)}{-6.45} + \underset{(3.52)}{0.007}\ \text{UNIVPAT} + \underset{(3.16)}{0.003}\ \text{SLOPE81};$$
$$N = 21,\ R^2 = 0.85,\ \text{D.W.} = 1.79.$$

Evidently, multinational enterprises reacted to the transition in Italian patent law somewhat more slowly than Italian-based firms. Since they were undoubtedly obtaining patents in other jurisdictions, it seems unlikely that their behavioral change merely reflects an increased propensity to patent in the United States inventions flowing from research in Italy. Whether the significant increase beginning in 1981 reflects an increasing multinational presence in Italy while the R&D operations of Italian firms acquired by MNEs remained essentially unchanged, or whether multinationals intensified their Italian R&D operations, cannot be determined, since only aggregated R&D data were available.

New Product Introductions

The regression analysis evidence suggests that Italian drug manufacturers altered their U.S. patenting behavior following the change in Italian patent law, but did not break the historical trend of their R&D expenditures, which had been rising at nearly the same rate as those of non-Italian firms. This inference is consistent with a tabulation of Italian firms' drug innovations compiled by Etienne Barral.[21] Barral divided

21 P. Etienne Barral, "Research, Medicines, Competition, Communication: Fifteen Years of Pharmaceutical Research Results Throughout the World (1975–1989)" 56 (Foundation Rhône-Poulenc Santé, 1990). Pablo M. Challu reports a declining trend in the number of new products invented by Italian firms after 1980. "Effects of the Monopolistic Patenting of Medicine in Italy Since 1978," 10 International Journal of Technology Management 237 (1995).

newly introduced drugs into four categories based upon their therapeutic contribution and novelty:

Type A: Therapeutic advance, new chemical configuration
Type B: Therapeutic advance, known chemical structure.
Type C: No therapeutic advance, new chemical structure.
Type D: No therapeutic advance, known chemical structure.

Italian manufacturers' new drug introductions break down by time period and novelty type as follows:

	1975–79	1980–84	1985–89
Type A	0	0	1
Type B	1	2	2
Type C	5	2	4
Type D	20	15	20
Total	26	19	27

Both before and after the change in patent regimes, most of the Italian firms' new products conform to the pattern expected of an industry specializing in imitative preparations. Although the single Type A product introduced during the late 1980s could conceivably presage a more innovative orientation, we are unable to reject the hypothesis that it was a chance event.[22]

International Trade Balance

Figure 6 traces trends in Italy's balance of international trade in pharmaceuticals.[23] Exports include those from both domestic and multinational companies. The trade balance index used is BALANCE = (Exports − Imports)/(Exports + Imports). It is calibrated in percentage terms. Prior to the change in patent law, a modest trade surplus prevailed. Following the change of patent regimes, both exports and imports rose, but imports increased more rapidly, so that the trade balance turned negative, attaining substantial negative values during the second half of the 1980s. Although alternative explanations cannot be ruled out, the data are consistent with the conjecture that importers become more competitive in Italy after gaining patent protection, while Italian manufacturers, unable to copy the newest drugs developed in other nations, saw their ability to sustain relatively strong export sales ebb.

22 In a test of differences among the three time periods, $F(2,9) = 0.067$, which falls far short of statistical significance.

23 The source is "Indicatori Farmaceutici," *supra* note 13.

Figure 6
Trends in Italian Pharmaceutical Trade Balance

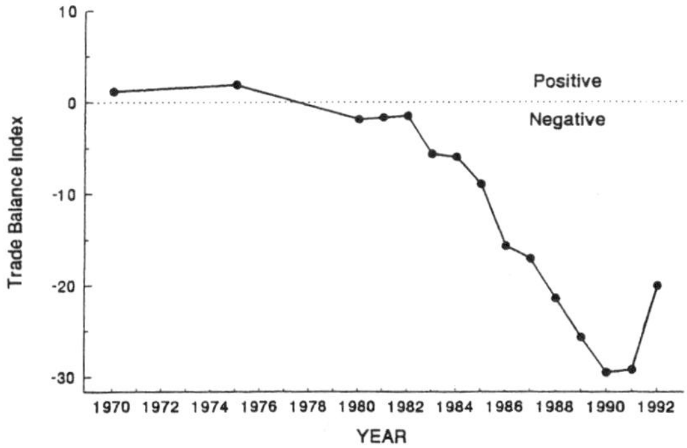

IV. Conclusion

The evidence available to us supports a conclusion that the legitimization of drug product patents in Italy did not induce a marked shift in Italian pharmaceutical manufacturers' strategic emphasis from emulating drugs developed elsewhere to developing innovative drugs. Drug R&D expenditure growth in Italy did not accelerate after the patent regime transition, the number and character of new product launches did not change significantly, and Italian firms' increased patenting of drug chemical entities in the United States appears mainly to have reflected a rise in the propensity to patent.

It does not necessarily follow that granting product patents could not under some conditions stimulate innovative drug development. The Italian experience was probably affected by the existence of an important confounding variable. The Italian government, like the governments of most nations, controls the prices at which pharmaceutical preparations are marketed.[24] Price controls were stringent in Italy during the period surrounding the shift in patent regimes. Between 1976 and 1992, the consumer price index for pharmaceutical products fell more in Italy relative to general price trends than in other European Community nations such as the United Kingdom, France, and Germany.[25] Tight price controls could have inhibited Italian firms' willingness to invest in innovative products and overwhelmed the stimulative incentive effects of drug product patent protection. Disentangling the

24 *See* BALLANCE, ET AL., "The World's Pharmaceutical Industries," *supra* note 4, at 142–145.

25 "Indicatori Farmaceutici," *supra* note 13.

conflicting chains of causation would require data much richer than those available to us.

Nevertheless, from our research, one must remain skeptical whether significantly increased new drug development efforts are likely in the nations required under the Treaty of Marrakech to offer drug product patents for the first time. Most of those nations are less-developed. Most have scientific infrastructures less conducive to state-of-the-art pharmaceutical research and development than Italy's. (India may be an exception.) Like their counterparts in industrialized nations, most of their governments impose more or less comprehensive price controls on pharmaceutical products. Thus, the factors that inhibited dramatic changes in the Italian environment are likely to operate even more strongly in the nations whose patent policies must be adapted to satisfy new World Trade Organization mandates.

Journal of International Economic Law (2002) 913–939

POST-TRIPS OPTIONS FOR ACCESS TO PATENTED MEDICINES IN DEVELOPING NATIONS

F. M. Scherer and Jayashree Watal***

ABSTRACT

This article explores the tension between granting patent protection under the TRIPS Agreements and the availability of medicines at affordable prices to developing countries. A crucial consideration under the TRIPS compulsory licensing option is the 'adequate remuneration' paid. A theoretical and empirical analysis shows that the royalties set under past compulsory licenses have been much lower than those that would be established under the 'foregone profits' standard of US patent law. To respect comparative advantage in the supply of licensed drugs, the TRIPS language requiring that compulsory licensing be predominantly for domestic supply needs clarification. The multinational drug pricing strategy that best combines equity with coverage of R&D costs is a variant of Ramsey pricing, under which prices are much lower in nations with low ability to pay and/or high price elasticities of demand than in wealthy nations. Statistical evidence on the prices of 15 AIDS drugs in 18 low- and medium-income nations reveals that tendencies toward Ramsey pricing were at best weak. To encourage Ramsey pricing, parallel exports should be barred from low-income nations, and price controls should not benchmark the prices charged in low-income nations. Outright donation can also enhance the supply of drugs to low-income nations. A quantitative analysis shows that when the marginal cost of production is low relative to 'inventoriable' average cost, donations can actually enhance a drug producer's after-tax profits under US tax laws. Minor tax law changes to enhance donation incentives are suggested.

INTRODUCTION

With the creation of the World Trade Organization in January 1995, a wide-ranging international agreement on Trade-Related Aspects of Intellectual

* Harvard University emeritus, and **Intellectual Property Division, World Trade Organization, respectively. A longer version of this article was written for Working Group 4 of World Health Organization's Commission for Macroeconomics and Health when co-author Watal was visiting scholar at Harvard University. It is available at http://www3.who.int/whois/cmh/cmh_papers/e/pdf/wg4_paper01.pdf.

This paper does not engage the responsibility of the WTO Secretariat or WTO Members, individu-

Property (TRIPS) entered into force. Its obligatory standards for patent protection on new medicines are closer to the pre-Uruguay Round norms existing in the United States and the European Community than those prevailing in many major developing nations. This has precipitated fears that nations previously choosing not to grant drug product patents might face higher prices for patented medicines. To be sure, drugs that were unpatented before 1995 will not be affected. But for nations such as India, Brazil, and Argentina, which before 1995 granted no drug product patents and hosted thriving generic drug industries, the ability to supply generic versions of new drugs will be curbed. And for other developing nations that relied upon such nations for generic drugs, the ability to import the newest medicines could be limited.

How the cost of the newest drugs will change depends upon the pricing policies pursued by research-oriented multinational drug companies prior to the implementation of TRIPS, how those policies evolve under the new legal regime, and the extent to which nations can take advantage of exceptions allowed under TRIPS. This article explores some means by which less-developed nations might maintain or even improve their access to the newest drugs at prices consistent with their limited purchasing power. In particular, we analyze three main possibilities: compulsory licensing, the expanded use of discriminatory pricing by drug suppliers (taking as a benchmark the actual pricing of AIDS drugs during the late 1990s), and increased donation of essential medicines by their producers.[1]

I. COMPULSORY LICENSING

One option available under TRIPS is the issuance of compulsory licenses, which authorize a third party to make, use, or sell a patented invention without the patent owner's consent. The TRIPS Agreement distinguishes between two situations in which patented inventions are subjected to compulsory licensing.

Article 31 allows compulsory licensing after a proposed user has been unsuccessful within a reasonable period of time in negotiating to obtain from the patent holder authorization to use the patented invention 'on reasonable commercial terms and conditions'. The failed negotiations clause can be waived in case of national emergency or extreme urgency or for non-commercial public use. Any use under Article 31 compulsory licenses must be 'predominantly for the supply of the domestic market' of the authorizing nation, and the user must pay to the patent holder 'adequate remuneration', taking into account 'the value of the authorization'.

---

ally or collectively, and does not present authoritative interpretations of WTO provisions, which can only be done by WTO Members jointly.

[1] For a complementary analysis of other measures, including price controls, see Scherer and Watal, 'The Economics of TRIPS Options for Access to Medicines' in Brigitte Granville (ed) *The Economics of Essential Medicines* (London: Royal Institute of International Affairs 2002) 32–56.

Article 40 permits WTO member nations to take appropriate measures, including compulsory licensing of patented inventions, when judicial or administrative procedures have identified conditions that constitute 'an abuse of intellectual property rights having an adverse effect on competition in the relevant market'. The article offers a non-exhaustive list of cases under which compulsory licensing might be authorized, including insertion of exclusive grantback conditions into patent licenses, conditions preventing challenges to a licensed patent's validity, and coercing a licensee to pay for packages including unwanted patents. The article's language appears to track in a general way the 'abuse' doctrine of US patent antitrust law, although the article as a whole can be reconciled with European legal traditions holding that failure to supply or license a patented product at all, or supplying the product at unreasonably high prices, might be deemed abusive. Article 31(k) states that the need to correct anti-competitive practices may be taken into account in determining the amount of remuneration paid to the patent holder in Article 40 cases, i.e. that payments might be less than in Article 31 compulsory licensing cases.

Both articles embody language sufficiently vague, no doubt as a result of compromises made during the Uruguay Round negotiations, that some key provisions will be clarified only through amendments, authoritative interpretations, or dispute settlement findings. For example in relation to Article 31, the Doha Declaration on the TRIPS Agreement and Public health has already clarified that each Member has the right to grant compulsory licences and the freedom to determine the grounds upon which such licences are granted and that each Member has the right to determine what constitutes a national emergency or other circumstances of extreme urgency. The following discussion offers historical background for understanding the implementation of compulsory licensing and an analysis of issues likely to be controversial in the foreseeable future.

A. Actual compulsory licensing regimes

Many nations, industrialized and developing, have included in their patent laws provisions permitting compulsory licensing of patents under specified conditions.[2] The most common ground for compulsory licensing has been non-working of the patented invention within the patent-granting nation, i.e. supply of the relevant product through importation rather than domestic production. Since Article 27 of TRIPS requires that 'patents shall be available and patent rights enjoyable without discrimination as to . . . whether products are imported or locally produced', this historical basis for compulsory licensing

[2] See, e.g. Fredrik Neumeyer, *Compulsory Licensing of Patents under Some Non-American Systems*, Study No. 19 of the Subcommittee on Patents, Trademarks, and Copyrights, US Senate Committee on the Judiciary (Washington: 1959).

is ignored here. The available evidence suggests that it has been invoked infrequently even when it was available statutorily.

Under US patent law, the federal government may utilize technology patented in the United States when such use serves the national interest and reasonable compensation is paid.[3] In effect, the government issues itself or its designated private sector agent a compulsory license. During the 1950s and early 1960s, the US Department of Defense procured patented pharmaceutical products at substantially reduced prices from sources other than the patent holder – in most cases, from producers in nations such as Italy that provided no patent protection for pharmaceutical products.[4] The practice was ended after a clause prohibiting importation of patented pharmaceutical products was attached as a rider to a foreign economic assistance bill in August 1961.[5] In 1999, the US Supreme Court ruled that individual states could emulate the federal government in infringing valid US patents for governmental purposes if they followed appropriate legal procedures, including the determination of reasonable compensation.[6]

The United States has led the world in issuing compulsory licenses to restore competition when violations of the antitrust laws have been found, or in the negotiated settlement of antitrust cases before full adjudication has occurred.[7] By the late 1950s, compulsory licenses had been issued in roughly 100 antitrust cases covering an estimated 40,000–50,000 patents, including AT&T's basic transistor concept patents, IBM's computer and tabulating card machine patents, General Electric's fluorescent and incandescent lamp patents, du Pont's nylon patents, and Eastman Kodak's color film processing patents. Additional cases since then have led to the licensing of Xerox's plain paper copying machine patents, the tranquilizer Meprobamate, synthetic steroids, the antibiotic Griseofulvin, Cytokine biopharmaceutical patents owned by Novartis and Chiron, and the 9-AC cancer drug patent rights assembled under the merger of Pharmacia AB with Upjohn. Some of the US antitrust

[3] 28 US Code 1498.

[4] See 'Pentagon Discloses Policy on Drug Buying Abroad; Announces 3 Purchases', *Wall Street Journal* (20 January 1961) 3. Savings on the order of 65% were claimed. In one case, however, antibiotics procured abroad may have been produced using cultures and processes stolen from the Lederle Laboratories Division of American Cyanamid Corp. See *US* v *Bottone, Salb, and Sharff*, 365 F 2d 389 (1966).

[5] Public Law 87–195, sec 606(c). The amendment was offered on the floor of the House of Representatives by Rep. Richard Roudebush of Indianapolis, Indiana, and, after a brief debate, accepted for inclusion by a vote of 87 to 65. *Congressional Record*, 18 August 1961, 16283–16285. The amendment's language was amended insignificantly in conference committee. US House of Representatives, Report No 1088, Foreign Assistance Act of 1961 (Washington: USGPO: 1961).

[6] *Florida Prepaid Postsecondary Education Expense Board* v *College Savings Bank*, 119 S Ct 2199 (1999).

[7] See Marcus A. Hollabaugh and Robert Wright, *Compulsory Licensing under Antitrust Judgments*, staff report, Subcommittee on Patents, Trademarks, and Copyrights, US Senate Committee on the Judiciary (Washington: 1960); and F. M. Scherer, *The Economic Effects of Compulsory Patent Licensing*, New York University Monograph Series in Finance and Economics (New York: 1977).

decrees, such as those covering General Electric's incandescent lamp patents and the 8,600 patents in AT&T's portfolio, required licensing at zero royalty rates. Most provided for 'reasonable' royalties, whose more precise meaning will be considered subsequently.

The circumstances under which compulsory licensing has been ordered in fully litigated US antitrust cases include the use of patents as a basis for price-fixing or entry-restricting cartels, the consummation of market-concentrating mergers in which patents played an important role, and practices that extended the scope of patent restrictions beyond the bounds of the patented subject matter. Charging high prices has not in its own right been deemed actionable under US precedents. Nor has the accumulation of monopoly power through pyramided patents gained through internal inventive efforts been found in litigated cases to warrant compulsory licensing.[8]

The competition policy precedents of leading European nations and the European Community are in this respect more expansive than those of the United States. A company that controlled patented processes used to produce a key chemical intermediate for a drug effective against tuberculosis was found under Article 86 of the European Community treaty to be abusing its monopoly power when, after entering into production of the drug through its subsidiary, it subsequently refused to sell or license the intermediate to an independent pharmaceutical manufacturer.[9] The patent holder was ordered to resume third-party intermediate sales. Reacting to the 'stagflation' tendencies of the 1970s, the German Federal Cartel Office instituted a series of actions asserting that enterprises with dominant market positions, based in some cases on patent rights, had abused their monopoly power by effecting unjustified price increases.[10] The most thoroughly litigated case focused on the tranquilizers Valium (at the time, the most-prescribed drug in Germany) and Librium. In the end, Hoffmann-LaRoche, the products' manufacturer, was not required to adjust its prices downward because no adequate basis for comparing HLR's prices against those of an 'as if effectively competitive' supplier could be found. It seems clear from the various higher German court decisions, however, that high monopolistic prices would have been ruled to be abusive if competitive benchmarks could be established.

The United Kingdom and Canada provide the leading examples of compulsory licensing of drug patents without a finding that the anti-monopoly

[8] But see *US* v *General Electric Co. et al.*, 115 F Supp 835, 844 (1953).

[9] *ICI and Commercial Solvents Corporation* v *Commission of the European Communities*, 1974 ECR 223, 250 (1974). The European Court's decision is abstracted in 'Refusal by a Dominant Firm to Sell Raw Materials,' 19 Antitrust Bulletin (Fall 1974) 605–18.

[10] See Erich Kaufer, 'The Control of the Abuse of Market Power by Market-Dominant Firms under the German Law against Restraints of Competition', *Zeitschrift fuer die gesamte Staatswissenschaft*, Band 137 (September 1980) 510–32; and Ingo Schmidt, 'Different Approaches and Problems in Dealing with Control of Market Power: A Comparison of German, European, and US Policy Towards Market-Dominant Enterprises', 28 Antitrust Bulletin (Summer 1983) 417–69.

laws have been violated. In the United Kingdom, Section 41 of the Patents Act of 1949 distinguished foods, medicines, and surgical devices from other patent-protected products by articulating a rebuttable presumption in favor of compulsory licensing to ensure that the products are 'available to the public at the lowest prices consistent with the patentees' deriving a reasonable advantage from their patent rights.' Between 1953 and 1971, a total of 20 compulsory licenses were granted in response to 54 applications, covering, *inter alia*, such important products as Chloromycetin, Librium, and Valium.[11]

With the United Kingdom's ratification of the Uruguay Round treaty, the UK laws on compulsory licensing were amended.[12] For UK patents held by residents of WTO signatory nations, provisions authorizing compulsory licensing of inventions not worked within the UK were eliminated. Also removed were the provisions singling out food, drug, and surgical device patents as subject to especially strong presumptions in favor of compulsory licensing. However, compulsory licensing of WTO-member nationals' patents can be ordered under Section 48 of the Patents Act when the UK demand for a patented invention is not being met 'on reasonable terms', or when the patent owner has refused to grant a license 'on reasonable terms', or (under Section 51) when a monopoly found by the UK competition policy authorities to be operating against the public interest has refused to make patent licenses available 'on reasonable terms'.

Canada's experience has been more far-reaching. Since 1923 Canada had a law providing for compulsory licensing of the right to manufacture within Canada drugs (and also food products) protected by patents (usually process patents, since product patents were not available at the time). The law saw little use, at first because few important drugs were covered by patents up to the time of World War II, and later because the Canadian market was considered too small to realize all economies of scale in the production of bulk therapeutic ingredients and because long delays in the granting of licenses left little or no profit to be realized by small-scale domestic generic producers.[13] Recognizing that importation of bulk ingredients was virtually essential if Canadian consumers were to receive the intended benefit of medicines 'available . . . at the lowest possible price consistent with giving to the patentee due reward for the research leading to the invention', the Canadian Parliament amended the law in 1969 to permit licenses for importation. The new law also required the responsible Commissioner of Patents to approve or disapprove a license application within 18 months of its receipt.

[11] Scherer, above n 7, 40–41 and 44–45.

[12] The amended provisions can be found at the United Kingdom Patent Office's web site, http://www.patent.gov.uk.

[13] See Paul K. Gorecki, *Regulating the Price of Prescription Drugs in Canada: Compulsory Licensing, Product Selection, and Government Reimbursement Programs*, Technical Report No 8, Economic Council of Canada (Ottawa: May 1981) 28–34.

Between 1969 and 1977, 227 licenses were issued, only eleven of them calling for domestic production alone without the right of importation.[14] The most typical approach has been for the active ingredients to be imported in bulk, with encapsulation and packaging occurring in Canada. Among 47 drugs for which licenses were issued between 1970 and 1978, the average number of licensees was 3.0, with a range of from 1–11. Although some license recipients did not follow through by actually supplying the drug in Canada, in many cases, and especially for the drugs with substantial sales volume, competition was secured in the generic provision of drugs that would otherwise have been monopolized by the patent owner. On average, generic drugs supplied under compulsory license captured roughly 19% of the total sales of the product lines in which they competed, with penetration rates varying widely across Canadian provinces, depending upon the extent to which provincial drug reimbursement rules encouraged or discouraged generic substitution.[15] In Ontario, where the rules were most conducive to substitution, penetration rates were as high as 55–64% at the retail level.[16] Gorecki estimates that for drugs on which competition through compulsory licensing occurred, prices during the late 1970s would have been 20% higher in the absence of such competition.[17] A later study of 29 drugs subjected to compulsory licensing in Canada but patented in the United States revealed that the Canadian prices were on average 47% lower than their US counterparts in 1982.[18] For Valium, one of the world's best-selling drugs during the 1970s and the licensed drug sales leader in Canada, the price to hospitals fell from $42 per 1,000 units before licensing began to $4.10 by the end of the 1970s.

In anticipation of proposed free trade treaties, the Canadian compulsory licensing law was weakened in 1987 with the imposition of a 7–10 year exclusivity period for drug patent holders. It was eliminated altogether in 1992. As quid pro quos, the multinational drug companies agreed to locate in Canada drug research and development activities proportional to Canada's share of their world sales and to accept a new regime of 'reasonable price' controls by a Canadian Patented Medicines Review Board.

[14] Gorecki, above n 13, at 46. Some companies received multiple licenses to import and/or produce a given drug, so the total number of non-duplicating licenses out of 227 approvals was 152.

[15] Gorecki, above n 13, at 86.

[16] Gorecki, above n 13, at 89. See also James J. McRae and Francis Tapon, 'Some Empirical Evidence on Post-Patent Barriers to Entry in the Canadian Pharmaceutical Industry', 4 Journal of Health Economics (March 1985) 43–61.

[17] Gorecki, above n 13, at 149.

[18] On this and other price comparison studies, see the submission of Lawson A. W. Hunter, Director of Investigation and Research under the Combines Investigation Act, to the Commission of Inquiry on the Pharmaceutical Industry, 14 August 1984, p. 6. The study cited in the text was by Tom Brogan, Mario Deschamps, and Guy Roberge, 'Drug Cost Differential Between Canada and USA' (Ottawa: Consumer and Corporate Affairs: 1983).

B. Determining license compensation rates: the theory

Third-party use of patents under compulsory licensing mandates or infringement for government purposes is usually accompanied by a condition that the user pay 'adequate remuneration' or 'reasonable compensation' for its use of the patented technology. The extent to which price-reducing competition can arise through compulsory licensing depends critically upon the criteria imposed to determine the magnitude of the compensation paid.

With the creation of a new Appellate Court for the Federal Circuit in 1982, the United States has evolved particularly generous criteria for compensating patent holders when infringement of their patents is proven. There have been record-breaking awards in a number of important cases – for example, $900 million for Kodak's infringement of Polaroid patents, $550 million to the Lemelson trust for bar coding, $212 million from Steelcase to Haworth for movable office panels, and $171 million from Mobil to Exxon for a plastic catalyst.

The US Patent Code, 35 US Code 284, provides that in cases of patent infringement, the damages awarded shall be:

> ... *adequate to compensate* for the infringement but in no event less than a *reasonable royalty* for the use made of the invention by the infringer, together with interest and costs. [emphasis added]

One implication, confirmed by court decisions, is that there are two potentially different standards, with the 'reasonable royalty' standard tending to be less generous and/or second-best.

Figure 1 shows how the new Appellate Court for the Federal Circuit has interpreted the 'adequate to compensate' provisions of the US statutory law.

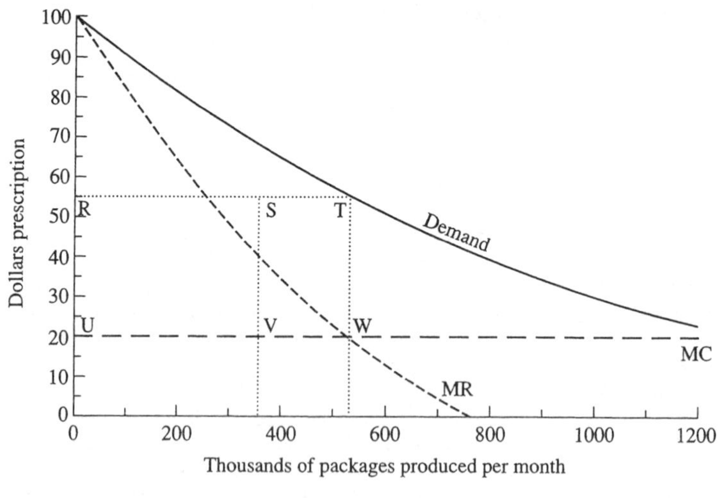

Figure 1. Royalty determination under profits lost standard.

Suppose the demand curve for a drug product is as marked in Figure 1 and marginal production costs are $20 per standard prescription package. Absent competition, the patent holder is in effect a monopolist in the sale of its product.[19] It derives its marginal revenue function (dashed line marked MR), equates marginal revenue with marginal cost, and sets a price of $55 per package, producing 530,000 packages monthly to satisfy the demand. Its contribution to profits and the repayment of research and development costs is 530,000 × (55–20) = $ 18.55 million per month. If another firm enters and supplies VW packages (= 175,000), inducing the patent holder to reduce its output to 355,000 packages, the infringer deprives the patent holder of profits measured by the area rectangle STWV, or 175,000 × (55–20) = $6.125 million per month, and under the 'adequate to compensate' law as interpreted, that is the initial measure on the basis of which damages are computed. If assessed on an *ad valorem* basis, the royalty would approximate 64% of the infringer's sales.

Suppose, however, that because it lacks a well-known brand name or other first-mover advantages, the infringer can only obtain a price of $40 per package for its version of the drug. Then the highest royalty it could agree to in arms-length negotiations would be $20 per package. If the courts accepted those facts as a basis for awarding only a reasonable royalty, the royalty rate would be reduced to 50% *ad valorem* on the infringer's lower-priced sales (which is 36% relative to the patent holder's price). If the infringer's lower price forced the patent holder to reduce its price (which is unlikely, given evidence that branded drug sellers tend not to reduce their prices in the face of generic competition), the court under a 'profits lost' standard would count as additional damages to be paid by the infringer the profit loss per package still sold by the patent holder due to price suppression times the number of packages sold by the patent holder. Given a downward-sloping demand curve, however, the lower price will lead to higher unit sales by the patent holder, complicating the damages assessment problem in ways beyond the value of further exploration here.[20]

Two things are evident from this elementary exposition. First, if 'adequate remuneration' were construed under the simplest of the 'profits lost' tests analyzed here, compulsory licensing would impose such high royalty payments on the licensed producer that there could be no price reduction and hence no expansion of drug availability at all. Since the purpose of virtually all known compulsory licensing schemes is to increase competitive supply and reduce prices, the 'profits lost' test cannot logically be the standard to be met

[19] This does not mean that there are no other chemically different products with similar therapeutic effects; it only means that the products are sufficiently differentiated that the drug's patent holder faces a downward-sloping demand curve taking into account the availability of imperfect substitutes.

[20] For an analysis showing that price suppression and loss of output by a dominant patent holder tend not to coincide, see the declaration of F. M. Scherer submitted in the matter of Mahurkar Double Lumen Hemodialysis Catheter Patent Litigation, MDL-853, US Federal District Court for the Northern District of Illinois, September 1992.

in determining compensation for compulsory licensing. Second, it is evident from a large number of cases that the royalties awarded in actual compulsory licensing cases have been much lower than the 50–64% *ad valorem* rates derived in our simple 'profits lost' example. On this second point, we now examine relevant evidence, dealing first with patent infringements by the US government, then with compulsory licensing orders used as antitrust remedies, and finally with compulsory licensing directed specifically toward pharmaceuticals.

C. Actual compensation determination experience

For US government use of Enrico Fermi's patent covering plutonium production, a payment of $300,000 was made – less than 1% of the government's World War II investment in the Hanford plutonium extraction facilities. The heirs of Robert S. Goddard were paid $1 million for the government's use of Goddard's rocket engine patents – about 0.01% of the value of the liquid-propelled rockets produced during the life of the patents.

In what was initially described as the largest patent compensation case in history, Hughes Aircraft claimed a 15% royalty, or $3.3 billion in total, on the value of 81 government satellites using Hughes' geostationary orbit technology. The US government argued for a 1% royalty.[21] The government's 1% proposal was accepted by the US Court of Claims and sustained on appeal.[22] According to a 1991 survey, the highest royalty rate paid by the US government in compensation for the use of a portfolio of pioneering private patents was 10%.[23] Rates of 6% were said to be applied 'as a general rule' in the absence of contrary evidence. In one such case, a relatively small firm, Tektronix, asked for recovery of its profits, said to be in the range of 24–28% on commercial sales, when the government authorized another company to use the Tektronix patent portfolio in delivering at substantially lower quoted prices special-design oscilloscopes.[24] The court of appeals observed that had Tektronix earned such high profits on sales to the government, they might have been reduced under the profit renegotiation procedures applicable at the time. It went on to observe that under precedents interpreting claims against the US government under 28 US Code 1498:[25]

> [The] goal of 'complete justice' implies that only a reasonable, not an excessive, royalty should be allowed where the United States is the user – even though the patentee, as a monopolist, might be able to extract excessive gains

[21] See 'Patent Case May Cost US Billions', *New York Times* (22 April 1989), at D1.

[22] *Hughes Aircraft Co.* v *United States*, 86 F. 3rd 1566 (June 1996), affirming 31 Fed Cl 481 (1994).

[23] Richard J. McGrath, 'The Unauthorized Use of Patents by the United States Government or its Contractors', 18 AIPLA (American Intellectual Property Law Association) Quarterly Journal (1991) 349, 359.

[24] *Tektronix Inc.* v *United States*, 552 F 2nd 343 (1977).

[25] 552 F 2d 343, 351 (1977). See also *Leesona Corp.* v *United States*, 599 F 2nd 958 (1969).

from private users. Much of the content of the constitutional requirement of just compensation derives from the equitable principles of fairness as between the Government and its citizens.

Some important US antitrust judgments, as noted earlier, have required that patent portfolios be licensed at zero royalty rates. In the more typical cases, royalty rates have been modest. For example, licensees were required to pay 0.5% *ad valorem* for the first Xerox plain paper copying machine patent they used, an additional 0.5% for the second patent, and then an additional 0.5% (implying a maximum royalty of 1.5%) for the remainder of Xerox's vast patent portfolio.[26] At the time, Xerox was devoting 5.6% of its sales revenue to research and development. In a decision later overturned on procedural grounds unrelated to the compensation question, the Federal Trade Commission ordered that the patent covering the antibiotic tetracycline be licensed at an *ad valorem* royalty rate of 2.5%. Before generic competition began, tetracycline was sold at wholesale for a price of $30.60 per 100 capsules. Production costs were on the order of $3.00, so a 'profits lost' royalty rate would have been on the order of 90%.[27] In most compulsory licensing cases, royalties were left to be settled through negotiations by the parties, so no public record exists. But in the minority of cases requiring the courts to step in and settle disputes, royalties of from 0.2 to 3.0% have been reported.[28]

The United Kingdom Comptroller of Patents pursued an essentially cost- and profit-based approach to setting royalties for compulsory licenses to drugs under Section 41 of the UK Patents Act. To research, development, and testing costs averaged over the licensing firm's pharmaceutical operations, a fairly generous profit margin was added to arrive at the royalty per kilogram. In an early case, this led to an *ad valorem* royalty rate of 18%.[29] A similar approach led to a fixed royalty per kilogram of the tranquilizer Librium that approximated 18% of the average price received by Hoffmann-LaRoche on its UK sales, but a higher percentage rate on the lower sales price attainable by generic producers. On Librium's more potent sister drug Valium, the per-kilogram royalty was set at roughly 22% of the selling price received by Hoffmann-LaRoche.[30] These royalties, although much less than the marginal profit rates realized by the patent-holding drug manufacturers, were sufficiently high to have impaired significantly the market inroads of compulsory-licensed substitute drugs.

During the 1970s and much of the 1980s, Canada had the world's most far-reaching compulsory drug licensing program, at least in part because of

[26] In the matter of Xerox Corporation, 86 FTC 364 (1975).

[27] F. M. Scherer, *Industrial Market Structure and Economic Performance* (Boston: Houghton-Mifflin 1980) 517–18.

[28] Scherer, above n 7, 49–50.

[29] In the matter of J. R. Geigy S.A.'s Patent, 1964 R.P.C. 391, discussed in Scherer, above n 7, at 45.

[30] See again Scherer, above n 7, 44–45.

the royalty determination approach adopted. The enabling statute declared that:[31]

> . . . in . . . fixing the amount of royalty or other consideration available, the Commissioner shall have regard to the desirability of making the medicine available to the public at the lowest possible price consistent with giving to the patentee due reward for the research leading to the invention.

In an early test case, the Commissioner of Patents rejected a fixed per-kilogram royalty proposal by Valium patent holder Hoffmann-LaRoche, which would have amounted to 30% of HLR's selling price and a substantially higher percentage of a generic substitute's price. Instead, an *ad valorem* rate (against the licensee's price, not the licensor's) of 4.0% was set. On appeal, the Exchequer Court affirmed the Commissioner's 4.0% rate, among other things rejecting the suggestion that royalties on licensed sales should reimburse a pro-rated share of the patent holder's research and development program outlays.[32] The 4.0% royalty rate was applied almost uniformly in subsequent compulsory licensing orders, among other things avoiding a detailed inquiry into unique cost factors by the Commissioner of Patents and the reviewing courts and hence facilitating procedures that expedited the entry of generic substitute drugs into the Canadian market.

To sum up, there is wide variation in the way responsible government agencies and courts have set the amount of compensation awarded to patent holders when patents have been subjected to compulsory licensing. The United Kingdom has provided the most generous compensation in its drug patent licensing decisions; the United States the least generous compensation in key antitrust case orders. None of the royalty determinations on which information is available have established rates approaching those that would emerge under a 'lost profits' criterion. The choices made in industrialized nations provide ample precedent for royalty-setting on the modest side of the range of possibilities.

D. Other obstacles

The longer the issuance of compulsory licenses is delayed after patented drugs enter the marketplace, the less time licensees have to recover their start-up costs and the more difficult it is to achieve effective competition among multiple generic substitute suppliers. Thus, if compulsory licensing is to be successful, expeditious licensing procedures are a necessity. TRIPS Article 31 requires judicial or other independent review of the decisions taken by the licensing authority. Here the experience in Canada is relevant. The licensing

[31] Section 41(4) of the Canadian Patent Act, as amended in 1969.

[32] *Hoffmann-LaRoche Ltd.* v *Frank W. Horner Ltd.*, 61 CPR 243 (1969), 64 CPR 93 (1970). In 1985 the so-called Eastman Commission recommended that the royalty rate be increased to 14%, compensating estimated R&D costs amounting to 10% of sales and promotional outlays of 4%. *Report of the Commission of Inquiry on the Pharmaceutical Industry* (Ottawa: 1985). The recommendation was not implemented.

authority there was required to reach its decisions within 18 months of a license application. In fact, the median time to decision was 10 months for applications filed between 1969 and 1977.[33] It will also be essential for the designated judicial authorities to establish clear and transparent precedents in early cases, as was done in Canada, so that they can perform subsequent reviews efficiently.

One of the most important unresolved challenges is the requirement under TRIPS Article 31(f) that any compulsory drug patent license be 'authorized predominantly for the supply of the domestic market of the Member authorizing such use'. Many and perhaps most less-developed nations lack the infrastructure and technical capabilities to build a domestic industry able reliably to supply modern pharmaceutical products. Even Canada, with high income per capita, excellent universities, and a population during the 1970s of roughly 22 million, found it necessary to import most of the bulk pharmaceuticals ultimately supplied under compulsory licenses. Thus, smaller less-developed nations will have to issue their compulsory licenses mainly for importation rather than domestic production. This in turn requires that competitive world market supply sources exist. The 'predominantly' term in Article 31(f) clearly implies that some exportation under compulsory license in the exporting nation will be allowed. At the Doha ministerial meeting in November 2001, the World Trade Organization's TRIPS Council was given a mandate to recommend a solution to the problem faced by countries with insufficient or no manufacturing capacities in the pharmaceutical sector by the end of the year 2002. A compilation of member nation proposals released in July 2002 (WTO document IP/C/W/363) revealed differences of detail unlikely to bar a compromise permitting importation by nations lacking domestic production capabilities.

II. DISCRIMINATORY PRICING?

Even with full patent protection, drugs might be sold in low-income nations at relatively low prices if multinational pharmaceutical companies engaged in a certain form of price discrimination, often called 'differentiated pricing', 'tiered pricing', or 'equity pricing' by non-economists.

Figures 2a and b tell the basic theoretical story. They assume two nations A and B with roughly equal numbers of cases in which use of a particular drug product might be indicated. At a zero price then, equal quantities of the drug – 1 million prescriptions (Rx) per month – would be demanded. However, Nation A is assumed to have high average income per capita while Nation B has low income per capita. This 'income effect' leads to different demand curves, assumed for illustrative purposes to be straight lines, for the two different nations, with the demand curve for Nation A being higher and

[33] Gorecki, above n 13, at 41.

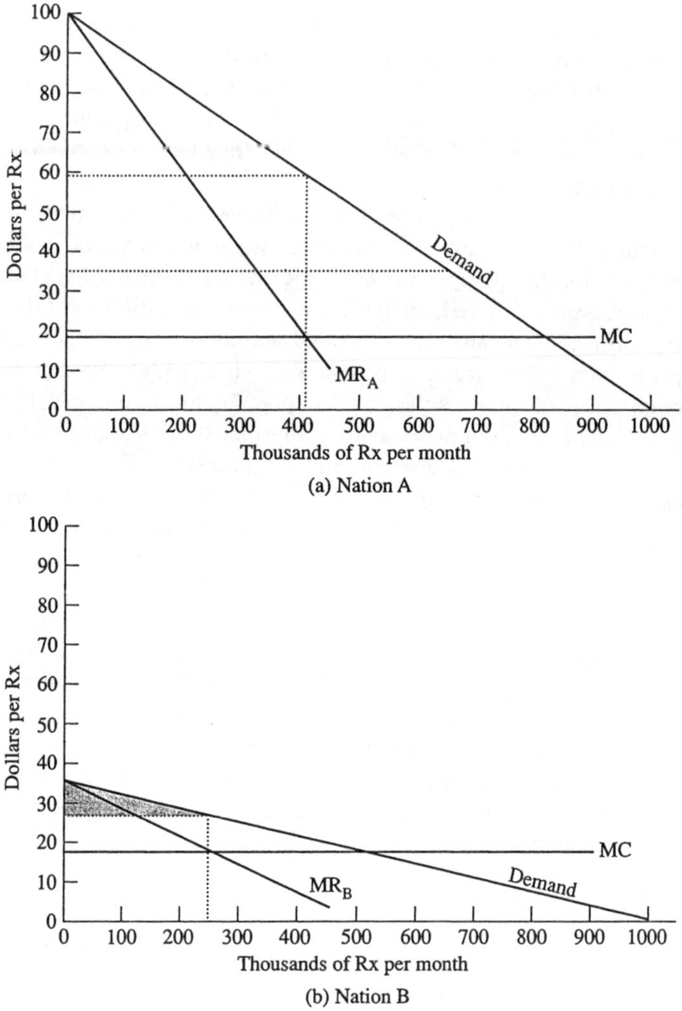

Figure 2. Price discrimination between markets of differing wealth.

(at any given positive price) less price-elastic than the demand curve for Nation B.[34] We assume also that the drug can be produced and distributed at a constant marginal cost of $18 per Rx, shown by the horizontal lines marked MC.[35] Patent protection permits the drug's producer to maximize its

[34] Where Q is quantity consumed per month and P is the price, the inverse demand curve for Nation A is assumed to be $P_A = 100 - 0.1\ Q_A$; and the curve for Nation B $P_B = 35 - 0.035\ Q_B$.

[35] Economies of scale or cost savings through learning by doing might invalidate this assumption, in some cases leading to price reductions in nations paying high prices.

profits, given demand and cost, in each market separately or in both markets together. In wealthy Nation A, the firm will derive its marginal revenue curve MR_A, equate marginal cost with marginal revenue at an output of 410(000) Rx, and set the corresponding price at $59 per Rx, earning a contribution to the repayment of its research and development outlays and to its profits measured by the rectangular area (59–18)(410,000) = $16.8 million per month.

If it must charge a uniform price in every national market, the drug's producer recognizes that at the price maximizing profits in Nation A, it can sell nothing in low-income Nation B, since the $59 price is higher than the maximum $35 price any consumer in Nation B is able and willing to pay. To sell anything at all in Nation B under a uniform price policy, the firm must reduce its price in Nation A to less than $35, entailing a profit and R&D cost recovery sacrifice in Nation A of at least (59–35)(410,000) less (35–18)(240,000) (i.e. the surplus of a $35 price less marginal cost on additional sales of 240,000 Rx) = $5.76 million. This sacrifice is larger than the zero profit the firm could make at a $35 price in Nation B. For prices lower than $35, it can be shown, the sacrifice in Nation A also exceeds the gain in Nation B, given the assumed demand functions.[36] Thus, if forced to charge a uniform price, the firm will not sell in Nation B.

If, however, it can engage in price discrimination, the firm will find it profitable to sell in both markets. It will derive its marginal revenue curve MR_B in national market B, equate marginal cost with marginal revenue at an output of approximately 243,000 Rx, and set a much lower $26.50 price, which maximizes profits in Nation B. Relative to selling only in Nation A, this is clearly profitable for the drug's producer. It contributes incrementally to the firm's profits and R&D cost recoupment by (26.50–18)(243,000) = $2.07 million. Relative to the no-sales-in-B case, it is also beneficial to the citizens of Nation B, adding consumers' surplus measured by the dot-shaded triangular area, or approximately $1.04 million per month. Ignoring any fixed costs that might be incurred setting up a sales outlet in Nation B, the drug producer will find it worthwhile to sell at a relatively low price in Nation B as long as at least part of Nation B's demand curve lies above the marginal cost line – a condition, to be sure, that may not be satisfied in the very poorest nations.

Thus, under the not-implausible conditions assumed here, both LDC consumers and the drug producer are better off with price discrimination than under uniform pricing. Although not valid under all plausible conditions, this case is typical of a broader range of economic situations conducive to what is

[36] If demand for the drug at relatively low prices in Nation B were many times higher than demand at the same prices in Nation A, an exception to this uniform price case could arise, and profits could be maximized by selling in both markets at a price much lower than the price that maximizes profits in Nation A.

called Ramsey-Baumol-Bradford pricing, or more simply, Ramsey pricing.[37] In the classic formulation, it is necessary to recover a substantial block of fixed costs (e.g. for research and development) by setting prices in a diversity of markets with differing demand elasticities. It can be shown that the most efficient solution is one in which prices are elevated above the marginal costs of production more, the less elastic demand is in any given market. 'Most efficient' in this sense means that the fixed costs are recovered and the sum of producer's surplus (e.g. contributions to fixed costs and profits) plus consumers' surplus (i.e. the amount consumers are able and willing to pay, less what they actually pay) is maximized.[38] Discriminatory pricing along Ramsey lines approaches as closely as one can reasonably hope to an ideal price-setting method in an intrinsically imperfect world.

A. Parallel trade and other obstacles

There are a number of reasons why pricing may not favor low-income nations in the way suggested by the basic theory. When prices are higher in one nation than in others, there is a tendency for arbitrage to occur through what is known as 'parallel trade'. That is, wholesalers in a low-price country divert supplies through international trade channels to nations in which the manufacturer is attempting to maintain high prices, undermining the high prices (and their contribution to research and development expenditures) in the wealthier nations and, if quantitatively substantial, inhibiting the manufacturer's willingness to supply at low prices in the low-income nation.[39]

At the negotiations that led to the Uruguay Round treaty, the participants essentially agreed to disagree, establishing in Article 6 of TRIPS language excluding from the WTO's dispute resolution jurisdiction the question of whether patent rights could block parallel trade. Widely varying national policies (under what is called the rights exhaustion principle) exist across countries and between different intellectual property regimes within countries.

Seeking to create a true 'common market', one feature of which is a tend-

[37] The relevant theory, attributable to Frank P. Ramsey (1903–1930), was brought back to the forefront of interest after long neglect by William J. Baumol and David Bradford, 'Optimal Departures from Marginal Cost Pricing,' 60 American Economic Review (June 1970) 265–83. Its application to drug pricing in international markets has been proposed, *inter alia*, by George Yarrow, 'CEC and EC Member State Industrial Policy and the Pharmaceutical Industry' in Adrian Towse (ed), *Industrial Policy and the Pharmaceutical Industry* (London: Office of Health Economics, 1995) 1–11; and Patricia Danzon, *Pharmaceutical Price Regulation: National Policies versus Global Interests* (Washington: AEI Press 1997), Chapter 7.

[38] Under the form of Ramsey pricing proposed for regulated utility price-setting by Baumol and Bradford, the regulator can squeeze the producer's profits down to the point at which only normal returns on investment are realized. When profits are subjected to such a squeeze, the so-called Ramsey number has a value less than unity. When firms are allowed to engage in price discrimination of the type illustrated here with no profit constraint, the Ramsey number has a value of unity.

[39] For a more thorough exploration of the theoretical possibilities and the underlying international law, see Scherer and Watal, above n 1.

ency for differences in product prices to be arbitraged away, the European Community has sharply discouraged impediments to parallel trade within the Community's jurisdiction. In pharmaceuticals, upon which (along with automobiles) EC authorities have focused with special vigor,[40] Bayer AG was fined 3 million ECUs in 1996 for attempting to restrict reshipment of the cardiovascular drug Adalat by wholesalers in Spain and France, where the wholesale price was relatively low, to the United Kingdom, where prices were 53–94% higher.[41] However, Community rules prevent unauthorized parties from importing drugs enjoying patent protection within the Community from nations outside the Community.

Although many nations share the European Community's ban on parallel imports of patented products from outside their borders, some less-developed nations, including Argentina, Thailand, and South Africa, have enacted laws permitting parallel imports – in the case of drugs for South Africa, at the discretion of the Minister of Health. The South African law was challenged in a national court by multinational pharmaceutical companies, but the suit was abandoned in April 2001. Concerned over the likelihood that governmental drug reimbursement programs would be extended to cover senior citizens and the high cost of pharmaceutical products, the US Congress also swerved from its long-standing antipathy toward parallel imports in September of 2000, passing bills that would permit the reimportation of patented drugs from sources such as Canada and Mexico, where prices charged by the patent-holding manufacturers have tended to be lower than in the United States. However, the law was declared by both the Clinton and Bush administrations to be unworkable because of difficulties in establishing quality control and labelling standards.[42]

Pharmaceutical manufacturers may also choose not to offer much lower prices in less-developed nations when the richer nations subject drugs to price controls, as many nations do, and when the controls emphasize what is called 'external reference pricing'. That is, prices in the home market are set at a level not exceeding the prices charged for the same drug in other countries. When the outside references include less-developed nations, drug manufacturers' incentive to offer low prices in those nations is undermined. Drug

[40] See Sir Leon Brittan, 'Making a Reality of the Single Market: Pharmaceutical Pricing and the EEC', address before the IEA Health and Welfare Unit, 1 December 1992, anticipating future EC actions against barriers to parallel trade. In 1990, parallel pharmaceutical imports amounted to an estimated 8% of domestic consumption in the United Kingdom, 5 to 10% in the Netherlands, and 1% in Germany—nations with relatively high drug prices. See Gernot Klepper, 'Pharmaceuticals' in Pierre Buigues, Alexis Jacquemin, and Andre Sapir (eds), *European Policies on Competition, Trade and Industry* (Aldershot, UK: Edward Elgar 1995) 335.

[41] Commission decision in re Adalat, Case IV/34.279/F3, decided 10 January 1996.

[42] See 'In a Turnaround, White House Kills Drug-Import Plan', *New York Times* (27 December 2000) 1; and 'Plan to Import Drugs from Canada Passes in Senate, but Bush Declines to Carry It Out', *New York Times* (18 July 2002).

manufacturers may also engage in niche pricing, finding it more profitable to sell their drugs at high prices to the wealthiest consumers in a low-income nation and forgo the possibility of achieving much broader sales at low prices to poor consumers.

B. Evidence on the pricing of AIDS drugs

It is useful to pause and ask whether multinational pharmaceutical firms have in fact engaged in Ramsey pricing strategies across the various nations, rich versus poor, of the world. That they have done so in the pricing of vaccines is well documented.[43] To shed new light on this question we obtained detailed information on sales revenues and quantities sold for 15 AIDS anti-retroviral drugs in 18 nations or national groups, all with low or intermediate per-capita incomes, over the years 1995 through 1999.[44] The nations or national groupings comprise Argentina, Brazil, Central America, Chile, Colombia, the Dominican Republic, Ecuador, French West Africa, India, Indonesia, Malaysia, Mexico, Peru, the Philippines, South Africa, Thailand, Uruguay, and Venezuela. The data cover wholesale transactions, usually on sales to retail outlets, although for four nations (Indonesia, the Philippines, Thailand, and South Africa), sales to hospitals are also included. The reported wholesale sales revenue of each product sold in each national market in any given year was divided by quantity sold to obtain average price realizations for standardized dosage forms – expressed here as the price paid for a standard daily dose. For the analysis that follows, these standardized prices were expressed as a ratio of *Red Book* wholesale list prices for the same products in the United States. These will be called *US price relatives*. The published *Red Book* prices, it must be emphasized, do not necessarily reflect the prices at which actual transactions occurred. During the past decade there has been extensive discounting off wholesale list prices on drug sales to US hospitals and health maintenance organizations as well as on transactions reimbursed under the government's Medicaid program and by pharmaceutical benefit management intermediaries. Discounts in the range of 15–25% are not unusual. Thus, a US price relative of 0.80 should be interpreted as implying rough parity with actual transaction prices prevailing in the United States.

Data were available for a total of 586 nation-product-year triplets. Of these, 461 were supplied by clearly identifiable multinational pharmaceutical companies. Some of the remaining 125 US price relatives may have come from companies that were acting as agents to multinational enterprises; others were

[43] See John F. Kennedy School of Government case study 14–98–1450, 'Vaccines for the Developing World' (1998).

[44] The data were obtained from IMS, the leading collector of information on pharmaceutical company sales. Research assistance was provided by Joan-Ramon Borrel.

known to be locally owned companies. Our analysis here focuses on the pricing policies of the known multinationals. However, a preliminary analysis was conducted to determine whether there were discernible differences between known multinational and other company prices. In a regression equation, the price relatives (again, the ratio of a sample company's price to the comparable US list price) were found to be about 22% higher for multinational enterprises than those of companies not identified as MNEs, controlling also for the general molecule type, whether the drug was dispensed as a tablet or in liquid form, and whether hospital sales were included in the transaction data. Neither the relatively new protease inhibitors nor the non-nucleoside reverse transcriptase inhibitors (RTIs) had significantly higher price relatives than older (RTI) drugs – the latter group including AZT, the first effective anti-AIDS drug.

Our basic question is, do the prices of multinational pharmaceutical companies exhibit patterns suggesting Ramsey pricing? If they do, we should expect to find systematically lower prices, or price relatives, the lower GNP per capita is in a sample nation. Since even the most affluent nation in our sample had GNP per capita (measured in purchasing power parity (PPP) terms at prevailing dollar exchange rates) of less than half the comparable figure for our benchmark price nation, the United States, and since the average for all sample nations was roughly 18% of US GNP per capita, we should expect the price relatives in our sample to be less than the 0.80 value that would (taking into account off-list discounts) imply parity with US pricing.

Figure 3 provides preliminary insight. It arrays all 461 price relative data points for the multinational companies against the sample nations' GNP per

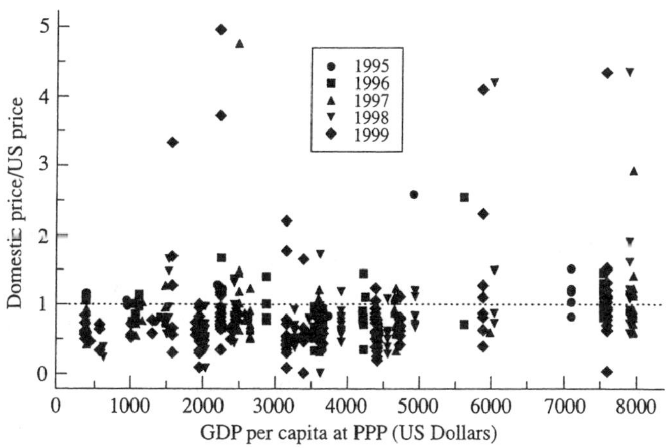

Figure 3. Scatter diagram relating MNEs' AIDS drug prices divided by US list prices to national GDP per capita.

capita, expressed in contemporary purchasing power parity terms.[45] There is only a faint indication of a systematic income-correlated pattern. The average price relative is 0.850, indicating that prices in our sample of nations were approximately equal on average to presumed US transaction prices. However, in 89 out of the 465 cases, they were higher than the US list price parity value of 1.0, sometimes very substantially. The simple correlation between the price relatives and GNP per capita (at PPP) was +0.127, which is significantly different from zero at the 99% confidence level. Pricing did conform in a crude way to the Ramsey predictions, but with a great deal of variation about central tendencies.

Figure 4 provides a more aggregated view of the same data. It correlates against GNP per capita the average price relatives experienced by the various nations during 1997 and 1998. The price relatives for a given year are weighted averages of individual product price relatives, with the weights being the number of daily doses sold for each product. A simple average was taken between the weighted price relatives for the two years. Here the tendency for prices to rise with GNP per capita is more visible, although the simple correlation is only +0.33. Relatively wealthy Uruguay is found to have paid the highest average prices; Indonesia and India had the lowest prices.

Further insight into the reasons for the extensive variability of price relatives was secured through multiple regression analysis. The details are reported elsewhere.[46] In brief, consistent with the Ramsey pricing hypothesis, price relatives increased with GNP per capita. The relationship was statistically significant at the 95% confidence level, although it cannot be said to be strong. An extra $1,000 in income per capita added 0.01 to the price relatives (whose average value, again, was 0.85). A more detailed analysis revealed that the correlation between GNP per capita was higher in early years; indeed, in the last sample year, the correlation was essentially zero. But as time passed, prices fell on average for all of the low- and middle-income nations in the sample, i.e. from an average of 0.974 in 1995 to 0.679 in 1999, suggesting that pharmaceutical companies were moving toward a cruder version of Ramsey pricing. There were lower prices on average in nations with large numbers of HIV cases. This could be attributable to economies of scale in providing drugs to those markets, the tendency for generic competition to be more vigorous in larger markets, or to drug companies' compassion in providing economical therapy for severely impacted nations. A variable indicating whether a drug was patentable in the sample nation or eligible for exclusive marketing rights under the 'letter box' provisions of TRIPS was statistically insignificant, apparently because multinational companies pursued widely

[45] The source is the World Bank's World Development Indicators Database, available on the worldwide web.

[46] http://www3.who.int/whosis/cmh/cmh_papers/e/pdf/wg4_paper01.pdf.

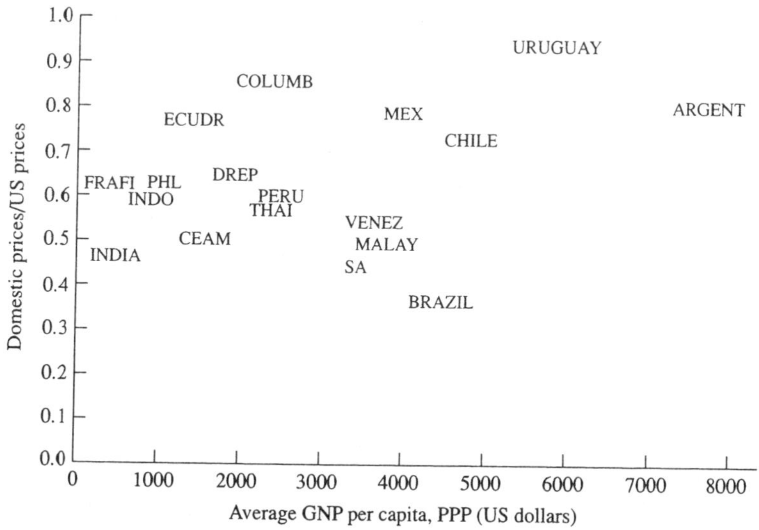

Figure 4. Scatter diagram correlating aggregated national price relatives with GNP per capita in 1997 and 1998.

varying policies toward seeking patent protection even in nations that granted full rights.

In the years 2000 and 2001, which are not covered by our statistical sample, multinational pharmaceutical companies began offering very substantial AIDS drug price discounts to some low-income nations.[47] Whether these recent developments will lead to a world price structure more closely resembling what would emerge under Ramsey pricing remains to be seen.

C. Implications

To sum up, there is much to be said for price discrimination in multinational drug markets. Setting prices lower in low-income nations than in high-income, low price elasticity markets achieves two desirable ends – it helps low-income nations' consumers obtain vital drug supplies, and it enhances drug producers' net revenues, which, if accurately foreseen, stimulates investment in research and the development of new drugs.[48] To the extent that

[47] See 'Africa's AIDS War', *New York Times* (10 March 2001) 1; 'South Africa May Cite Crisis to Lower Cost of AIDS Drugs', *New York Times* (12 March 2001) A3; 'Maker Yielding Patent in Africa for AIDS Drug', *New York Times* (15 March 2001) 1; '6 Companies in New AIDS Pact', *New York Times* (6 April 2001) C12; and 'Lifting the Curtain on the Real Costs of Making AIDS Drugs', *New York Times* (24 April 2001) C1 and C10.

[48] See F. M. Scherer, 'The Link Between Gross Profitability and Pharmaceutical R&D Spending', 20 Health Affairs (September/October 2001) 216–20, showing that R&D spending and drug companies' net cash flow are highly correlated over time.

parallel trade or other obstacles interfere with the attainment of these results, as evident in AIDS drug pricing during the late 1990s, there is reason to discourage them. A nuanced policy that makes the best of an inherently imperfect situation is likely to have the following characteristics:

(1) To encourage the low-price provision of drugs to low-income nations, low-income nations should be allowed to bar parallel exports of drugs received at preferential prices. Pharmaceutical manufacturers should be given the legal means to discourage parallel importation into high-income markets of the patented drugs they have sold at lower prices in nations identified as less-developed under United Nations criteria.

(2) To reduce the adverse consequences from multinational drug providers' niche-pricing strategies, parallel imports into low-income nations should be allowed.

(3) To reduce the product misallocations and impairment of research and development capacity caused by price controls in affluent nations, parallel exports would not be permitted from price-controlled jurisdictions. High-income nations should also agree not to base the prices they allow under their price control regimes on the prices observed in low-income nations, i.e. to limit the geographic scope of any external reference price-based controls. Since foregoing external reference pricing may not be in the interest of high-income nations, an international covenant may be required to achieve this desirable result.

III. DRUG DONATIONS

When the potential recipients of drug therapy have too little income to purchase a needed drug even at discriminatory prices close to the marginal cost of production, pharmaceutical companies sometimes provide their products through outright donations. In 1998, four drug companies led the list of leading corporate philanthropists in the United States.[49] Many of their contributions were for purposes other than providing medicines in kind, and most went to domestic rather than international causes. However, more detailed data reveal an impressive history of in-kind medicine donations to less affluent nations. Merck set an example in 1987 by announcing that it would donate, rather than sell, its Ivermectin drug, effective against the worms that cause river blindness, for use in needy nations. Between then and 1998, nearly 25 million individuals were treated with the donated drug.[50] Over the years 1970 to 1999, Merck reported total product donations (including other drugs)

[49] 'The List: Corporate Giving', *Business Week* (24 January 2000) 8.

[50] See Peter Wehrwein, 'Pharmacophilanthropy,' www.hsph.harvard.edu/review/summer_pharmaco.shtml.

valued at more than $235 million.[51] Other important drug donation programs include the provision of the anti-malaria drug Malarone by Glaxo Wellcome, the antibiotic Zithromax by Pfizer, and the anti-elephantisis drug Albendazole by SmithKline Beecham.[52]

Under the tax laws of the United States, which is the only nation for which we have detailed information, donations sometimes permit sufficient tax savings to entail little or no out-of-pocket cost to the drug manufacturers. For charitable donations in general, corporations can take the accounting cost of the good donated as a deduction against net income on which the corporate income tax is levied. Suppose, for example, a company donates to an appropriately accredited charitable organization drugs whose accounting cost is $100,000. Taxable income is reduced by the amount of that cost. Since the marginal corporate income tax rate for a large and profitable corporation is 34%, the deduction entails a tax saving of 0.34 x $100,000 = $34,000, which only partly offsets the cost of the products donated.

There is an alternative tax provision, however, with more interesting implications. Under Section 170(e)(3) of the US Internal Revenue Code, special deductions apply for donations to qualified charitable organizations when the donated property is used by the donee 'solely for the care of the ill, the needy, or infants'. To qualify under Section 170(e)(3), drug products must also satisfy applicable regulations issued by the Food and Drug Administration, which means, *inter alia*, that their remaining shelf life must not be too short.[53] For drugs subject to these special provisions, the donor can take a deduction against taxable income equal to the accounting cost of the product donated *plus* one-half of the so-called 'stepup', i.e. the difference between the fair market value of the product and its accounting cost, *provided* that the deduction cannot be more than twice the accounting cost of the product.[54] The total of such deductions is limited to 10% of total taxable corporate income.

To illustrate, suppose a pharmaceutical manufacturer donates drugs whose accounting cost is $1,000 and which are normally sold in the US market at a net wholesale price of $4,000. The amount deductible from pre-tax corporate income is $1000 + 0.5 × ($4000 – $1000) = $2500. Ignoring the two-times

[51] From the Taft Group, 'Top 10 Givers in the Top 9 Categories – 1999', www.taftgroup.com/taft/topnews, 27 October 2000.

[52] Wehrwein, above n 50.

[53] Michael Reich et al., 'International Strategies for Tropical Disease Treatments', http://www.hsph.harvard.edu/takemi/pzq/pzq.html (1999), report that pharmaceutical companies sometimes use their donation programs as a method of inventory management. Roughly 30% of the donated drugs surveyed by a Harvard University team had expiration dates less than one year from the time of shipment.

[54] The 'one-half' fraction will be called a stepup factor in our subsequent analysis. For fairly understandable explanations of a complex subject, see *Merten's Law of Federal Income Taxation* (Deerfield, IL: Clark Boardman Callaghan 2000) 31–194; and *CCH Federal Income Tax Code and Regulations*, 1998/1999 loose-leaf version, para 11,675.0204.

cost cap, the tax saving would be 0.34 × \$2500 = \$850. However, the cap applies here, so the deduction can only be \$2000, and hence the tax saving is 0.34 × \$2000 = \$680. With tax savings of \$680 and a product cost of \$1,000, it would appear that the company incurs net out-of-pocket costs of \$320 in making its donation.

However, this mildly pessimistic conclusion ignores that fact that costs as computed by tax accountants are not the same as marginal or incremental costs as computed by economists. In producing the extra units of product needed for its donation, the company incurs only the marginal costs of production, which are almost surely lower than the costs company accounts tally in valuing the product for inventory record-keeping purposes. The 'inventoriable' costs normally include in-factory costs that are more or less fixed, regardless of the production volume, and perhaps also allocations of corporate overhead costs which are invariant over appreciable changes in production volume. In the specific example at hand, the company would actually increase its net after-tax profits if true marginal costs were less than 68% (i.e. twice the corporate income tax rate) of inventoriable accounting costs. In other words, if true marginal cost in the above example were only 67% of the \$1,000 accounting cost, the company would expend \$670 out-of-pocket to produce the donated drugs and obtain a tax saving of \$680.

To glean insight into the structure of drug production costs, aggregated data published by the US Census Bureau for 1992 on SIC industry 2834, 'Pharmaceutical Preparations', were analyzed.[55] Assuming all materials and production worker labor costs to be marginal and other reported in-plant costs to be fixed, the breakdown is as follows, in millions of dollars:

Materials costs	13,545	
Production worker wages	1,867	
Production worker fringe benefits	299	
Subtotal: incremental costs		15,709
Salaried employee compensation	3,083	
Salaried employee fringe benefits	493	
Depreciation	934	
Rentals	186	
Total reported costs		20,405

By these tabulations, incremental costs are roughly 77% of total in-plant costs – nine percentage points too high for tax deductions to reach the level of marginal production costs. However, if companies' inventoriable costs include additional corporate overhead loadings, and especially if the drug makers produce intermediate materials in other plants and transfer them to

[55] US Bureau of the Census, *1992 Census of Manufactures*, Industry Series report MC92-I-28C, 'Drugs' (June 1995).

the reporting plants at values exceeding marginal costs, which must happen frequently, true marginal costs could very easily be less than 68% of the costs reported for tax purposes. In such cases, pharmaceutical companies can do well, i.e. increase their after-tax profits, by doing good.

The relationships among prices, inventoriable costs, and true marginal cost leading to a zero net out-of-pocket cost in making drug donations entail moderately complex nonlinearities. Figure 5 plots break-even lines under existing US tax policies and plausible changes in the tax parameters. When inventoriable costs are one-third or less of the product's normal market price, the company gains in the net whenever its marginal costs are less than 68% of costs used in valuing inventories for tax purposes. When inventoriable cost is higher than one third of the normal market price, lower ratios of marginal to inventoriable costs are required to break even. All price-cost combinations southwest of the lines marked 'Existing Policy' and 'Stepup Factor = 0.5' yield tax savings in excess of marginal production costs. If the law were changed to allow deductions up to three times inventoriable cost, the break-even line shifts upward so that, at high product price/cost values, donors gain in the net if their marginal costs are even slightly below book cost values. Raising the stepup factor, that is, the fraction applied to the difference between price and inventoriable cost, to 0.67 instead of the currently applicable 0.5 expands further the set of donation possibilities profitable in the net. To encourage substantially increased donation programs, the tax laws would also have to be revised to allow corporations to deduct more than 10% of their taxable income. Thus, by modifying the parameters of the tax deduction provisions, the government can increase the profitability of donations by drug companies.

To be sure, strengthening incentives for product donations by revising tax

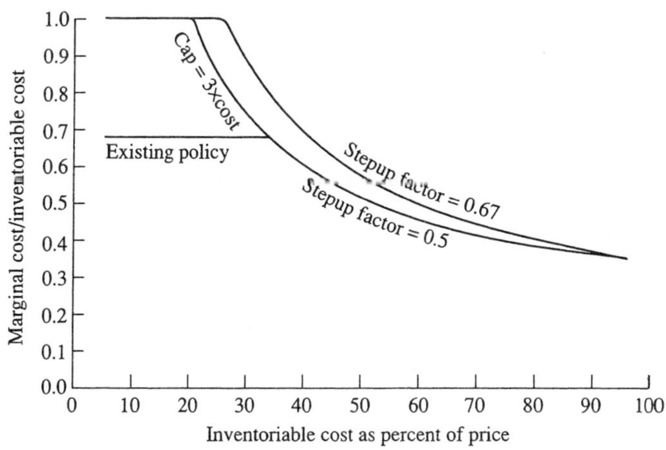

Figure 5. Break-even lines for alternative tax deduction policies.

rules implies a sacrifice of tax revenue by the US Treasury. Why should the US government be willing to forgo tax revenues when politicians are clamoring to channel any budget surpluses into tax relief and increased 'pork barrel' spending? Ignoring moral justifications as a domain into which it is hazardous for us to intrude, we can cite three other reasons. Reducing the spread of infectious diseases such as AIDS and tuberculosis on other continents has spillover benefits in reducing disease risks at home (e.g. as fewer overseas air travellers ingest resistant tuberculosis bacilli); healthier citizens in developing nations provide better markets for US high-technology exports; and better health in the third world may lead to a lower incidence of civil strife requiring costly intervention by first world military forces.

Certainly, there are precedents for using the tax code to promote economic development abroad – notably, the decision of the US government in 1950 to treat crude oil severance payments to the governments of crude oil-producing nations (initially, those located in the Middle East) as income taxes rather than royalties, allowing the multinational petroleum companies to take tax credits for them rather than treating them as deductible costs.[56] This permitted the oil companies to negotiate more generous revenue-sharing formulas with the oil-exporting nations, enhancing those nations' prosperity, encouraging their affiliation with the United States rather than the Soviet Union, and delaying expropriation initiatives until the 1970s. As a result of the tax credit treatment for severance payments, US-based multinational oil companies paid little or no US corporate income tax on their multi-billion-dollar profits from overseas crude oil operations during the 1950s and 1960s.

CONCLUSION

How the TRIPS agreement is interpreted and implemented could have life or death consequences for the citizens of less well-developed nations. Some nations – especially those that in the past have actively encouraged generic substitution for drugs protected elsewhere by patents – will experience a substantial economic shock, compensated to an unknown degree by an increase in pharmaceutical innovation.[57] But consequences will redound throughout the world health system. This article has tried to find policies consistent with the TRIPS agreement that minimize the adverse consequences for the world's least affluent inhabitants.

[56] See US Senate, Committee on Foreign Relations, report, *Multinational Oil Corporations and US Foreign Policy* (Washington: 1975), especially Chapter IV.

[57] For a theoretical analysis of how the number of new drugs is affected by changes in patent protection, see F. M. Scherer, 'The Economics of Human Genome Patents', forthcoming in *Academic Medicine*. For evidence on how drug innovation in Italy changed with a Supreme Court decision requiring the issuance of drug product patents, see F. M. Scherer and Sandy Weisburst, 'Economic Effects of Strengthening Pharmaceutical Patent Protection in Italy', 6 International Review of Industrial Property and Copyright Law (1995) 1009–24.

In any set of future policies, generic drugs must play a crucial role. Thousands of effective medicines are available without the protection of expired patents. Many less-developed nations have hurt themselves by not taking full advantage of the opportunities for encouraging generic substitution. For the smaller of these nations and those less well-endowed with technically trained personnel, vigorous international trade in generic drugs must provide a solution.

To make life-saving new drugs available at affordable prices and to strengthen the development of internationally competitive drug industries, the compulsory licensing opportunities opened up by TRIPS should be seized selectively and imaginatively. This will require, *inter alia*, the creation of national institutions ensuring expeditiously that TRIPS rules are respected. To avoid total free-riding on the research and development efforts of research-oriented pharmaceutical companies, appropriate compensation must be set for patents subjected to licensing. From precedents established in industrialized nations, it seems clear that a 'reasonable' royalty is one that is higher than zero, but much less than the royalty that would compensate a patent holder fully for the loss of whatever monopoly position it might enjoy by virtue of the patent. Permissive clarification of the 'predominantly for the supply of the domestic market' provision in TRIPS is urgently needed. Comparative advantage requires that compulsory licensing orders issued by importing nations be applicable in the nations from which the generic drugs are exported.

In an ideal world, pharmaceutical manufacturers would pursue Ramsey-Baumol-Bradford price discrimination, setting relatively high prices for their patented products to recover R&D investments in the most affluent nations while selling drugs at only a modest markup above marginal production and distribution cost in nations with weak ability to pay. For a variety of reasons, as our analysis of AIDS drug prices has shown, the situation is less than ideal. We have advanced several proposals to minimize the retarding effects of parallel trade, niche pricing, and external reference price controls.

At the end of the day, it must be recognized that the poorer residents of the world's least affluent nations cannot pay even the marginal cost of drugs that might save their lives or permit them to become productive workers. Here the only alternative to death or debility is charity. Charity, like high-technology drugs, is often in short supply. We have shown that it could be facilitated if corporate income tax laws, at least in the United States, are interpreted so that outright donations of essential drugs confer tax advantages sufficiently large as to impose no net cost on the donor, with the burden falling upon the tax collector. Charity through non-transparent tax expenditures is often more feasible politically than outright governmental gifts and grants. It should be encouraged energetically.

[8]

A Note on Global Welfare in Pharmaceutical Patenting

F. M. Scherer

Harvard University and Haverford College

1. INTRODUCTION

THE Uruguay Round TRIPS[1] provisions requiring the extension of first-world patent protection standards to third world nations, including especially the mandate that patents be granted on pharmaceutical products, have been enormously controversial. The ensuing debate led among other things to a decision at the Doha WTO conference to delay the requirement that the least-developed nations grant pharmaceutical product patents by at least a decade, to the year 2016.

It is reasonably well established in the economics literature that, especially in a world of AIDS and resistant tuberculosis epidemics, low-income nations enjoy higher economic welfare when they can free-ride on pharmaceutical innovations made and patented in the first world than when they must pay monopolistic prices for the newest and most effective drugs.[2] Less settled is the question of whether total world welfare is higher under uniform pharmaceutical patent standards or with free-riding. This paper provides what I believe are some fresh insights into the global welfare problem.

A foundation is laid by revisiting diagrams used in my 1996 analysis of the pharmaceutical patent question,[3] which in turn was based in part upon an analysis by Alan Deardorff (1990 and 1992). Figure 1 views the welfare implications from the perspective of an LDC, whose demand curve for a particular drug is given by D. If the LDC grants patents, the drug is assumed to be sold at the monopoly profit-maximising price OP_M. In that case, rectangle H measures the producers' surplus realised by the drug's patent holder, presumably a multinational company, and consumers' surplus triangle B is realised by the citizens of the LDC. If, however, patents are not granted and the drug is supplied to the

[1] Trade-related aspects of intellectual property rights.

[2] See e.g. Commission on Intellectual Property Rights (2002), Maskus (1997 and 2000, Ch. 5) and Scherer and Watal (2002).

[3] Scherer (1996, Ch. 9).

FIGURE 1
Impact of Patent Policy on LDC Welfare

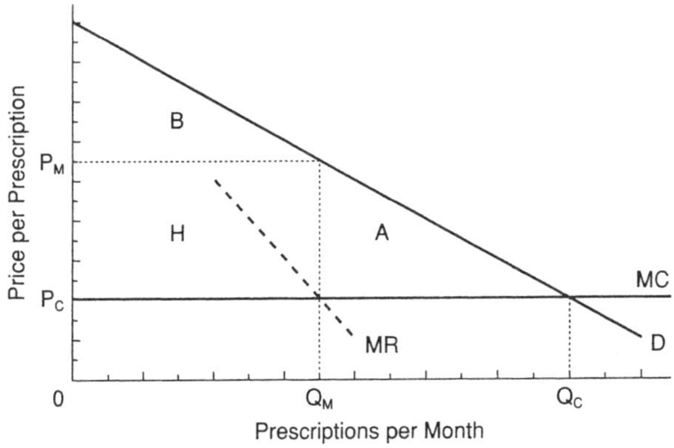

LDC's citizens competitively (a strong assumption not questioned here),[4] the citizens of the LDC consume more at the lower price OP_C and gain the much larger consumers' surplus of B + H + A. Given the linearity of demand and cost assumed in my example, B + H + A is four times as large as consumers' surplus B under monopoly. I argue that the LDC is better off only if granting patent rights increases the development and supply of comparable new drugs by at least three times.

Figure 2 asks whether such an increase in the number of new drugs is likely. Extending concepts originally articulated by William Nordhaus (1969), the solid line RD(NCE) shows how the number of new chemical entities (NCEs) developed on average per year (vertical axis) can be expected to vary with the amount spent on research and development (horizontal axis). Broken lines Q_1 and Q_2 show how appropriately discounted quasi-rents captured by innovators (horizontal axis) vary with the number of NCEs introduced into the market. Both have a curvature reflecting diminishing marginal returns; the most lucrative and/or least costly possibilities are explored first.[5] Both are hypothetical but with intuitively plausible values. Econometric estimates are unavailable, but much to be desired. Q_2 is shifted to the right by 20 per cent relative to Q_1 to reflect my assumption that extending patent protection to LDCs would increase rents by 20 per cent, or low-income nations' approximate share of world GDP. Given these functions and

[4] The post-Doha resolution allowing drug exports to least-developed nations under compulsory licences is likely to facilitate world-wide generic competition that will help to make reality approximate the assumption.

[5] The diminishing returns are roughly quadratic. Specific functions used in constructing Figure 2 were $RD(NCE) = (0.205\ N)^{1.8}$ (for convenience, the inverse of the likely causal relation) and $Q_1 = (100\ N)^{0.51}$.

FIGURE 2
Impact of Quasi-rent Volume on NCE Development

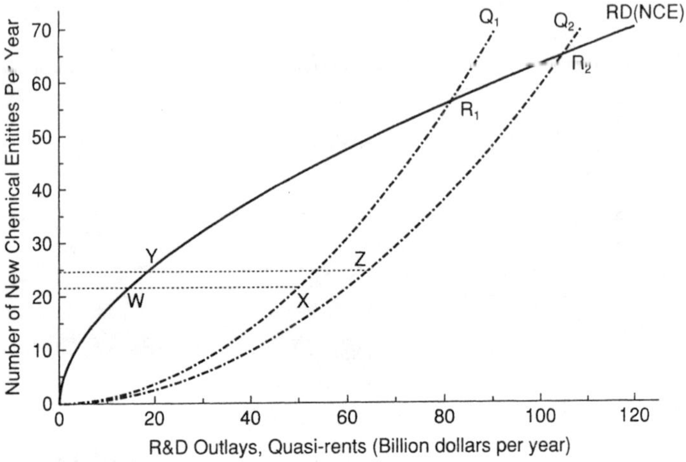

the profit maximisation approach postulated by Nordhaus, which I question in my book and will question later but accept tentatively, the equilibrium number of new chemical entities is found by maximising the horizontal distance between RD(NCE) and a quasi-rent function, leading to the development of 21.25 new chemical entities per year on average if LDCs do not offer patent rights and 24.4 if they do. The 15.5 per cent change falls far short of the three-fold increase required to render LDCs better off in granting patents.[6]

2. GLOBAL WELFARE: NORDHAUS ASSUMPTIONS

Analysing the global welfare question requires a more complex model. The one presented here is deliberately austere, attempting to focus on three fundamental variables: the relative increase in producers' surplus achieved through patented pharmaceutical sales in the third world, the number of additional pharmaceutical NCE products induced (each assumed tentatively to have identical demand functions), and – a key variable that cannot be ignored in comparative welfare analyses – the average difference in the marginal utility of income for third world as compared to first world consumers. Another variable will be held constant for the sake of simplicity – the number of consumers in the third world relative to the

[6] If alternatively the world quasi-rent function is shifted to the right by 40 per cent or 100 per cent, as in subsequent sensitivity tests, the increases in the number of NCEs are 29.4 per cent and 72.2 per cent respectively. Again, diminishing returns are evident.

FIGURE 3
(A) Demand Conditions in the First World; (B) Demand Conditions in the Third World

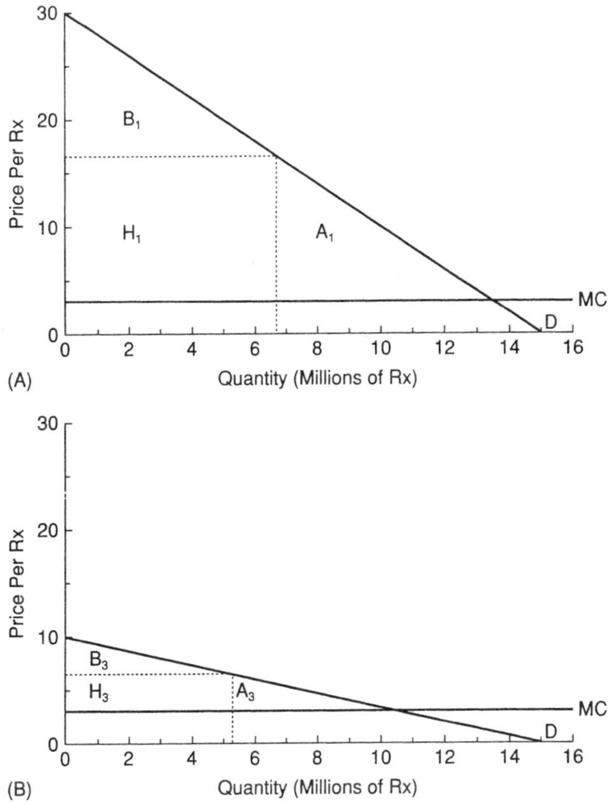

number in the first world. The relevant data suggest setting the first- and third-world population shares at $S_1 = S_3 = 0.5$. Complications such as time patterns, price controls, parallel trade, information lags, non-patent barriers to entry and demand non-linearities are for the most part neglected; other caveats are introduced in a final section.

Figures 3(A) and 3(B), which are conceptually identical to demand curves used by Jayashree Watal and myself to illustrate the benefits of Ramsey-Baumol-Bradford discrimination for the pricing of pharmaceutical products in the first and third worlds,[7] assume that at a zero price, the same number of prescriptions (15 million) would be demanded monthly in the first world and in the third world.[8]

[7] Watal and Scherer (2002) and Scherer and Watal (2002).

[8] The demand equation is $P = 30 - 2Q$ for the rich nations and $P = 10 - 0.667Q$ for the poor nations.

However, income effects cause the demand function to be flatter in poor nations than in rich nations. Long-run marginal production and distribution cost (excluding R&D outlays) is assumed to be $3 per Rx. As the diagram is drawn, a firm enjoying patent protection in both the rich and poor nations will set a price of about $16.50 per Rx in the rich countries, realising a contribution to profits and the recoupment of R&D costs (quasi-rents) of roughly $91 million per month there. In the poor nations its price will be $6.50 and its contribution to profits will be approximately $18.4 million, or roughly 20 per cent of first-world profits and 16.67 per cent of first *plus* third world quasi-rent potential with universal patent protection.

To keep the numbers simpler and more memorable, we recalibrate the rich-nation producer's surplus to be 100. Using the welfare gain notation of Figure 1, and assuming linearity of demand functions, the initial tally of surpluses per new product, assuming patent protection in both jurisdictions, is as follows:

	Rich Nations	*Poor Nations*
H Producer's surplus	100	20
B Consumers' surplus	50	10

It will be subjected in a moment to sensitivity analysis. The producers' surplus in third world nations, it should be noted, is assumed provisionally to accrue to rich-nation shareholders of multinational corporations. If, on the other hand, the third world nations offer no patent protection and free-ride at competitive prices on pharmaceutical products that would be developed in any event in response to rich nation consumers' demand, the within-patent-period welfare gain to poor (i.e., third world) nation consumers is $A_3 + B_3 + H_3 = 40$ and producers forgo a surplus of 20.

From the surplus accruing to producers, R&D costs must be subtracted for a correct net benefits accounting. Calculations using the functions specified in footnote 5 revealed that they approximated 28 to 29 per cent of quasi-rents in the cases to be examined. We round the deduction to 30 per cent.

A common assumption in benefit/cost analysis is that one party's surplus of given magnitude in some numéraire currency is equivalent to another party's surplus. To accept that assumption would be to miss much of what the debate over pharmaceutical product patent rights in the third world is all about. Roughly half of the world's population live in nations where income per capita is less than one-tenth that of the United States or western Europe. If one accepts the notion dating back at least to Alfred Marshall that 'the richer a man becomes the less is the marginal utility of money to him',[9] one needs to assign greater weight to the benefits realised by poor-nation citizens than to those of rich-nation inhabitants.

[9] Marshall (1948, p. 96). An even older source is the widow's mite parable in the Gospel according to Mark 12: 41–44.

I do this through the weighting factor U, which measures the ratio of the marginal utility of income for the median poor-nation inhabitant to that of the median rich-nation citizen.[10]

We must now tally the appropriately weighted sum of surplus for two different intellectual property regimes: Case 1, in which pharmaceutical products receive full patent protection in all nations, and Case 2, in which patent rights are only conferred in the first world and LDCs free-ride on the inventions made in the first world. Plainly, with larger producer rents in Case 1, there will be more inventions, as implied by Figure 2. How many more successful new chemical entities there are is a key variable. We assume in base Case 1 that with worldwide patent rights, the average number of new chemical entities approved and introduced into the market per year, each with the same demand characteristics, is 25. In Case 2, the number is a variable N whose value is less than 25.

After patents expire, the new chemical entities will be available to both first and third world consumers at competitive prices, yielding surpluses of 25 or N times (A + B + H). Since patents are applied for at about the time human subject tests begin and since clinical testing periods average 8 years, we assume 12 years of patent life. Assuming level quasi-rents and a discount rate of 10 per cent, the discounted present value multiplier for within-patent-period benefits is $[1 - e^{-0.1 \times 12}]/0.1 = 6.99$, and the post-patent benefit multiplier (with infinite horizon) $[e^{-0.1 \times 12}]/0.1 = 3.01$. Thus, we assume that discounted post-patent benefits are 100(3.01/6.99) = 43 per cent of within-patent-period benefits.

The question is, over what configurations of variables U and N is worldwide welfare higher under Case 1, and when is it higher under Case 2? Using the notation of Figure 3, and disaggregating the accounts so that within-patent-period benefits are on the first line and post-patent benefits on the second line, and within a line, benefits in the first world are in the first set of brackets, the producer's surpluses realised by first world firms in third world markets in the second set of brackets, and benefits to third world consumers in the third set of brackets, the accounting is as follows:

$$\begin{aligned}\text{Case 1: } & [25S_1(B_1 + H_1)] + [25S_3H_3] + [25S_3B_3U] \\ & + 0.43[25S_1(A_1 + B_1 + H_1)] + [0] + 0.43[25S_3(A_3 + B_3H_3)U] \\ & = 1875 + 250 + 125U + 1075 + 215U = 3200 + 340U.\end{aligned}$$

$$\begin{aligned}\text{Case 2: } & [NS_1(B_1 + H_1)] + [0] + [NS_3(A_3 + B_3 + H_3)U] \\ & + 0.43[NS_1(A_1 + B_1 + H_1)] + [0] + 0.43[NS_3(A_3+B_3+H_3)U] \\ & = 75N + 20NU + 43N + 8.6NU = 118N + 28.6NU.\end{aligned}$$

10 When new drug products must be imported in bulk or finished form, U might alternatively or in addition reflect the higher Lagrangian multiplier on foreign exchange budgets in less affluent nations.

FIGURE 4
Total Welfare Tradeoff: NCE Numbers vs. Utility Ratio
Nordhaus Case; Monopoly Period = 12 Years; Discount Rate = 0.10

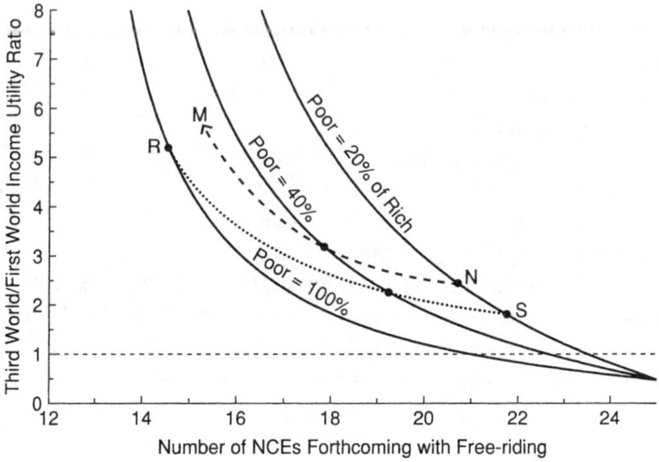

When these two equations are set equal, global welfare is identical under either policy; equality identifies the breakeven case. The solid line marked 'Poor = 20% of Rich' in Figure 4 shows the breakeven line for the assumptions accepted thus far. Variable pairs above (north of) the solid line delineate situations in which global welfare is higher under Case 2, i.e., with LDC free-riding. If extending patent protection world-wide leads to a loss of only one NCE per year relative to the base case, global welfare is maximised under free-riding even if the marginal utility of income is the same for poor as compared to rich nation citizens. But if (ignoring the diminishing marginal returns tendency, and hence erring on the side of favouring patent protection) the sacrifice from eliminating 20 per cent of rich-nation rents or 16.67 per cent of the world rent potential is 16.67 per cent of base Case 1 NCEs, i.e., 4.17 NCEs, global welfare is higher if the marginal utility of income in the third world exceeds that of the first world by more than a multiplier of 2.40. With diminishing returns in the inducement of NCEs as the quasi-rent potential increases (see Figure 2), free-riding becomes globally optimal with even smaller utility differentials. A key implication is that strong value judgements – on the magnitude of that utility differential – are unavoidable in determining which policy is globally optimal.

Less-developed nations might conceivably have more than 20 per cent of wealthy nations' pharmaceutical rent potential.[11] Suppose the rent potential is 40 per cent rather than 20 per cent – e.g., because the poor nation demand curve's

[11] This criticism of my 1996 assumptions is advanced in Sykes (2002). A potential smaller than 20 per cent was suggested by a referee. The implications of such a downward variation will be transparent.

vertical intercept is raised to 15 instead of 10, or because a higher fraction of the poor nation's citizens are afflicted. Then holding the rich nation quasi-rent potential constant at 100, consumers' surplus in Case 1 rises to 20 for the poor nation and so also does avoided deadweight loss triangle A_3 in Figure 3(B). The higher surpluses in LDCs shift the breakeven curve downward (marked 'Poor = 40%' in Figure 4), increasing the likelihood that free-riding is globally optimal for any given retardation of new chemical entity development. To be sure, the sacrifice of larger potential quasi-rents due to free-riding makes it likely that the number of new chemical entities will be smaller than in the 20 per cent scenario. With a proportional 28.6 per cent (40/140) diminution to 17.9 NCEs, breakeven occurs when the income utility ratio is 3.20. Again, diminishing marginal returns are more probable, so free-riding is globally optimal at utility differentials less than 3.2.

A third case relating to diseases especially prevalent in the third world is illustrated by the breakeven curve marked 'Poor = 100%', where demand in the third world yields a rent potential equal to that of the first world, or one-half of the total world potential. This enlargement of the third world's relative rent potential shifts the breakeven curve even more toward the origin. But of course, the absence of patent rights in half the world market also implies a greater sacrifice of new chemical entities – in the proportional case, to 12.5 NCEs. In this instance, the breakeven utility ratio soars to 24.7 – outside the range of plausible value judgements.

More generally, the larger the third world rent potential is relative to the first world potential, the more the breakeven curve shifts toward the origin and the smaller is the breakeven utility differential for any given impairment of incremental NCE supply due to third world free-riding. But given the probable diminution of NCE supply, the larger is the breakeven utility ratio required for free-riding to be globally optimal. Accepting the proportional diminution assumption most favourable to third-world patent grants, the dashed line MN in Figure 4 interpolates the locus of breakeven cases for diverse third- as compared to first-world quasi-rent potentials. With diminishing returns in the quasi-rent–NCE relationship, a lower breakeven locus exists. For example, with the parameter assumptions of Figure 2, the locus shifts to the dotted line RS, with breakeven utility ratios of 1.80 when third world quasi-rents are 20 per cent of those in the first world and 5.35 in the equal quasi-rents scenario.

For the limiting case in which the whole potential resides in the third world – e.g., for drugs combating diseases existing only in LDCs – quasi-rents converge on zero in Case 2. Absent inducements other than those provided by the patent system, there would be no incentive to perform drug R&D and hence no likely marketing of NCEs.[12] We explore those special conditions further after considering some complications.

[12] But see Bond and Lean (1977) for a discussion of non-patent first mover advantages in pharmaceuticals.

3. ALTERNATIVE ASSUMPTIONS

In addition to the customary reluctance of economists to make the kind of income utility comparisons I have advanced, there are two broad criticisms of this model.

a. The Rent-seeking Model

One questions whether the producer's surplus rectangles H in my model are preponderantly welfare gains. Those surpluses are a stimulus to investment in research, development and testing. My service as chair of the US Office of Technology Assessment advisory committee for its study of pharmaceutical profits and R&D during the early 1990s led me to wonder how quasi-rent margins in the pharmaceutical industry can be so high, stimulating R&D spending, while correctly computed returns on industry investment exceed risk-adjusted norms by only two or three percentage points on average.[13] A new analysis (Scherer, 2001) suggests an answer. Extracting from US Census reports data on pharmaceutical industry gross margins and from Pharmaceutical Research and Manufacturers of America reports data on R&D spending for the years 1962–96, I computed percentage deviations of those variables (adjusted to 1992 price levels) from their exponential trends. Figure 5 shows the plotted percentage deviations. When gross margins

FIGURE 5
Percentage Deviations from Trend for US Pharmaceutical Industry Gross Margins and R&D Outlays, 1962–1996
(Constant 1992 dollars)

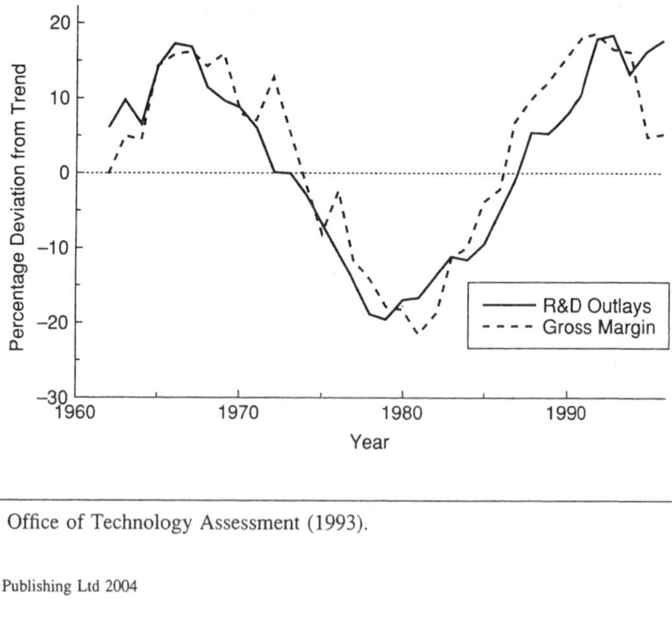

[13] See US Office of Technology Assessment (1993).

rise, so also do R&D outlays; when margins fall, R&D outlays fall virtually in tandem. Although the exact adaptive expectations structure remains uncertain, it seems probable that a competitive rent-seeking model best describes the relationship of pharmaceutical R&D spending to cash flows. That is, rising quasi-rent potentials are almost fully exhausted by the competitive escalation of costs – for R&D, marketing and implicit returns on R&D investment – leaving only a small pure surplus for the stockholders of pharmaceutical companies.

One implication of this alternative behavioural model is that industry equilibrium occurs at points R_1 and R_2 rather than W and Y in Figure 2, and so, depending upon the strength of the diminishing returns phenomenon, far more new products are forthcoming in a given year. Another implication follows from the largely forgotten insights advanced in a debate two decades ago, showing that when innovators are competing rent-seekers rather than secure Nordhaus-type profit maximisers, the welfare-maximising life of invention patents is drastically shortened.[14]

In the model presented here, the logic of rent-seeking requires that none of the H rectangles can be counted as social welfare gains during the within-patent period. Thus, assuming that with full world patent rights the average number of NCEs developed per year under rent-seeking rises to 50,[15] the comparison of gains when LDC quasi-rent potential is 20 per cent of the rich-nation potential is altered to:

$$\begin{aligned}\text{Case 1: } & [50S_1(B_1)] + [0] + [50S_3(B_3)U] \\ & + 0.43[50S_1(A_1 + B_1 + H_1)] \\ & + [0] + 0.43[50S_3(A_3 + B_3 + H_3)U] \\ & = 1250 + 250U + 2150 + 430U = 3400 + 680U.\end{aligned}$$

$$\begin{aligned}\text{Case 2: } & [NS_1(B_1)] + [0] + [NS_3(A_3 + B_3 + H_3)U] \\ & + 0.43[NS_1(A_1 + B_1H_1)] + [0] + 0.43[NS_3(A_3 + B_3 + H_3)U] \\ & = 25N + 20NU + 43N + 8.6NU = 68N + 28.6NU.\end{aligned}$$

With this change of assumptions, the breakeven curves shift dramatically toward the origin. This is illustrated in Figure 6. If a loss of LDC quasi-rent potential amounting to 20 per cent of rich-nation rents in Case 2 implies a proportional 16.7 per cent reduction in the number of new products from 50 to 41.7 per year, global welfare (with breakeven given by the solid line) is higher under free-riding if the ratio of LDC to rich-nation income utility exceeds 1.13.[16] With 40 per cent LDC rent potentials relative to those of the high-income nations,

---

[14] McFetridge and Rafiquzzaman (1986), with a comment by Roger Beck.

[15] The breakeven analysis is unaffected by rescaling the base number of new chemical entities.

[16] If the relationship between quasi-rents and N were continuously proportional, no stable rent-seeking equilibrium would exist. Diminishing returns ensure equilibrium.

FIGURE 6
Total Welfare Tradeoff: NCE Numbers vs. Utility Ratio
Rent-seeking Case; Monopoly Period = 12 Years; Discount Rate = 0.10

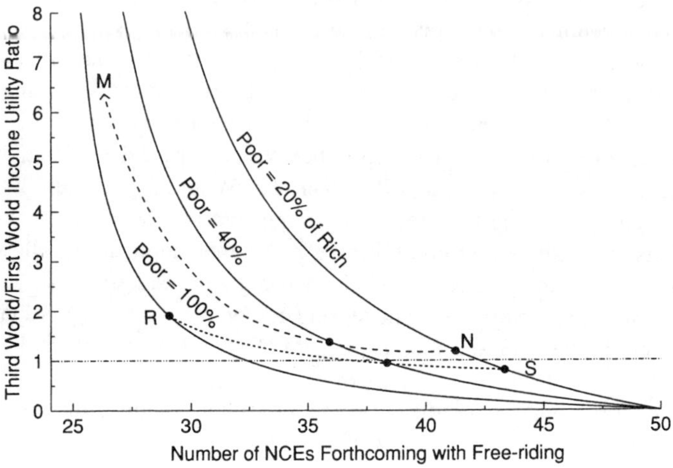

the downward shift of the breakeven curve is even more, and so global welfare is maximised under free-riding and a proportional diminution in NCEs with a utility ratio of 1.42.

As in Figure 4, the dashed line MN shows the interpolated locus of breakeven utility ratios under the assumption of proportional NCE losses and the dotted line RS the locus using the parameter assumptions of Figure 2. These rent-seeking assumptions, I am convinced, are more realistic than the assumption that producers' surplus is a pure social gain, accepted in my initial model. If one believes that competitive rent-seeking dissipates much or all of producers' surpluses, free-riding policies become all the more justifiable, even from a global perspective.

b. Variable Demand Conditions

Another debatable point in the models presented here is the assumption that each product has the same demand curve and hence the same constellation of surpluses. This is at odds with reality in pharmaceuticals and indeed in virtually every field of technological innovation. The distribution of returns to innovation is highly skewed. Blockbuster innovations always capture a share of quasi-rents or profits disproportionate to their numbers.[17]

It might be argued that consumers in less-developed countries will generate quantitatively large quasi-rents because there are vast numbers of them, they

[17] See Scherer, Harhoff and Kukies (2000).

suffer from diseases not prevalent in the rich nations, and (arguably) the granting of patent rights will induce pharmaceutical companies (presumably, multinational pharmaceutical companies)[18] to do research they would otherwise not undertake on diseases found mainly in poor nations and discover new drugs with blockbuster potential. This is indeed the strongest argument for extending pharmaceutical product patent rights throughout the world. Recognising it, Jean O. Lanjouw (2001) has advocated a bifurcated patent system, letting the companies achieving pharmaceutical product innovations seek patent rights in *either* the first world *or* in the third world, but not in both. This would allow third-world nations to free-ride on drugs that would have been developed in any event for the first world but maximises the stimulus for private investment in the development of drugs targeted solely toward third-world diseases. It is an attractive argument, but four doubts intrude.

First, in the nations where the so-called 'tropical diseases' abound, most potential consumers are very poor, with annual incomes measured in the low hundreds of dollars. Those nations also tend to have at best primitive public health systems, and most drugs must be purchased with consumers' own funds, for which a myriad of life-sustaining uses compete. In other words, demand functions are pressed even closer to marginal cost functions than implied in Figure 3(B). Even when they are aggregated over hundreds of millions of consumers, it is not clear that there are quasi-rent potentials anywhere near those associated with medicines targeted toward coronary problems, common cancers, gastritis, depression and inadequate sexual function in rich nations. If the quasi-rent potential is weak, not much rent-seeking research and testing will be induced.

Second, there is much uncertainty in predicting during R&D phases how large the market for a particular therapeutic molecule will be. Molecules often turn out to have therapeutic uses quite different from those initially contemplated. Viagra is an example of such serendipitous discovery, as was the discovery that Bayer's praziquantil is effective against schistosomiasis and that the German anti-cancer drug miltefosine is effective against black fever.[19] And at the clinical testing stage, it remains uncertain what fraction of a target audience will respond favourably to the drug's use. Even as simple a matter as the way the drug is introduced into the body – e.g., injection, three-per-day oral delivery, or once-per-week time-release delivery – can significantly affect the extent of use in less-developed nations, where systematic care by nurses or physicians is the exception rather than the rule. *Ex ante*, therefore, the expected distribution of returns from innovation

[18] Note that to the extent quasi-rents are appropriated within third world markets, e.g., by indigenous companies, they must be multiplied by U, shifting breakeven curves in Figure 4 (but not Figure 6) outward and strengthening, all else equal, the case for third-world patents.
[19] See 'Drug Proves Able to Cure Disease Borne by Parasite', *New York Times* (28 November, 2002, p. A6).

entails more uniformity of quasi-rents than it does after full technological and marketing experience has unfolded.[20]

Third, interest in developing new tropical disease cures may be keener *in situ* than in pharmaceutical laboratories thousands of miles away from the disease locus.[21] But in the nations most afflicted by tropical diseases, the technological capabilities needed to do state-of-the-art research and product development are scarce. Advancing from incentive to breakthrough innovation is not automatic. Research by Sandy Weisburst and mentored by me, showed, for example, that Italy, with a vibrant generic drug industry, did not achieve any significant increase in the discovery of innovative drugs during the first decade after the Italian Supreme Court mandated the issue of pharmaceutical product patents.[22]

If there are exceptions to the Italian experience, India, having replaced Italy as the world's leading generic drug source, is the most likely candidate. But this raises my fourth caveat. In preliminary interviews on the probable response of Indian pharmaceutical firms to a new regime in which pharmaceutical product patents can be received in India, Jean O. Lanjouw (1997) found that the leading Indian firms seemed more interested in developing new drugs that will be blockbusters in the first world than in targeting tropical disease remedies. More recent public statements convey the same implication.[23]

It is nevertheless possible that multinational pharmaceutical companies will reorient their R&D portfolios to place more emphasis on third-world diseases. Certainly, in the past, that has not been the case. A study for Medicins sans Frontières (2001) revealed that among the 1,393 new drug chemical entities introduced into world markets between 1975 and 1999, only 13 (or 15 including tuberculosis drugs) were indicated for so-called tropical diseases. Among the 26 new chemical and biological entities approved in 2002 for marketing in the United States, only one, the anti-diarrhoeal nitazoxanide, had a larger potential market in third world nations than in the United States, where the orphan drug's case incidence was estimated to be roughly 4,000.[24] Since the Uruguay Round intellectual property agreements had been a reality for eight years, one might expect some results to have materialised by 2002, but again, the evidence is meagre. The principal two exceptions to this gloomy prognosis involve the establishment by GlaxoSmithKline of a laboratory in Spain to seek drugs effective against tuberculosis and malaria and the creation of a Novartis laboratory in Singapore to work on drugs for resistant tuberculosis and dengue fever.[25]

[20] This is shown in simulations by Scherer, Harhoff and Kukies (2000).
[21] Cf. note 18 *supra*.
[22] Weisburst and Scherer (1995). See also Challu (1995).
[23] 'Patently Ambitious', *The Economist* (6 September, 2003, p. 56).
[24] Pharmaceutical Research and Manufacturers of America, 'New Drug Approvals in 2002 (Washington, January 2003).
[25] 'Exotic Pursuits', *The Economist* (1 February, 2003, p. 52).

In the meantime, I believe, an opportunity for encouraging first-world cross-subsidisation of third-world disease research has been ignored. The debate over drug patent rights under TRIPS would have provided an ideal opportunity for someone like Kofi Annan to say to the multinational pharmaceutical companies, 'We will support your demand for strong patent rights throughout the world if you will commit 20 percent of your research, development, and testing budgets to diseases specific to less-developed nations'. The multinational pharmaceutical companies made an analogous commitment in persuading Canada to abandon its vigorously enforced drug patent compulsory licensing laws. As a *quid pro quo*, the pharmaceutical companies agreed to locate in Canada R&D activities proportional to Canada's share of the companies' drug sales. If such a commitment were forthcoming, my caveats would be less persuasive.

Finally, we must recognise a quite different rationale for third world free-riding not expressly modelled here. India's most powerful pharmaceutical companies grew from an environment in which drug product patent rights were not issued, permitting Indian companies to build their technological capabilities initially by concentrating on process innovations to produce and then export still-patented drugs to nations similarly denying product patent rights and then, after relevant patents expired, becoming first-moving generic drug suppliers to first-world nations. Similarly, the United States developed its early industrial capabilities by free-riding upon the technology pioneered in the leading European nations – e.g., England for textiles, steam engines and machinery, and France for gunpowder. Until 1836, foreigners were not allowed to obtain US patents unless they had resided at least two years in the United States and declared an intent to become US citizens. After that, between 1836 and 1861, aliens paid higher registration fees than citizens. For example, in 1844 the patent fee for US citizens was $30. For citizens of Great Britain it was $500; for residents of other nations $300. In the 1836–61 period, foreigners received fewer than two per cent of all US patents issued. By the 1850s, English technicians visiting the United States observed with astonishment, US enterprises were producing a wide range of products with methods more mechanised than those prevailing in England, the cradle of the Industrial Revolution.[26] Analogous strategies of building industrial prowess first by free-riding or cheap-riding on other nations' technology and only then strengthening patent mechanisms can be observed in the histories of Switzerland, the Netherlands and Japan.[27] Although the issue cannot be pursued here, it is at least arguable that industrial development can proceed faster when more advanced foreign technology can initially be appropriated at low cost.

---

[26] See Habakkuk (1962, p. 4).
[27] See Schiff (1971), Ordover (1991) and Granstrand (1999).

4. CONCLUSION

To sum up, my analysis suggests that global welfare is maximised by letting low-income nations free-ride on the patented inventions of first-world nations over a wide range of negative new product development impacts if one accepts the reasonable premise that the marginal utility of income is appreciably higher in poor nations than in rich nations. The Doha round of negotiations appears to have gravitated toward a proper solution, deferring implementation of the TRIPS provisions on pharmaceuticals in the least-developed nations for a decade.[28] In the interim, we will be able to observe the response of pharmaceutical companies to the limited grants of exclusivity already implemented under the Treaty of Marrakech. And there will be time for commitments to be extracted that could change the conditions under which tropical medicines are supplied.

REFERENCES

Bond, R. S. and D. F. Lean (1977), *Sales, Promotion, and Product Differentiation in Two Prescription Drug Markets* (Federal Trade Commission Staff Report, Washington).

Challu, P. M. (1995), 'Effects of the Monopolistic Patenting of Medicines in Italy Since 1978', *International Journal of Technology Management*, **10**, 2/3, 237–50.

Commission on Intellectual Property Rights (2002), *Integrating Intellectual Property Rights and Development Policy* (London).

Deardorff, A. V. (1990), 'Should Patent Protection be Extended to All Developing Countries?', *The World Economy*, **13**, 4, 497–507.

Deardorff, A. V. (1992), 'Welfare Effects of Global Patent Protection', *Economica*, **59**, 35–51.

Granstrand, O. (1999), *The Economics and Management of Intellectual Property* (Edward Elgar, Cheltenham, UK).

Lanjouw, J. O. (1997), 'The Introduction of Pharmaceutical Patents in India', National Bureau of Economic Research Working Paper 6366.

Lanjouw, J. O. (2001), 'A Patent Policy Proposal for Global Diseases' (Brookings Institution Policy Brief, Washington).

Marshall, A. (1948), *Economics* (8th ed., Macmillan, New York).

Maskus, K. E. (1997), 'Implications of Regional and Multilateral Agreements for Intellectual Property Rights', *The World Economy*, **20**, 5, 681–94.

Maskus, K. E. (2000), *Intellectual Property Rights in the Global Economy* (Institute for International Economics, Washington).

McFetridge, D. G. and M. Rafiquzzaman (1986), 'The Scope and Duration of the Patent Right and the Nature of Research Rivalry', in J. Palmer (ed.), *Research in Law and Economics*, **8** (JAI Press, Greenwich, CT), 91–129.

Nordhaus, W. D. (1969), *Invention, Growth, and Welfare: A Theoretical Treatment of Technological Change and Economic Welfare* (MIT Press, Cambridge, MA).

Ordover, J. A. (1991), 'A Patent System for Both Diffusion and Exclusion', *Journal of Economic Perspectives*, **5**, 1, 43–60.

[28] The UK Commission on Intellectual Property Rights (2002, p. 162) recommended extension of this delay for all subject matter covered by TRIPS along with flexible interpretation of its provisions after the year 2016.

Scherer, F. M. (1996), *Industry Structure, Strategy, and Public Policy* (HarperCollins, New York).
Scherer, F. M. (2001), 'The Link Between Gross Profitability and Pharmaceutical R&D Spending', *Health Affairs*, **20**, 5, 216–20.
Scherer, F. M. and J. Watal (2002), 'Post-TRIPS Options for Access to Patented Medicines in Developing Nations', *Journal of International Economic Law*, **5**, 4, 913–40.
Scherer, F. M., D. Harhoff and J. Kukies (2000), 'Uncertainty and the Size Distribution of Rewards from Innovation', *Journal of Evolutionary Economics*, **10**, 1–2, 175–200.
Schiff, E. (1971), *Industrialization without National Patents* (Princeton University Press, Princeton).
Sykes, A. (2002), 'TRIPS, Pharmaceuticals, Developing Countries and the Doha "Solution"', *Chicago Journal of International Law*, **3**, 1, 47–68.
US Office of Technology Assessment (1993), *Pharmaceutical R&D: Costs, Risks and Rewards* (Washington: USGPO).
Watal, J. and F. M. Scherer (2002), 'The Economics of TRIPS Options for Access to Medicines', in B. Granville (ed.), *The Economics of Essential Medicines* (Royal Institute of International Affairs, London).
Weisburst, S. and F. M. Scherer (1995), 'Economic Effects of Strengthening Pharmaceutical Patent Protection in Italy', *International Review of Industrial Property and Copyright Law*, **26**, 6, 1009–24.

PART II

USING PATENT DATA TO MEASURE TECHNOLOGICAL INNOVATION

[9]

FIRM SIZE, MARKET STRUCTURE, OPPORTUNITY, AND THE OUTPUT OF PATENTED INVENTIONS

By F. M. SCHERER*

During the past three decades hypotheses concerning the most favorable industrial environment for technological progress have proliferated like foliage in the Vietnam jungles.[1] Only recently have we begun to penetrate the rich theoretical growth with empirical insights. This paper contributes to the defoliation program, reporting on a statistical study of the relationships between inventive activity and technological opportunity, firm size, product-line diversification, and monopoly power.

The main sample consisted of 448 firms on *Fortune's* list of the 500 largest U.S. industrial corporations for the base year 1955.[2] Independent variables included three measures of firm size for 1955, profits for 1955 through 1960, liquid assets for 1955, an index of diversification, dummy variables differentiating industry and technology classes, and (for a special sample to be described later) four-firm concentration ratios. The principal dependent variable is the number of U.S. invention patents issued to the sampled firms in 1959 [33]. The year of patent issue is lagged four years from the base year 1955, whose inventive output is to be measured, because on the average nine months pass between the conception of an industrial invention and the filing of a patent application [17, p. 71] and because during the 1950's three and one-half years were required for the Patent Office to process an average application to the point of issue.

* The author is assistant professor of economics at Princeton University. The research was supported by a grant from the Inter-University Committee on the Microeconomics of Technological Change, sponsored in turn by the Ford Foundation. Use was also made of computer facilities supported in part by National Science Foundation Grant NSF-GP 579. I am indebted to Jacob Schmookler and to numerous participants in the Princeton seminar on economic research in progress, and especially to Jesse Markham, for critical comments; to Roald Buhler and Insley Pyne for computer programming assistance; to E. C. Taylor for advice on the classification of organic chemical patents; and to Mrs. Dorothy Register, Mrs. Noah Meltz, and my wife for help in collecting data.

[1] For surveys of the literature, see [7] [11] [13] [14].

[2] Fifteen firms on *Fortune's* 1955 list merged with other surviving firms between 1955 and 1959. Their data were consolidated, and so the sample includes 463 firms independent in 1955. Twelve firms were excluded because they were not primarily engaged in manufacturing; 16 because they could not be fitted into industries with sufficient observations; two because of inadequate data; two because they were liquidated by 1959; four because their merger records were too complex to unscramble; and one inadvertently.

I. *Patent Statistics as an Index of Inventive Output*

Unfortunately, no completely satisfactory index of inventive output is available to the researcher. A straight count of patents has two well-known limitations: (1) the propensity to patent an invention of given quality may vary from firm to firm and from industry to industry; and (2) the quality of the underlying inventions varies widely from patent to patent.

The most important countermeasure is to recognize the existence of these deficiencies, but other steps are also available. Interindustry differences in the propensity to patent will be analyzed through the use of dummy variables. Within a given major industry or field, interfirm differences in the propensity to patent can be viewed as a random disturbance which, unless correlated with some independent variable, imparts no bias to regression estimates of inventive output, only increasing the unexplained variance.

The underlying economic (or technological) significance of any sampled patent can also be interpreted as a random variable with some probability distribution. Here a more serious problem arises, however. Fragmentary data on the profitability of patents—one indicator of economic significance—reveal a distribution highly skewed toward the low private value side, with a very long tail into the high value side [18] [20]. A graphic test suggested the existence of a Pareto-type distribution of profits with an α coefficient of less than .5. Asymptotically such a distribution possesses neither a finite mean nor a finite variance, and so one cannot be sure that the mean economic value of any particular sample of patents converges (under the weak law of large numbers) toward the true population mean value if large enough samples are drawn [10]. This forces us to acknowledge that patent statistics are likely to measure run-of-the-mill industrial inventive output much more accurately than they reflect the occasional strategic inventions which open up new markets and new technologies. The latter must probably remain the domain of economic historians.[3]

Given the limitations of patent data, many of the hypotheses explored in this paper were also tested with respect to 1955 research and development employment (including supporting staff members) for a subsample of 352 firms on which data were available. One finding of this double-barreled attack deserves immediate mention. The number of patents received in 1959 by the 352 firms was (by cross-section

[3] In accepting this limitation, we in effect cut off the most valuable tail of the invention value distribution, eliminating the cause of the infinite mean and variance problem. This is analogous to limiting the number of plays in defining the value of a St. Petersburg game. See [3, pp. 228-37].

standards) fairly strongly correlated with the number of 1955 R&D employees, with an r^2 of .72.[4] More generally, as we shall see, the results of the analyses were similar whether patents or R&D employment was taken as the index of inventive activity.

II. *Demand-Pull and Technology-Push Hypotheses*

Let us consider now the relationship between patenting and the gross size of the 448 firms. Regression of the ith firm's patents P_i on its 1955 sales S_i (in billions of dollars) gave the equation:

$$P_i = 10.65 + \underset{(4.09)}{73.81}\, S_i; \tag{1}$$

with r^2 of .422. (Standard errors of regression coefficients will be given in parentheses.) In comparable regressions, the r^2 was .431 when 1955 assets were the scale variable and .480 with 1955 employment as the scale measure. Although least suitable in terms of a crude goodness-of-fit test, 1955 sales will be used as the scale variable in most of the analyses which follow.[5] This choice is taken for three main reasons. First, the sales variable is most likely to be responsive to short-run changes in demand. Second, the sales variable is essentially neutral with respect to factor proportions—an important attribute, given the fact that more than three-fourths of U.S. industrial research and development is directed toward new products, as opposed to new internal production processes [6]. Third, interview studies show that sales are the principal scale variable considered in company research and development budget decisions.

The correlation between patenting and firm size can be interpreted, along lines suggested by Schmookler and Brownlee [26] and extended by Schmookler and Griliches [27], as reflecting the response of inven-

[4] An even higher r^2 of .89 was obtained in a correlation with patents and R&D employment of the 352 firms aggregated into 14 two- and three-digit industry groups. This gain in explained variance is undoubtedly due to the canceling out of special, perhaps transitory, factors influencing individual firm patenting. When the ratio of government R&D support for the various industries in 1956 was added as a second independent variable, R^2 increased to .94 in this aggregated correlation. The more government support an industry received, the less patenting it did relative to R&D employment, no doubt because exclusive rights cannot be retained for patents received in connection with government contracts. Both regression coefficients were statistically significant at the .01 level.

[5] The reasons why different scale variables give different results and a more complete argument for using sales are presented in [22]. The differences in correlation coefficients are due to peculiarities in the industry distribution of firms. Notably, when each industry assumes its own best-fitting slope and intercept, the over-all fit is closer with sales than with employment.

Ideally we should like to use value added rather than sales to eliminate any biases due to differences in vertical integration, but value-added data for a large sample of firms are not available.

tive output to the over-all pull of demand. Presumably, the greater the sales of a firm in any given market, the more incentive and resources the firm has to generate patentable inventions related to that market.

Nevertheless, systematic differences between industries not related to mere sales volume also affect corporate patenting in an important manner. This can be seen in the effect of estimating separate linear regressions for each of the 14 two- and three-digit industries covered by my study. The results are summarized in Table 1 and in the Appendix.[6] By letting each industry assume its own best-fitting slope and intercept, we are able to "explain" 84.7 per cent of the total variance in corporate patenting about its grand mean. This represents an incremental gain of 42.5 percentage points over the simple regression of patents on sales for 448 firms together.[7] Thus, interindustry differences accounted for almost exactly the same amount of variance as interfirm variations in sales volume.

These observed interindustry differences in patenting, holding firm sales constant, can have a number of underlying causes. Perhaps most important is a set of influences best described under the heading "technological opportunity." Technological opportunity in this context could relate partly to industry traditions or to demand conditions not manifested in mere sales volume, but it seems most likely to be associated with dynamic supply conditions dependent in turn upon the broad advance of scientific and technological knowledge. Thus, the high sales regression coefficients for the electrical and general chemicals industries are undoubtedly due mainly to the vigorous scientific climate of those fields, which has assured a continuous supply of new technical possibilities exploitable to satisfy existing or latent demands. In effect, science and technology exert a push on inventive output in these fields. On the other hand, common observation suggests that the low-coefficient paper, food products, and conventional textile and clothing industries face a much more limited (although not empty) barrel of new technological possibilities.

At least part of the observed interindustry differences in patent out-

[6] A further implication of Table 1 is worth noting. The slopes of the industry regression equations are correlated positively with the r^2 for those equations, with a rank correlation coefficient of .69. The higher an industry's average patent output per sales dollar is, the less variable patenting tends to be relative to size. My interpretation of this result is that in technically progressive fields like electrical equipment and general chemicals, technological competition forces firms to match each others' inventive efforts. But in unprogressive fields like textiles, food products, paper, and fertilizers, soap, and cosmetics, invention is only a business strategy option.

[7] See A-7 of the Appendix, which provides a step-by-step analysis of covariance. Actually, as Appendix step A-12 shows, most of the patenting variance can be explained with even fewer variables, and 20 of the slope and intercept terms allowed to vary in Table 1 contribute little incremental variance reduction.

TABLE 1—LINEAR REGRESSIONS OF PATENTING ON SALES, BY INDUSTRY

Industry	Intercept	Regression Coefficient	N	R^2	Total Patents
Food and tobacco products	− .4	+ 18.05 (2.04)	75	.52	366
Textiles and apparel	2.8	− .48 (7.00)	25	.00	70
Paper and allied products	4.5	+ 7.11 (6.14)	21	.07	120
General chemicals[a]	8.1	+262.48 (25.68)	41	.73	3,316
Misc. chemicals[b]	13.0	+ 19.33 (20.50)	14	.07	231
Petroleum	4.5	+ 81.10 (10.50)	30	.68	2,194
Rubber products	7.3	+ 52.32 (11.29)	8	.78	303
Stone, clay, and glass	−12.4	+200.92 (25.03)	19	.79	434
Primary metals	.4	+ 23.21 (2.50)	50	.64	486
Fabricated metal products and miscellaneous[c]	5.9	+ 61.86 (16.56)	31	.32	516
Machinery	6.1	+ 90.40 (12.58)	46	.54	967
Electrical equipment and communications	22.5	+311.06 (17.61)	35	.90	5,036
Transportation equipment, except aircraft	2.7	+ 59.72 (5.28)	30	.82	1,685
Aircraft and parts	6.8	+ 70.38 (22.39)	23	.32	739
All industries combined	10.7	+ 73.81 (4.09)	448	.42	16,463

[a] Includes S.I.C. 281, 282, and 283 (inorganic, organic, and drugs).
[b] Includes S.I.C. 284, 285, 287, and 289 (soap, paints, fertilizer, and misc.).
[c] Includes, in addition to fabricated metal products, ordnance, watches and clocks, optical equipment, and the S.I.C. 39 miscellaneous category.

put is nonetheless attributable to a different set of causes: systematic variations in the propensity to patent inventions actually made. It is well known, for example, that aircraft makers and other defense specialists seldom seek patent protection on inventions made under government contract, since they must give their principal customer either exclusive rights or at least a royalty-free license in any event [24, pp. 80-82]. Conversely, patent attorneys reported in interviews that in such fields as organic chemicals and petrochemicals, the ease of gaining differentiated market positions through "manipulated molecule" patents encourages profuse patenting. A crude indicator of differences in the propensity to patent is differences in average patent output per unit of engineering input. As Table 2 demonstrates, the average number of 1959 patents received per thousand 1955 research and development employees varied from a low of 27 in the rubber goods industry to a high of 150 in the machinery industry.[8]

TABLE 2—1959 PATENTS PER THOUSAND 1955 R&D EMPLOYEES FOR 352 FIRMS, BY PRINCIPAL INDUSTRY

Industry	Mean P/R&D	Number of Firms
Machinery	150.0	34
Stone, clay, and glass	134.6	18
Petroleum	119.5	21
Fabricated metal products and misc.	117.9	25
General chemicals	111.1	37
Primary metals	95.6	39
Electrical equipment and communications	84.5	31
Paper and allied products	84.2	18
Textiles and apparel	81.3	17
Transportation equipment, exc. aircraft	72.1	18
Food and tobacco products	61.8	55
Miscellaneous chemicals	55.1	14
Aircraft and parts	33.4	19
Rubber products	26.6	6
All industries	83.4	352

Further analysis suggests that these differences in the propensity to patent account for only a minority of all systematic interindustry differences in patenting. To see this, we assume that varying propensities to patent should show up as differences in patent outputs per unit of R&D input. As indicated in Section I, a simple regression of patent outputs on R&D employment inputs for 352 firms yielded an r^2 of .72. If slope (not intercept) dummy variables for 14 industries are introduced into this regression, permitting each industry to have its own input-output coefficient, the multiple R^2 rises to .84. This 12-percentage point gain in patenting variance explained affords a maximum estimate of the effect of interindustry differences in the propensity to patent.[9]

We may consequently partition the over-all variance in corporate

[8] The extremes of this distribution deserve further comment. If we delete the 3,919 R&D employees of Goodyear Aircraft Company, a subsidiary specializing in defense work which accounted for 34 per cent of all rubber industry R&D personnel, the industry's *P/R&D* ratio rises to 40.5. Other rubber companies were also deeply involved in defense work. At the other extreme, I suspect that the craft-oriented machinery industry obtained many patented inventions from employees not in formal R&D organizations.

[9] It is a maximum estimate because other factors such as the type of R&D done also affect patenting per R&D employment input. For instance, the auto industry's preoccupation with styling change may lead to a low output of patentable inventions per engineering man-year.

patenting as follows: 42 per cent to interfirm differences in sales; less than 12 per cent to interindustry differences in the propensity to patent; more than 30 per cent to interindustry differences in technological opportunity, broadly defined; and 16 per cent as (thus far) unexplained residual. Even if this 12-30 split of the 42.5 percentage point interindustry component is inaccurate by a wide margin, it is impossible to avoid a conclusion that interindustry differences unrelated to mere sales volume account for a major proportion of the variance in corporate patenting.

III. *Corporate Bigness and Inventive Output*

We have seen that patenting is an increasing function of firm size, with or without technological opportunity taken into account. We turn next to a question more in the Schumpeterian vein: Does patenting increase more than proportionately with firm size, less than proportionately, or is the relationship essentially linear? Disciples of Schumpeter argue that inventive output ought to increase more than proportionately with firm size due to the scale economies and more effective incentives associated with bigness. Others have postulated the opposite relationship, pointing mainly to the stultifying effects of bigness on incentives and initiative.

Tests of the neo-Schumpeterian scale hypothesis encounter a number of methodological problems. One is that the results depend to some extent on the choice of a scale variable. For reasons analyzed in [22], the neo-Schumpeterian bigness contention receives greatest support when total employment is chosen as the scale measure and least when assets are chosen. Here the sales variable is emphasized partly because it compromises the extremes, but chiefly because it is considered most frequently in actual R&D budget decisions and because it is neutral with respect to factor proportions.

The relationship between sales and patenting or R&D employment can be studied by computing concentration measures or through nonlinear regression methods. The concentration approach is less elegant but more lucid, so let us use it first. Table 3 presents cumulative sales, patenting, and R&D employment shares for selected groups from my subsample of 352 firms on which fully comparable data were available. Firms are ranked on the basis of 1955 sales for all cumulations. Two findings stand out. First, sales volume is persistently more concentrated among the largest firms than R&D employment, which in turn tends to be slightly more concentrated than patenting. The top 20 firms by sales, for example, made 41.5 per cent of the sales of all 352 firms, but employed only 36.7 per cent of the R&D personnel and received only 32.9 per cent of the patents. Second, it follows by subtraction that smaller firms were responsible for a higher relative share of inventive activity than sales. The implication is that inventive inputs (R&D employment) and outputs (patents) increase less than proportionately with sales.

TABLE 3—CONCENTRATION OF SALES, PATENTS, R&D EMPLOYMENT, AND TOTAL EMPLOYMENT IN A SAMPLE OF 352 CORPORATIONS

Number of Firms Included, Ranked by 1955 Sales	Percentage of Total for All 352 Firms			
	1955 Sales	1959 Patents	1955 R&D Employment	1955 Total Employment
First 4	19.9	10.4	9.7	16.1
First 8	27.5	16.8	16.4	24.1
First 12	32.8	24.9	25.9	28.3
First 20	41.5	32.9	36.7	35.0
First 30	49.0	42.9	44.7	42.4
First 40	55.0	45.0	50.4	48.8
First 50	59.9	50.8	57.8	55.1
First 75	69.3	64.5	68.1	63.9
First 100	75.9	71.0	71.9	71.9
First 150	84.9	81.3	81.7	81.8
First 200	90.8	89.4	90.0	88.9
First 250	94.9	94.2	94.8	93.5
First 300	97.7	97.6	97.8	97.2
All 352	100.0	100.0	100.0	100.00

The last column of Table 3 gives cumulative total employment for the 352 companies, again ranked by sales volume. It illustrates the earlier suggestion that a somewhat different result is obtained when employment is taken as the scale variable, R&D employment in particular being more concentrated than total employment for firms ranked below the 12 sales leaders.

Further insight into the role of size can be gained by introducing outside data. The 448 firms in my full sample accounted for 56.9 per cent of the sales of all U.S. manufacturing corporations in 1955, as estimated by the Federal Trade Commission. They accounted for 55.1 per cent of the 29,888 patents issued to U.S. corporations in 1955.[10] These figures are not quite comparable, since the population of all U.S. corporations includes nonmanufacturing corporations. But the latter do very little patenting. It is estimated that nonmanufacturing corporations received no more than 600 patents in 1959.[11] Thus manufactur-

[10] This figure was provided by P. J. Federico, examiner-in-chief, U.S. Patent Office.

[11] Total patents issued in 1959 to the 150 corporations on *Fortune's* 1955 lists of leading retailers, transportation firms, and public utilities amounted to 55. The sampled firms accounted for about one-fifth of all nonmanufacturing corporate sales, suggesting an

ing corporations excluded from my sample—mostly firms with sales of less than $55 million in 1955—accounted for approximately 43.1 per cent of manufacturing sales and 43.9 per cent of patents issued to manufacturing corporations.

This interpolation is a bit surprising, in view of National Science Foundation data showing that industrial firms with fewer than 5,000 employees—a population closely analogous to the population of firms excluded from my sample[12]—accounted for a much smaller share of U.S. research and development expenditures (about 14 per cent) than their share of sales and employment [32, pp. 23-24]. At least three explanations for the disparity between inventive input and output shares among small firms seem relevant. First, government-contract R&D spending, which leads to few patents, is much more concentrated among large firms than private R&D expenditures. Second, small firms apparently obtain a higher proportion of their patented inventions from employees not assigned to formal R&D activities [21]. Finally, there is weak evidence that small firms conduct their R&D with greater cost-consciousness than large firms, and therefore they may tend to generate more inventions per unit of input [2]. In any event, the data suggest that smallness is not necessarily an impediment to the creation of patentable inventions and may well be an advantage.

Nonlinear Regression Analyses

For a more detailed analysis of the relationship between patenting and sales, nonlinear regression methods were employed. If, when P_i is regressed nonlinearly on S_i, the second derivative d^2P/dS^2 of the estimated function is positive, P_i is increasing at an increasing rate with S_i, and so patenting must generally be increasing more than proportionately with sales.[13] A negative second derivative implies the opposite relationship. For expositional convenience let us call the positive d^2P/dS^2 case the "increasing returns" case and the negative d^2P/dS^2 case

estimate of 300 patents for all such corporations. A count based upon several published lists indicated that 153 additional patents were issued to universities, nonprofit research institutions, and nonprofit patent management corporations. Doubling this last figure and adding the two estimates leads to the over-all estimate of roughly 600 patents issued to all nonmanufacturing corporations. This is consistent with National Science Foundation data showing that nonmanufacturing firms do 2 per cent of all industrial R&D [32, p. 19].

[12] Out of the 448 firms sampled, 341 had 5,000 or more employees in 1955, and only three had fewer than 1,000 employees. The firms with 5,000 or more employees had 95.8 per cent of all sample employees and 97.0 per cent of all sample patents.

[13] The last statement is strictly true only when the intercept term is equal to zero. In fact, all the intercept terms for regressions discussed in this section were insignificantly different from zero. The two statements are also not strictly consistent in the neighborhood of inflection points.

the "diminishing returns" case, taking care not to confuse this use of terms with the somewhat different usage accepted in the standard theory of production.

Considering first all 448 firms together, P_i was estimated as a function of S_i with squared and cubic S_i terms:

$$\text{(2)} \qquad P_i = -3.79 + \underset{(14.44)}{144.42\, S_i} - \underset{(4.83)}{23.86\, S_i^2} + \underset{(.316)}{1.457\, S_i^3};$$

with R^2 of .454. This is the nonlinear analogue of equation (1). Introduction of the two nonlinear terms permitted a statistically significant incremental variance reduction of 3.22 percentage points (see the Appendix). When the second derivative of (2) is taken, we find diminishing returns dominating up to a point of inflection at 1955 sales of $5.5 billion, after which increasing returns set in. Since only three firms in the sample of 448 (General Motors, Standard Oil of New Jersey, and Ford) had 1955 sales greater than $5.5 billion, and since the cubic (increasing returns) coefficient is of doubtful statistical significance,[14] the indication is that patent outputs generally increase less than proportionately with increases in sales among corporations large enough to appear on *Fortune's* 1955 list.

To determine whether this result was influenced by the industry distribution of firms, separate regressions of P_i on S_i with nonlinear terms were estimated for each industry with 19 or more observations. In 10 out of 12 industries, the equations were of the same form as (2), with negative squared-term coefficients and positive cubic-term coefficients. In the two nonconforming cases (paper and aircraft) diminishing returns set in with firms of intermediate size and continued all the way out to the largest firms. However, due to the small number of observations for many industries and the high intercorrelations among S_i, S_i^2, and S_i^3 terms, most of the nonlinear coefficients were not statistically significant by conventional standards.[15] Some regressions with few observations were also dominated by one or two extreme values.

[14] The standard error estimates in equations (1) and (2) are biased downward because of heteroscedasticity. When equation (2) was recomputed, letting each centered observation be weighted by the inverse of the standard error of estimate for firms of its size, the squared term's coefficient was significant in a two-tail test at the .09 level and the cubic coefficient at only the .25 level. The heteroscedasticity problem was not nearly as severe for the individual industry regressions, since standard errors of estimate were typically smaller and less correlated with firm size.

[15] Intercorrelation coefficients among the three sales variables ran from .930 to .995, causing the standard errors of all coefficients to explode. For example, the unsquared sales term's t-ratios fell by an average of 70 per cent from the values obtained for the regressions in Table 1. This is a common result of multicollinearity which cannot be avoided without getting into even more serious specification problems discussed in [22].

To combat these statistical problems while taking into account differences in technological opportunity, the original list of 14 industries was consolidated into four groups, depending upon the linear regression coefficients reported in Table 1. The group with coefficients from 0 to 23, including food, textiles, paper, miscellaneous chemicals, and primary metals, will be called the unprogressives. The group with coefficients from 52 to 90 will be called the moderates. Stone, clay, and glass was combined with general chemicals to form a third group, while the electrical industry stood alone in the fourth group. Separate regressions of patents on sales for these four groups explained 83.6 per cent of the total variance in patenting about its grand mean, closely approaching the 84.7 per cent reduction achieved with the 14 separate industry regressions of Table 1. This indicates that the four-group classification was able to allow for the most significant interindustry differences in patenting relative to sales.

Cubic regression equations estimated for the four groups were as follows:

(3) ELECTRICAL
$$P_i = -21.9 + \underset{(98.16)}{582.87}\, S_i - \underset{(95.88)}{177.41}\, S_i^2 + \underset{(22.13)}{26.67}\, S_i^3; \quad R^2 = .94;$$

(4) CHEMICAL
$$P_i = -23.1 + \underset{(131.45)}{432.04}\, S_i - \underset{(221.07)}{203.85}\, S_i^2 + \underset{(85.59)}{58.42}\, S_i^3; \quad R^2 = .74;$$

(5) MODERATES
$$P_i = 5.7 + \underset{(12.98)}{82.85}\, S_i - \underset{(3.78)}{10.63}\, S_i^2 + \underset{(.235)}{.789}\, S_i^3; \quad R^2 = .77;$$

(6) UNPROGRESSIVES
$$P_i = 1.9 + \underset{(6.53)}{21.04}\, S_i - \underset{(5.61)}{10.18}\, S_i^2 + \underset{(1.06)}{3.00}\, S_i^3; \quad R^2 = .55.$$

However, as Appendix steps A-9 and A-10 show, the introduction of the 12 squared and 12 cubic sales variables led to variance reductions each significant in F-ratio tests at the 1 per cent point.

All four equations, plotted in Figure 1, display the initially diminishing and then increasing returns pattern of equation (2).[16] Despite multicollinearity, three of the four squared (diminishing returns) terms' coefficients are significant at the .10 level or higher. In the increasing returns stage dominated by the positive cubic coefficients, the unprogressive group included five large firms, the moderates three, chemicals and stone-clay-glass two, and the electrical industry only one. Thus, the indication again is one of diminishing returns except for a few giant firms which led their two-digit industries in sales.

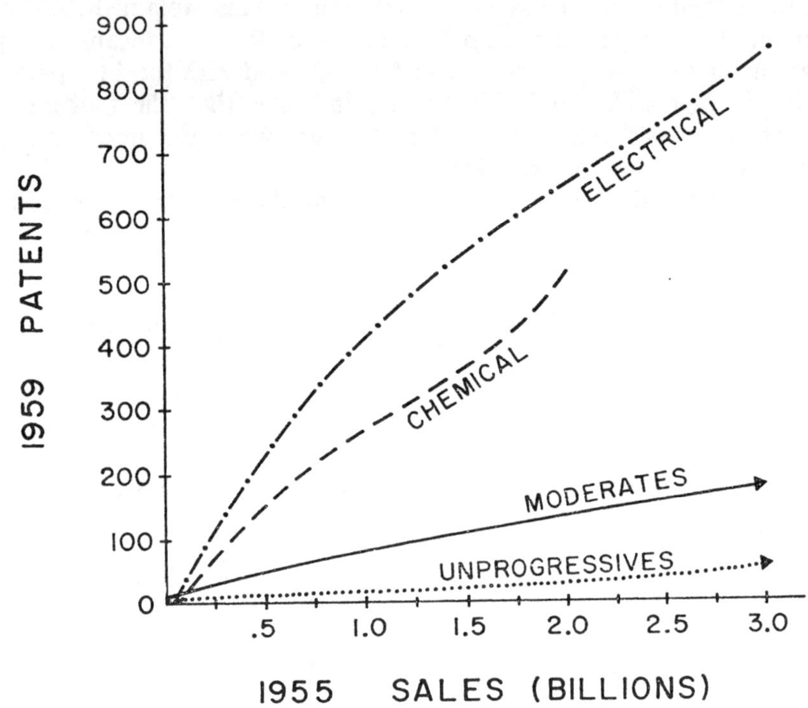

FIGURE 1

All of the regression equations discussed thus far are dominated to some extent by the observations for large firms. This is not wholly undesirable, since the largest firms account for such a substantial proportion of the full sample's sales and presumably had the market penetration and resources to support a similar lion's share of inventive activi-

[16] An alternative way of interpreting the figures is to sweep a straight line from the origin to points on the plotted curves. The slope of the line intersecting any point shows estimated patenting per billion dollars of sales. It is readily seen that P/S generally attains a maximum at sales levels below $500 million and a minimum at higher sales levels.

The upward bending part of the moderates' curve, including only General Motors, Jersey Standard, and Ford, is excluded from Figure 1 to maintain a larger drafting scale.

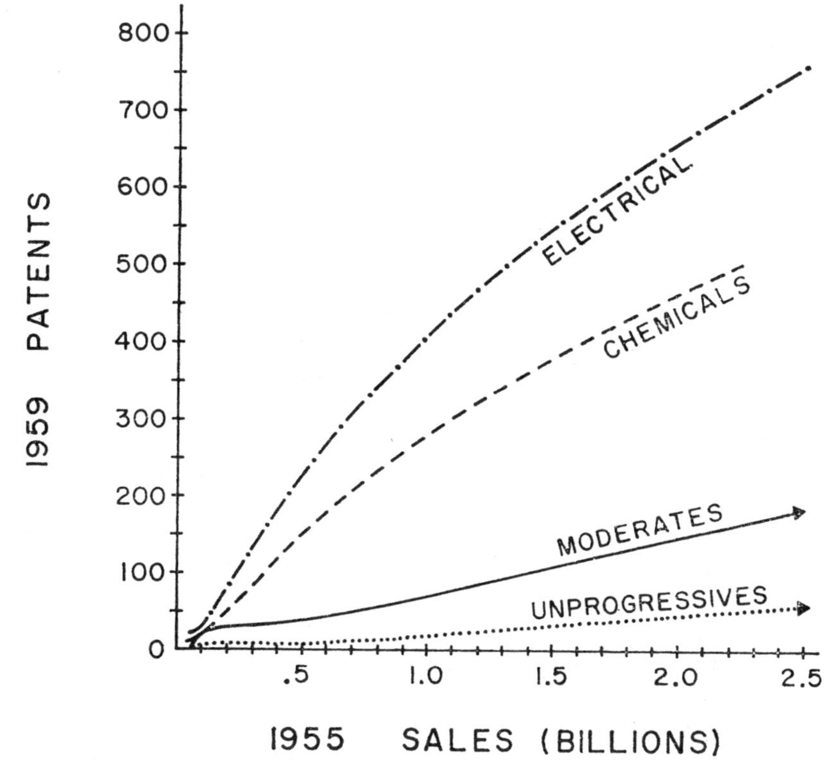

FIGURE 2

ty. Still for the sake of accuracy two further analyses were executed which either avoided large firm biases or erred in the opposite direction.

First, along lines suggested by Edwin Mansfield [11], logarithms of the sales variables were taken to compress the large-firm tail of the sales distribution, and then P_i was regressed on log S_i, $(\log S_i)^2$ and $(\log S_i)^3$ for each of the four groups. Because the equations are almost impossible to interpret intuitively, the results are presented graphically in Figure 2.[17] Diminishing returns in patenting are evident for both chemicals and electrical equipment—the most progressive groups. For the unprogressives there is a tendency toward increasing returns

[17] For the unprogressive and moderate groups, all three regression coefficients were statistically significant at the .01 level. None of the chemical and electrical regression coefficients was significant at even the .20 level, even though the three coefficients together permitted reductions in variance significant in F-ratio tests at well beyond the 1 per cent point. The latter result was again due to extreme collinearity among the three independent variables, with intercorrelation coefficients running between .980 and .996. Also, because of the logarithmic transformations, the more concavely nonlinear the functions appear when plotted on arithmetic grids, the more the regressions are likely to be linear on semi-log grids.

up to sales of about $1 billion, and then initially imperceptible diminishing returns set in. This relationship, however, is probably due in part to the greater diversification of the largest firms in unprogressive fields, as discussed in Section IV.[18] For the moderates slightly increasing returns prevail up to sales of $800 million, at which point a faint diminishing-returns effect begins to dominate.

Second, each group was broken down into from three to five size subgroups, and separate linear regressions were estimated for each subgroup. The results are shown in Figure 3. Although many of the coefficients were not statistically significant, the diverse pieces splice together well enough to suggest that something more than random phenomena was at work. For both chemicals (with stone, clay, and glass) and electrical equipment, there may be slightly increasing returns up to sales of roughly $500 million, but beyond that point a definite flattening out of the patenting function is evident. For the moderates very slightly decreasing returns appear to prevail continuously. The pattern for the unprogressives is so irregular that no conclusions can be drawn.

By way of synthesis, two possible interpretations can be attached to the full set of regression results, depending upon one's attitude toward statistical significance tests generally and the disutility of Type I as opposed to Type II inferential errors. Where significance is doubtful by traditional standards, one may incline toward the Scotch verdict that corporate patenting has not been shown to increase either more or less than proportionately with sales. But if the regressions are accepted as best estimates of some true behavioral pattern, it would appear that after a stage of slightly increasing returns extending to 1955 sales of approximately $500 million, corporate patenting tends to increase less than proportionately with sales, except in the case of a few giant firms which lead their two-digit sectors in sales. The least vigorous patent recipients relative to their size appear to be non-leader firms with sales over $500 million.

Conditions Underlying the Observed Relationships

The relevance of these observed relationships might be questioned if one could show that the propensity to patent is inversely correlated with firm size. It is conceivable that large firms receive proportionately fewer patents than smaller firms, not because they generate fewer inventions, but because they patent a smaller proportion of what they invent.[19] This could be so, for instance, because their inventive output

[18] Notably, in a regression of unprogressive group patents on S_i, S_i^2, S_i^3, and a diversification index, the diversification variable had by far the highest t-ratio and (positive) partial correlation coefficient and raised R^2 to .594 from the value of .545 for equation (6).

[19] It is also possible that large firms tackle more complex problems than small firms, and so their patents reflect greater inventive input. The former statement is probably

includes a higher fraction of unpatentable contributions to pure knowledge, or because patents afford less marginal benefit than they do for smaller and more vulnerable firms, or because government antitrust policies have been hostile to the exploitation of patent power by large firms [24]. But arguments for the opposite conclusion are also compelling. Notably, restrictive power may grow more than proportionately with the accumulation of patents, and for large firms with a staff of in-house patent attorneys, the short-run marginal cost of patenting may be less than for smaller firms dependent upon outside counsel.[20]

It is impossible to choose between these arguments on a priori grounds, so we must resort to two slender reeds of data. First, an intensive questionnaire survey of more than 500 randomly selected patents issued in 1938, 1948, and 1952 showed no tendency for the proportion of inventions actually produced or used in production to vary

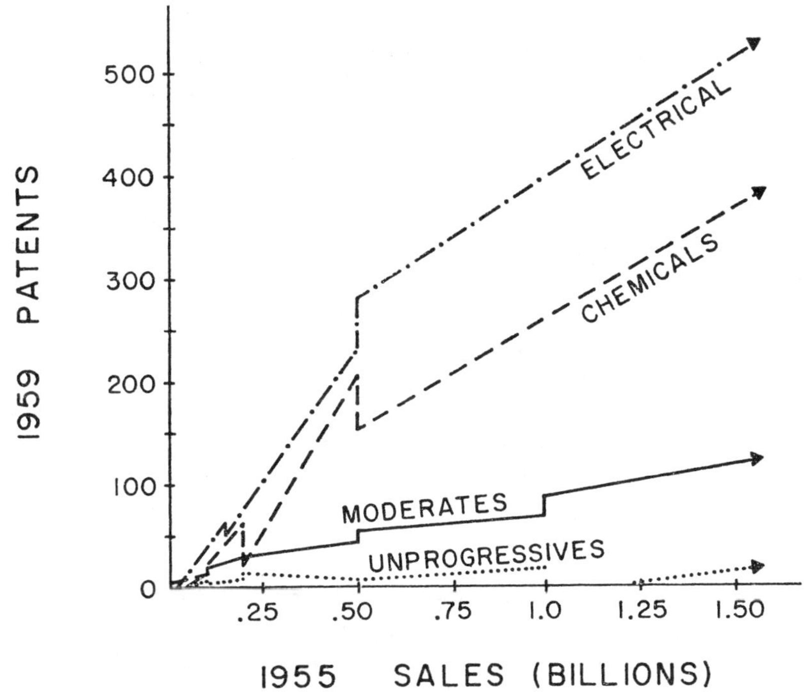

FIGURE 3

valid, but the latter is doubtful. Complex "system inventions" are usually broken down into many elementary parts for purposes of patenting, and I see no reason to expect the quality of these elementary inventions to vary with firm size. For example, RCA has well over a thousand patents on color television transmitting and receiving devices.

[20] Patents assigned in 1959 to the 448 firms in my sample were correlated with an incomplete count of in-house patent attorneys derived from the Martindale and Hubbell 1955 law directory, with a resultant r^2 of .78. Although 365 of the 448 firms received patents in 1959, only 161 were found to employ in-house patent attorneys.

with firm size for firms with 1949 sales exceeding $55 million [18]. There was, however, evidence that the percentage of patents in *current* use varied inversely with firm size.[21] One would expect firms which patent profusely to have a lower utilization rate than firms which patent only significant inventions, and so the survey results imply either a constant or (for current utilization) an increasing propensity to patent as firm size increases. Second, and in partial conflict with the survey results, the data underlying Table 3 show that the 40 largest corporations ranked by sales received an average of 74.7 patents per 1,000 R&D employees, compared to the over-all average of 83.4 patents for 352 firms (as in Table 2). Similar results are found for the 20 largest, 30 largest, and 50 largest. On the other hand, industry leaders Swift, New Jersey Standard Oil, U. S. Steel, Pittsburgh Plate Glass, International Harvester, and General Electric each received more than 120 patents per 1,000 R&D employees. These findings suggest that marked differences in the propensity to patent among the largest firms may in part explain both the relatively poor showing of large firms which did not lead their two-digit sectors in sales and the increasing returns associated with leading firms.

Variations in the propensity to patent clearly do not tell the whole story, however. Nonlinear regressions of R&D employment on sales presented in [22] generally paralleled those estimated here for patenting on sales. A global regression analogous to equation (2) had coefficients with the same signs as (2), but the nonlinearities were more pronounced. Cubic regressions for seven industries with 20 or more observations and grouped regressions similar to equations (3) through (6) revealed a general tendency toward diminishing returns, although the chemicals industry was an exception with R&D employment increasing more than proportionately with sales.[22] There was somewhat less evidence of an upturn in R&D employment for industry sales leaders (consistent with their high ratio of patents to R&D employment) and somewhat more evidence of increasing returns for firms with sales under $500 million.

Thus, one is led to believe that the diminishing returns observed in connection with patenting are related at least in part to similar tendencies in research and development employment, especially for firms with sales exceeding $500 million in 1955. It is not completely clear why this is so. My guess is that a firm's optimal competitive strategy tends to vary systematically but not monotonically with both absolute

[21] This may be the more relevant result due to lower response error and because of its closer proximity in time to my sample. The survey also showed that the largest corporations had a somewhat shorter mean duration of utilization for their patents.

[22] Recognition of this exception brings my results into full agreement with Mansfield's findings concerning major inventions, major innovations, and R&D spending in the petroleum, steel, chemicals, and glass industries [11] [12].

and relative size. Relatively small firms have strong incentives to do product-differentiating R&D and to patent the results in order to secure their positions against large rivals with economies of scale in promotion, distribution, and perhaps production. Giant firms in positions of industry sales leadership *may* also have effective incentives to support R&D—possibly on prestige grounds, or to guard against the hazard of organizational stagnation and decline, or because they inherited a research orientation which was partly responsible for their growth. At the same time such firms also tend to have a high propensity to patent. Large firms which do not lead their two-digit sectors in sales, on the other hand, appear to be least active relative to their size in both R&D employment and patenting, perhaps because they find it advantageous to focus their main attention on promotional activities and production, counting upon rapid imitation or licensing of others' inventions to guard against technological displacement.

Two further observations on the relationship between R&D inputs and patent outputs must be recorded. First, there was no evidence that patenting increased either more or less than proportionately with R&D employment. A regression of P_i on $R\&D_i$ with quadratic and cubic terms was very nearly linear. Second, there was some indication of diminishing returns to R&D input intensity, as suggested in the following regression, where S_i is scaled in billions of dollars and $R\&D_i$ in thousands of employees:

$$(7) \qquad P_i/S_i = (14 \text{ constants}) - \underset{(7.92)}{12.54}\, S_i + \underset{(7.87)}{61.33}\, R\&D_i/S_i - \underset{(.51)}{2.84}\, (R\&D_i/S_i)^2,$$

with R^2 of .54. The last two terms are of primary interest here. The more R&D employees per million dollars of sales a firm retained, the more patents per billion dollars of sales it received, but with diminishing returns. The function attains a maximum at 16.7 R&D employees per million dollars of sales—an input intensity exceeded only by the Raytheon Corporation in 1955. It would appear that for firms with sales under \$500 million, which tended to have the highest ratios of R&D employees to sales, patenting was damped by the intensity effect of equation (7), and so some of the increasing returns found in R&D employment did not show up in patent outputs. The diminishing returns relationship of equation (7) is in turn probably due to organizational problems associated with high R&D input intensity, although the precise mechanism of this phenomenon is not clear.[23]

[23] Also involved are some peculiarities in the distribution of propensities to patent. The six firms with the highest *R&D/S* ratios in 1955 were all predominately engaged in defense contracting, and so they undoubtedly had low propensities to patent. Ideally, the

In conclusion, the evidence does not support the hypothesis that corporate bigness is especially favorable to high inventive output. If anything, the results show that firms below the half-billion dollar sales mark generate more inventions relative to their size than do giant firms. Such a conclusion is not necessarily grounds for a campaign of breaking up large firms, since the observed tendencies are less than completely uniform. It is also possible that large size does confer advantages for the development and integration of complicated "systems"—activities less likely to yield patentable inventions. Small firms at the same time may enjoy a comparative advantage at inventing and developing the more readily patentable component parts for such systems. My results do suggest, however, that a heavy burden of proof must be sustained by firms emphasizing research and development potential as a justification (i.e., in merger cases) for bigness.

IV. *The Role of Diversification*

One of the most interesting neo-Schumpeterian hypotheses assigns a potentially important role to diversification as a stimulus to invention. The idea, as formulated by Richard Nelson [15], is as follows: Research, and especially basic research, is an uncertain activity, yielding inventions and discoveries in unexpected areas. The firm with interests in a diversity of fields will generally be able to produce and market a higher proportion of these unexpected inventions than a firm whose product line is narrow. Therefore, the expected profitability of speculative research is greater for highly diversified firms, and such firms will tend to support more research.

Even though patentable inventions are undoubtedly not the leading product of basic industrial research, uncertainty is an important component of all research, and so the diversification hypothesis seemed worth testing. The usual list of 447 S.I.C. four-digit manufacturing industries was consolidated into a more technologically meaningful list of roughly 200 industries. A diversification index was compiled for each of the 448 companies sampled by counting the number of consolidated industries in which the company operated.[24] The index varied from 43 (for General Electric) down to 1, with a mean of 5.7.

industry dummy variables should have compensated for this, but five of the six firms were defense specialists from the electrical industry, which on the whole was much less defense-oriented. Still when equation (7) was estimated for six separate industries (chemicals, food, petroleum, metals, machinery, and electrical), the diminishing returns phenomenon appeared for all but the chemicals industry.

[24] My index is analogous to, but because of consolidation not the same as, one of the indices developed for a smaller number of companies by M. Gort [5, pp. 155-57]. Our samples overlapped for 88 companies. The r^2 when his indices were correlated with mine was .71. An industry-by-industry comparison of his indices with mine yielded an average Spearman rank correlation coefficient of .83.

The results of the global 448-firm analysis were striking. Although the diversification index was correlated with sales, with r^2 of .20, it was much more strongly correlated with patents, with r^2 of .37. When the index was added to a regression of patents on sales, the variance in patenting was reduced by 13 percentage points. Similar results were obtained when the diversification variable was introduced into regressions of P_i/S_i on sales, $R\&D_i$ on sales, and P_i on $R\&D_i$. In every case the diversification variable permitted a significant reduction in unexplained variance, and the t-ratio of its regression coefficient was on the order of 5.0 or more.

The picture changed considerably when each of the 14 two- and three-digit industry groups was permitted to assume its own best-fitting sales and diversification coefficients. For all industries summed, inclusion of the diversification variable accounted for an incremental variance reduction of 1.05 percentage points beyond the contribution of the simple linear sales regressions in Table 1—an increment just significant at the 1 per cent point in an F-ratio test.[25] This decline in explanatory power suggests that the diversification index acted partly as a surrogate industry dummy variable, and that two- or three-digit industry groups with high patenting tended to be the home base of firms operating in a greater number of narrow fields, as defined in my consolidation procedure.

Within the two-digit groups, diversification played an uneven role. For five industries, including the three (electrical; general chemical; and stone, clay, and glass) with the highest patenting per billion dollars of sales (Table 1), the diversification coefficients were negative but statistically insignificant. For nine industries, patenting was positively correlated with diversification, in most cases with t-ratios exceeding 2.0; and the diversification variable accounted for an appreciable reduction in variance not explained by the simple regressions of Table 1. These included, with the incremental variance reduction percentage in parentheses, textiles and apparel (33), paper (26), aircraft (26), rubber (12), miscellaneous chemicals (10), petroleum (5), food products (4), machinery (4), and land transportation equipment (1). With the exception of machinery, petroleum, and perhaps aircraft, these were for the most part industries whose members obtained few patents per billion dollars of sales.

A similar result was experienced when the diversification variable was introduced into regressions of R&D employment on sales for seven

[25] See Appendix steps A-8 and A-11. When the diversification variables were added to the set of industry regressions with quadratic and cubic sales terms, the incremental variance reduction was 1.24 points. In total, the set of linear and nonlinear sales and diversification variables for 14 industries accounted for 91 per cent of the variance in company-patenting about its grand mean.

two-digit industry groups with 20 or more observations. R&D employment did not increase at all with diversification for the general chemicals and electrical groups, which were by far the most R&D-oriented of the seven. It increased most for food products, which ran a close second to primary metals in fewness of R&D personnel per million dollars of sales. (The metals diversification coefficient was also positive, but significant at only the .20 level.)

The main reason for these relationships becomes clear when one examines in detail the product structure of diversified companies classified in a parent two-digit industry with relatively little private research and development and patenting. The more diversified such a firm is, the higher is the probability that it will operate also in a more progressive industry—that is, one with ample opportunities for profitable private investment in new technology. For instance, of the food-products firms with sales of less than $500 million, the leading patent recipient was National Distillers Products, which manufactured basic chemicals in addition to its more thirst-quenching specialities. Sales leaders in the food products, textiles, paper, and miscellaneous chemicals groups tended to be a good deal more diversified than most of their followers, and despite some collinearity the diversification index for these firms picked up incremental variance in patenting. And for a final example from the government-oriented sector, aircraft-makers General Dynamics and North American, which were in 1955 either established in or moving aggressively into electronics, atomic energy, and chemicals, received many more patents than less diversified Douglas and Boeing, which had higher sales at the time. It would appear then that diversification was not per se a structural condition necessarily favorable to patentable invention. Rather, the diversification variable's impressive performance shows chiefly the effect of operaton in dynamic industries by firms whose home base is not conducive to patenting and (except for aircraft) research and development.

V. *Market Power and Patented Inventions*

We consider now market-power's effect on corporate inventive output. There are two broad hypotheses assigning a favorable role to monopoly. No attempt is made to test the older but still controversial hypothesis that the expectation of a future monopoly position (e.g., through a patent grant) is necessary before risk capital will be committed to invention or development.[26] We will examine only the assertion more original to Schumpeter that monopoly power already at-

[26] But see [24], which deals with this problem at length.

tained is a favorable base from which to make advances in modern technology.

Schumpeter's writings on this subject consist of little more than *dicta* demanding further interpretation. One construction is that profitable exploitation of market power is the easiest way to assemble the funds necessary for investment in new technology [28, p. 87]. This can be subdivided into two related hypotheses: that technological output (or input) increases with corporate profitability, or that technological output increases with corporate liquidity. The latter is not explicitly Schumpeterian, but it presupposes the same assumptions.

To test these hypotheses, data on base year 1955 profits and liquid assets were collected for the 448 corporations.[27] None of a variety of tests showed any noteworthy tendency for 1959 patenting or 1955 research and development to increase with either 1955 profitability or liquidity. The r^2 between patents per billion dollars of sales and 1955 profits as a percentage of sales was .001. Between P_i/S_i and liquid assets as a percentage of total assets it was .003. The 1955 profit rate, measured in relation to both sales and total assets, was included in some 26 regressions involving raw and deflated patents, R&D personnel, and an assortment of other variables for both the full sample and particular industries. Although there was a tendency for more of the coefficients to be positive than negative, in only five cases was a t-ratio of more than 1.0 achieved, and there were no profitability t-ratios of 2.0 or more.[28] Similarly negative results appeared when the liquid asset /total asset ratio was included in the same equations.

Rejection of the profitability and liquidity hypotheses does not, however, exhaust the list of proposals ascribing a progressive bias to market power. Schumpeter argued that the temporary security and longer-range perspective afforded by a monopoly position facilitated shooting at the rapidly and jerkily moving targets of new technology [28, pp. 88 and 103]. This implies that technological output should tend to increase with industrial concentration, other things being equal. More recently it has been proposed that oligopolists colluding overtly or tacitly on the price dimension have especially vivid incentives to engage in new product rivalry [4, p. 575]. This implies that technological output should tend to increase with concentration up to a point, but that it may decline if too much of an industry's output becomes concentrated in the hands of a single dominant seller.

Prior attempts to test these hypotheses quantitatively [1] [7] [8]

[27] The principal items included in the liquid-assets definition were cash, marketable securities, government securities, foreign currency, and bank acceptances.

[28] There *was* a tendency for lagged profits to increase with patenting. This is discussed in [23].

aggregated both the technological progressiveness indices and the concentration indices into 20 or fewer two- and three-digit industry categories. While the multitude of sins underlying such a procedure may well cancel out, one never really knows what has been cooked in the stew. Therefore, a completely different approach was tried.

The only published source which links firm names with concentration ratios for a large sample of industries is a Federal Trade Commission report covering the base year 1950 [31]. That report lists the four leading firms in four- and five-digit industries, along with industry concentration ratios and sales. From the report's listings a sample of 48 industries was drawn[29] based upon several selection criteria: (1) The industry definition had to be as meaningful as possible in terms of economic analysis. (2) The Census Bureau's coverage of industry producers had to be at least 75 per cent for industries with four-firm concentration ratios below 50 per cent and 85 per cent for industries with higher concentration ratios. (3) Primary-product sales had to comprise at least 75 per cent of reported sales for the industry. (4) The broad parent industry had to be reasonably well suited for making and patenting inventions. On the strength of this criterion the food and tobacco, apparel, conventional textile, lumbering, furniture, and publishing two-digit sectors and all primarily defense-oriented industries were excluded. (5) The industry's technology had to be such that classification of patents was feasible. (6) To the extent that all other criteria could be satisfied, a wide range of technologies, concentration ratios, and industry sizes was sought.

The index of inventive output P_{4i} employed in this analysis was the number of industry-related patents issued in 1954 to the leading four firms in the *i*th industry. A four-year lag from the 1950 base year was used to approximate the mean lags between invention, patent application, and patent issue. Each patent assigned in 1954 to each of 152 different firms occupying leadership positions in the 48 industries was

[29] The sample includes hard-surface floor coverings; pulp, paper, and paperboard; paperboard boxes; sanitary paper food containers; sanitary paper health products; plastics materials; synthetic fibers; soaps and organic detergents; paints and varnishes; inorganic color pigments; printing ink; dentifrices; carbon black; tires and inner tubes; leather; footwear (excluding house slippers and rubber footwear); flat and laminated glass; glass containers; vitreous plumbing fixtures; porcelain electrical supplies; gypsum products; abrasive products; iron, steel, and rolling mill products; primary, rolled, drawn, and extruded aluminum; tin cans; razors and razor blades; files and rasps; power boilers; wire springs; wheeled tractors and farm machinery; track-laying tractors; machine tools; textile machinery; industrial pumps; conveyors and conveying equipment; household mechanical washing machines; vacuum cleaners; valves and fittings; ball and roller bearings; motors, generators, and steam turbines; transformers; electrical welding apparatus; household radios, television sets, and phonographs; electron tubes and transistors; storage batteries; telephone and telegraph equipment; motor vehicles and parts; and locomotives.

separately examined to determine whether or not it pertained primarily to a sampled industry. Some 5,000 patents were examined in this manner; 2,202 patents were assigned to the sampled industries and the rest were thrown out. This procedure was fairly straightforward for most industries, but in a few cases (especially in those involving by-products and substantial vertical integration) a good deal of judgment had to be applied, and the count was subject to uncertainty.

Independent variables included the value of shipments S_{4i} of the leading four firms in the ith industry in 1950, four-firm concentration ratios M_i for 1950, and slope (not intercept) dummy variables to differentiate industries operating primarily in chemical and electrical technologies from those facing largely mechanical technologies.[30]

The analysis revealed a positive but very modest and statistically insignificant influence on patenting for the market share variable. The simple linear r^2 between patenting and M_i was .069, with a t-ratio of 1.85 for the corresponding regression coefficient. In a regression of the logarithms of P_{4i} on the logarithms of M_i, r^2 was .051 and the t-ratio 1.57. The t-ratios and partial correlation coefficients for market share declined when additional variables, especially the value of shipments S_{4i}, were introduced.[31] Because one would expect the effects of scale, market share, and technology to be multiplicative rather than additive, and in order to suppress the undesirable effects of heteroscedasticity, the regression model most suitable on theoretical and statistical grounds appeared to be:

$$P_{4i} = aS_{4i}^{\alpha}M_i^{\beta}C_i^{\gamma}E_i^{\delta}u_i; \tag{8}$$

where C_i is a chemical technology slope dummy variable, E_i is an electrical technology dummy variable, and u_i is the error term. The fitted regression equation in logarithmic form was:

$$\begin{aligned}(9)\qquad \log P_{4i} = 1.51 + \underset{(.133)}{.680} \log S_{4i} + \underset{(.302)}{.064} \log M_i + \underset{(.181)}{.192} \log C_i \\ + \underset{(.193)}{.498} \log E_i;\end{aligned}$$

[30] Another approach to the interindustry technological opportunity problem, using deviations of actual patenting from what adjusted equations based on those in Table 1 would predict, gave results essentially the same as those reported here.

[31] Since M_i was correlated with S_{4i}, with an r^2 of .065 in the linear version and .049 in the logarithmic version, it is possible that the higher simple correlations between patenting and market share were partly spurious. Notably, introduction of M_i into regressions of P_{4i} on S_{4i} with the two dummy variables permitted an incremental reduction in unexplained variance of only 1.2 percentage points in the additive version and 0.1 points in the logarithmic version.

with R^2 of .47. The partial correlation for market share was .032—far lower than any of the other variables' partial r's. If despite the low t-ratio for market share we use (9) as a best estimator, we find that for four industry leaders with total shipments valued at $100 million, an increase in market share from 15 per cent to 90 per cent brings an increase of one patent. For four firms with sales of $1 billion, the same market share differential implies a difference of five patents.

It must be noted that, although (9) was deemed most suitable on theoretical and statistical grounds, it assigned a weaker role to market share than the analogous additive (nonlogarithmic) equation (in which the M_i coefficient's t-ratio was 1.06). But further analysis showed this difference to be largely the result of patenting and market shares associated with the telephone and telegraph industry (with $M =$ 92 per cent). When the observations for that industry were deleted, the market share coefficient's t-ratio in the additive regression fell from 1.06 to .09.[32]

To test further the sensitivity of my results to potential biasing factors, regressions were run without the observations for eight industries with marked patent classification and assignment uncertainties.[33] The general effect was a distinct but far from statistically significant increase in the t-ratios and partial correlation coefficients associated with M_i. For example, the M_i t-ratio for equation (9) rose from .21 with $n = 48$ to .67 with $n = 40$. Deletion of telephone industry observations in addition to those of the uncertain industries caused t to fall in this case from .67 to .51.

Additional additive regression equations were estimated with a quadratic market share term in order to test for nonlinearities. In every such regression a U-shaped relationship was found, with minimum patenting occurring at four-firm concentration ratios between 44 and 71 per cent. Since the nonlinear coefficients made little incremental contribution to the reduction of variance and had no t-ratio exceeding 1.46, and since no compelling theoretical rationalization for such a relationship is known, it seems likely that the observed U-shape was the result of chance elements.

In sum, the analysis suggests that if structural market power has a beneficial effect on the output of patented inventions, it is a very modest effect indeed. One highly monopolistic industry—telephone and telegraph equipment—was responsible for much of the observed corre-

[32] Because more nearly homoscedastic logarithmic regression (9) assigned much less weight to the telephone industry's deviant patent value, it estimated a weaker role overall for the market share variable.

[33] They were plastic materials, synthetic fibers, printing ink, tracklaying tractors, transformers, radios and related products, motor vehicles, and locomotives.

lation between market share and patenting, and that industry, dominated as it is by quasi-regulated Western Electric, is clearly a special case.

This conclusion would have to be reconsidered if there were indications that the propensity to patent varied inversely with market power. But although nothing even approximating proof is available, conventional wisdom holds that the opposite is more apt to be true. In many of the most concentrated industries sampled (i.e., gypsum products, razors, synthetic fibers, flat glass, telephones, and aluminum) the market power of the leaders was built partly upon a patent base, and this fact could scarcely have been forgotten by company executives anxious to preserve that power.

It is also conceivable that existing monopoly power is a uniquely favorable environment only for investment in basic and the most fundamental applied research, either because these activities carry the highest risks or because their benefits are apt to be largely external to more competitive firms. Since knowledge—presumably an important output of such activities—cannot be patented, patent statistics do not permit a sensitive test of this narrower interpretation.[34]

As a final caveat, it must be pointed out that my data assign a weaker beneficial role to market power than do the studies of two-digit industry R&D spending reported by Hamberg [7] and Horowitz [8]. Still the fact that their analyses were unable to take into account differences in technological opportunity may have permitted the correlations to attribute a spuriously strong influence to market power.

VI. *Conclusions*

The principal conclusions of this study are as follows: (1) Inventive output increases with firm sales, but generally at a less than proportional rate. (2) Differences in technological opportunity—e.g., differences in technical investment possibilities unrelated to the mere volume of sales and typically opened up by the broad advance of knowledge—are a major factor responsible for interindustry differences in inventive output. (3) Inventive output does not appear to be systematically related to variations in market power, prior profitability, liquidity, or (when participation in fields with high technological opportunity is accounted for) degree of product line diversification.

[34] And a narrow interpretation it is, since basic research expenditures have comprised only about four per cent of all industrial R&D outlays in recent years, and since private industry has done only about 30 per cent of all U.S. basic research. Patentable inventions may also emerge from basic and certainly from applied research, even though knowledge is the main output. For instance, the early basic transistor patents were one output of Bell Laboratories' fundamental semiconductor research.

These findings among other things raise doubts whether the big, monopolistic, conglomerate corporation is as efficient an engine of technological change as disciples of Schumpeter (including myself) have supposed it to be. Perhaps a bevy of fact-mechanics can still rescue the Schumpeterian engine from disgrace, but at present the outlook seems pessimistic.

REFERENCES

1. Y. BROZEN, "R&D Differences Among Industries," in Richard Tybout, ed., *The Economics of Research and Development*, Columbus, Ohio, 1965.
2. A. C. COOPER, "R&D Is More Efficient in Small Companies," *Harvard Bus. Rev.*, May-June 1964, *42*, 75-83; and *ibid.*, Sept.-Oct. 1964, *42*, 44-51.
3. W. FELLER, *An Introduction to Probability Theory and Its Applications*, 2nd ed. New York 1957.
4. W. FELLNER, "The Influence of Market Structure on Technological Progress," *Quart. Jour. Econ.*, Nov. 1951, *55*, 556-77.
5. M. GORT, *Diversification and Integration in American Industry*. Princeton 1962.
6. W. E. GUSTAFSON, "Research and Development, New Products, and Productivity Change," *Am. Econ. Rev., Proc.*, May 1962, *52*, 177-85.
7. D. HAMBERG, "Size of Firm, Oligopoly, and Research: The Evidence," *Can. Jour. Econ.*, Feb. 1964, *30*, 62-75.
8. I. HOROWITZ, "Firm Size and Research Activity," *So. Econ. Jour.*, Jan. 1962, *28*, 298-301.
9. S. KUZNETS, "Inventive Activity: Problems of Definition and Measurement," in [16, pp. 19-43].
10. B. MANDELBROT, "New Methods in Statistical Economics," *Jour. Pol. Econ.*, Oct. 1963, *71*, 421-40.
11. E. MANSFIELD, "Size of Firm, Market Structure, and Innovation," *Jour. Pol. Econ.*, *71*, 556-76.
12. ———, "Industrial Research and Development Expenditures," *Jour. Pol. Econ.*, Aug. 1964, *72*, 319-40.
13. J. W. MARKHAM, "Market Structure, Business Conduct, and Innovation," *Am. Econ. Rev., Proc.*, May 1965, *55*, 323-32.
14. R. R. NELSON, "The Economics of Invention," *Jour. Bus.*, April 1959, *32*, 101-27.
15. ———, "The Simple Economics of Basic Scientific Research," *Jour. Pol. Econ.*, June 1959, *67*, 297-306.
16. ———, ed., *The Rate and Direction of Inventive Activity*, Nat. Bur. Econ. Research Conference Report. Princeton 1962.
17. B. SANDERS, "Some Difficulties in Measuring Inventive Activity," in [16, pp. 53-77].
18. ———, "Patterns of Commercial Exploitation of Patented Inventions by Large and Small Corporations," *Patent, Trademark, and Copyright Jour.*, Spring 1964, *8*, 51-93.

19. ———, "Speedy Entry of Patented Inventions into Commercial Use," *Patent, Trademark, and Copyright Jour.*, Spring 1962, *6,* 87-116.
20. ———, J. ROSSMAN, AND L. J. HARRIS, "The Economic Impact of Patents," *Patent, Trademark, and Copyright Jour.*, Sept. 1958, *2,* 340-63.
21. ———, "Patent Acquisition By Corporations," *Patent, Trademark, and Copyright Jour.*, Fall, 1959, *3,* 217-61.
22. F. M. SCHERER, "Size of Firm, Oligopoly, and Research: Comment," *Can. Jour. Econ.*, May 1965, *31,* 256-66.
23. ———, "Corporate Inventive Output, Profits, and Growth," *Jour. Pol. Econ.*, June 1965, *73,* 290-97.
24. ———, S. E. HERZSTEIN, *et al.*, *Patents and the Corporation,* 2nd ed. Boston 1959.
25. J. SCHMOOKLER, "Changes in Industry and in the State of Knowledge as Determinants of Industrial Invention," in [16, pp. 195-232].
26. ———AND O. BROWNLEE, "Determinants of Inventive Activity," *Am. Econ. Rev., Proc.,* May 1962, *52,* 165-76.
27. ———, AND Z. GRILICHES, "Inventing and Maximizing," *Am. Econ. Rev.*, Sept. 1963, *53,* 725-29.
28. J. A. SCHUMPETER, *Capitalism, Socialism, and Democracy,* 3rd ed. New York 1950.
29. J. S. WORLEY, "Industrial Research and the New Competition," *Jour. Pol. Econ.*, April 1961, *69,* 183-86.
30. NATIONAL ACADEMY OF SCIENCES, National Research Council, *Industrial Research Laboratories of the United States,* 10th ed. Washington 1956.
31. U.S. FEDERAL TRADE COMMISSION, *Report on Industrial Concentration and Product Diversification in the 1,000 Largest Manufacturing Companies: 1950.* Washington 1957.
32. U.S. NATIONAL SCIENCE FOUNDATION, *Research and Development in Industry: 1961.* Washington 1964.
33. U.S. PATENT OFFICE, *Index of Patents* (annually).

Please turn page for Appendix

Appendix

Incremental Analysis of Covariance: 448-Firm Patenting

Step	Description		Sum of Squared Deviations	SSD as Per Cent of A-1 SSD	D.F.	F Ratio
A-1	No independent variables		3,928,423	100.00	1	
A-2	Global regression of P_i on S_i (Eq. (1))	Reduction from A-1	1,657,710	42.20	1	326
		Residual	2,270,713	57.80	446	
A-3	Global regression on S_i and Diversification	Reduction from A-2	498,012	12.68	1	125
		Residual	1,772,701	45.12	445	
A-4	Global regression on S_i and S_i^2	Reduction from A-2	23,440	.60	1	4.64**
		Residual	2,247,273	57.20	445	
A-5	Global regression on S_i, S_i^2, and S_i^3	Reduction from A-4	102,985	2.62	1	21.32
		Residual	2,144,288	54.58	444	
A-6	Global regression on S_i, S_i^2, S_i^3, and Diversification	Reduction from A-5	400,700	10.20	1	102
		Residual	1,743,588	44.38	443	
A-7	14 Industry regressions on S_i (Table 1)	Reduction from A-2	1,667,626	42.45	26	44.66
		Residual	603,087	15.35	420	
A-8	14 Industry regressions on S_i and Diversification	Reduction from A-7	41,314	1.05	14	2.13*
		Residual	561,773	14.30	406	

APPENDIX (*Continued*)

Step	Description		Sum of Squared Deviations	SSD as Per Cent of A-1 SSD	D.F.	F Ratio
A-9	14 Industry regressions on S_i and 12 S_i^2	Reduction from A-7 Residual	170,788 432,299	4.35 11.00	12 408	13.43
A-10	14 Industry regressions on S_i, 12 S_i^2, and 12 S_i^3	Reduction from A-9 Residual	30,494 401,805	.77 10.23	12 396	2.50*
A-11	14 Industry regressions on S_i, 12 S_i^2, 12 S_i^3, and Diversification	Reduction from A-10 Residual	48,700 353,105	1.24 8.99	14 382	3.76*
A-12	4 Technology Group regressions on S_i	Surplus over A-7 Residual	42,211 645,298	1.08 16.43	−20 440	1.44***
A-13	4 Technology Group regressions on S_i, S_i^2, and S_i^3 [Eqs. (3)–(6)]	Reduction from A-12 Residual	94,471 550,827	2.41 14.02	8 432	9.26
A-14	Global regression of P_i on $R\&D_i$ ($n=352$)	Original variance Reduction Residual	3,784,738 2,732,927 1,051,811	100.00 72.21 27.79	1 1 350	909
A-15	Regression of P_i on 14 R&D slope dummies	Reduction from A-14 Residual	458,905 592,906	12.12 15.67	13 337	20.03

*** Significant at the 10 per cent point, ** at the 5 per cent point, * at the 1 per cent point. *F* ratios with no asterisk are significant at well above the 1 per cent point.

[10]

CORPORATE INVENTIVE OUTPUT, PROFITS, AND GROWTH[1]

F. M. SCHERER
Princeton University

BUSINESS firms presumably invest in research and development because they find it profitable to do so. Successful invention and innovation can increase profits in two principal ways: by widening the profit margin earned on a dollar of sales (or in the case of capital-saving changes, invested capital) and by opening up new opportunities for profitable sales growth. This paper reports the results of a study of inventive output, profitability, and sales growth in 448 firms appearing on *Fortune*'s list of the 500 largest U.S. industrial corporations in 1955.[2]

The index of inventive output employed here is the number of U.S. invention patents issued in 1959 to firms in the sample. The year of patent issue is lagged four years from the base year 1955, whose inventive output is to be measured, because on the average nine months pass between the conception of an industrial invention and the filing of a patent application and because during the 1950's three and one-half years were required for the Patent Office to process an average application to the point of issue.[3]

I. THE MODEL AND ITS LIMITATIONS

The main hypothesis proposed in this paper is that the dollar volume of profits π_{ti} of the ith firm in period t is a function of the number of patented inventions P_{si} conceived within that firm during some earlier base period s. Of course, a firm's profits are related not only to its inventive output. We are really interested in the effect of patented inventions, holding firm size (as measured by base period sales S_{si}) and industry structure constant. Differences in industry structure can be taken into account by means of industry dummy variables. Since the profits of a firm with zero sales should be zero, the dummy variables in an analysis of undeflated profits must pertain to the slope rather than the intercept of the relevant regression equation.[4] We may therefore write as our initial regression model:

$$\pi_{ti} = a + \left(b_1 + \sum_{j=2}^{m} b_j d_{ji}\right) S_{si} + cP_{si} + u_i\,, \qquad (1)$$

where d_{ji} is an industry dummy variable having a value of either 0 or 1 and u_i is an error term.

Several limitations in this model deserve note. First, the profits associated with new products and processes in period t arise from

[1] This research was supported by a grant from the Inter-University Committee on the Microeconomics of Technological Change, sponsored in turn by the Ford Foundation. Use was also made of computer facilities supported in part by National Science Foundation Grant GP579. I am indebted to Robert Chandross for help in collecting data.

[2] Actually, the sample included 463 firms listed as independent by *Fortune* in 1955, since 15 consolidations were made to accommodate subsequent mergers among listed firms. The remaining 37 firms were excluded mainly because they were not engaged primarily in manufacturing or because they could not be fitted into industry groups with sufficient observations.

[3] This mean lag has a variance, and so when 1959 patents are referred to as an index of 1955 invention, the context is probabilistic. The mean lag between application and issue was estimated through a random sample of 111 patents. The standard deviation is 1.6 years. Regarding the invention-to-application lag, see Barkev Sanders, "Some Difficulties in Measuring Inventive Activity," in *The Rate and Direction of Inventive Activity* (Princeton, N.J.: Princeton University Press, 1962), p. 71.

[4] Cf. D. B. Suits, "Use of Dummy Variables in Regression Equations," *Journal of the American Statistical Association*, December 1957, pp. 548–51.

inventions conceived in periods other than base period *s* as well as in *s*. At the same time the profits from period *s* inventions are distributed over a number of periods. If a firm's inventive output were constant from year to year and if each invention had the same time pattern of returns, the inventive profits realized during any particular year by that firm, while related to several years' inventions, would equal the total profits over time of a given year's inventions. In such a case the *c* coefficient of equation (1) would, *ceteris paribus*, be an unbiased estimator of the total profits associated with an incremental patented invention.

However, an analysis of patenting by twenty-one firms over nine years showed that company patenting adjusted for national trends varies significantly from year to year. Therefore a count of base year *s* inventions estimates only imperfectly the number of inventions conceived in other years but affecting year *t* profits. This condition is analogous to the presence of errors in measuring an independent variate, imparting an asymptotic bias toward zero in estimating the least-squares regression coefficient for that variate if the errors are random.[5] The most likely effect of this intertemporal variation problem is to bias profitability estimates downward, although exceptions might occur in special cases of non-random measurement error.

A second problem arises because profits from different years' inventions may be synchronized differently with respect to the business cycle. On this point the empirical results will speak best for themselves.

A third problem stems from limitations in a straight count of patents as an index of inventive output. No accessible set of data is fully satisfactory. Patent statistics have two principal advantages: greater availability than any other index of R & D inputs or outputs, and less susceptibility than either R & D spending or employment data to "mass engineering" biases unrelated to real inventive activity. Their main weakness involves the wide variation in quality of the underlying inventions, and hence in the economic significance and profitability of those inventions.[6] We can, of course, view the profitability of any sampled patented invention as a random variable with some probability distribution. Nevertheless, available evidence suggests that the distribution of profits associated with patented inventions has theoretically infinite first and second moments, and so one cannot be certain that convergence of a sample mean toward some true population mean of the sort implied by the weak law of large numbers will occur.[7] A peculiar distribution of spectacularly successful inventions could bias profitability estimates either upward or downward.

Fourth, estimates of inventive profits must take into account not only returns but also inventive costs. Since research and development employment inputs are fairly highly correlated with lagged patent outputs,[8] one should as a first approximation expect the profits associated with patents in a regression analysis to be net of inventive costs. But because returns lag inputs by several years, because R & D costs are typically written off annually, and because R & D costs were growing at a rate of roughly 10 per cent per annum during the 1950's, the R & D costs offsetting year *t* inventive returns should tend to exceed the actual

[5] Cf. J. Johnston, *Econometric Methods* (New York: McGraw-Hill Book Co., 1963), pp. 148–50.

[6] It is also possible that the propensity to patent and hence the mean value of a patent varies systematically with firm size. But the evidence I have examined elsewhere lends little support to this hypothesis (see my "Firm Size, Market Structure, Opportunity, and the Output of Patented Inventions," *American Economic Review* [forthcoming]).

[7] Analysis of the data discussed in n. 16 below suggested a Pareto-type distribution with an α coefficient of less than .5. For a summary of the problems connected with Pareto-type distributions see Benoit Mandelbrot, "New Methods in Statistical Economics," *Journal of Political Economy*, LXXI (October, 1963), pp. 421–40.

[8] For 352 firms on which data were available, the r^2 between 1955 R & D employment and 1959 patents was .72. Addition of thirteen industry slope dummy variables to account for interindustry differences in the propensity to patent raised the multiple R^2 to .84.

(earlier) R & D costs leading to those returns. This condition would also bias profitability estimates on the low side.

Finally, equation (1) is subject to statistical difficulties. The P_{si} and S_{si} variables are, by accepted standards, harmfully collinear. Introduction of either variable into a regression of profits on the other variable causes a Frisch bunch map to explode.[9] The estimates given by equation (1) might therefore be strongly influenced by relatively few deviant observations.

II. THE RESULTS: FIRST APPROXIMATION

Keeping these limitations in mind, let us turn to the empirical results. Two sets of profit data were collected. For all 448 firms in my basic sample, data were obtained on 1955 and 1959 after-tax profits, adjusted for all sizable mergers occurring during the 1955–59 interval. The initial year 1955 was chosen because one would not expect profits in that year to be augmented much by the returns from 1955 inventions. If anything, causality might run from 1955 profits to 1955 inventions under a Schumpeterian view of the research and development process. On the other hand, it was expected that by 1959 most of the 1955 inventions would be approaching their maximum contribution to corporate profits and growth. This hypothesis was suggested by independent survey evidence showing that about 90 per cent of patented industrial inventions commercially utilized at all were brought into initial use before the time of patent issue.[10]

For a more complete analysis it was desirable to have profit data covering several consecutive years. This posed practical problems, because comparability over time deteriorates with the number and size of intervening corporate mergers. To expand the periodic coverage without laborious iterative reconsolidation of company data, 83 firms involved in sizable or numerous mergers between 1955 and 1960 were dropped from the sample, and annual profit and sales data for the remaining 365 firms were gathered. Results with the reduced sample for 1955 and 1959 did not vary appreciably from results for the full 448-firm sample, and so only the 365 firm subsample estimates will be presented here.

Regressions similar to equation (1) were estimated for each year's profits from 1955 through 1960, both with and without industry-slope dummy variables. In each regression 1955 sales was used as the scale variable. The results are summarized in Table 1. Considering first the relationships without dummy variables (second and third columns), we see that the partial correlation and regression coefficients for patents, holding 1955 sales constant, rose steadily from the base year 1955 through 1958 and then declined. The results are roughly similar, although more irregular, when dummy variables are introduced (last two columns), letting each firm's profits be measured from the mean for firms of comparable size in its own home industry.

The dummy and non-dummy estimates diverge by far most sharply for 1959—a year characterized by strong recovery from the 1958 recession. Analysis of industry dummy-variable correlation coefficients suggested that the marked growth in profits associated with the technically vigorous electrical, chemical, and petroleum industries as a whole held much less consistently in 1959 for individual firms within those industries. It is conceivable, as we shall see, that the sharp drop in profits estimated by the 1959 dummy variable regression and the rise in the patent coefficient's standard error (numbers in parentheses) was a delayed effect of the 1958 recession.

More generally, the regression estimates indicate that corporate profits are related positively in normal times to inventive output. There is a suggestion of both growing inventive profits and a growing correlation between profits and 1955 inventions as the base-period lag increases, perhaps pointing

[9] For a concise discussion of the bunch map technique see E. F. Beach, *Economic Models* (New York: John Wiley & Sons, 1957), pp. 171–75.

[10] Cf. Barkev S. Sanders, "Speedy Entry of Patented Inventions into Commercial Use," *Patent, Trademark, and Copyright Journal*, Spring, 1962, pp. 87–116.

toward a peak in either 1958 or (had the 1958 recession not intervened) 1959. It is difficult to be certain about the exact time pattern, however, since potential business-cycle effects cannot be disentangled completely.

III. DECOMPOSITION OF THE PROFIT GROWTH EFFECT

Let us proceed further and attempt to decompose the observed increase in profits associated with inventive output into two components: a profit margin effect and a sales growth effect.

We might expect profit margins to be larger, the more patented inventions a firm makes relative to its scale. This would be true if patent protection or other barriers to rapid imitation permitted firms to enjoy significant temporary monopolies of product and process inventions. However, the results of tests of this hypothesis were uniformly negative. In correlations between π_{ti}/S_{ti} and P_i/S_{55i} both with and without industry dummy variables for each year from 1955 through 1960, the correlation coefficients ranged between +.047 and −.062. None was significantly different from zero even at the .25 level in a two-tail test. Similarly negative results were obtained when the profit rate was measured relative to assets instead of sales.

One possible rationalization for this surprising finding is that firms with high rates of patenting relative to sales were caught on a treadmill of qualitative competition, with imitation occurring so rapidly that monopolistic profit margins could be maintained only in exceptional cases (to which gross patent statistics are apt to be insensitive). The utterances of businessmen are not inconsistent with such a hypothesis.

TABLE 1

RESULTS OF REGRESSIONS: LAGGED PROFITS ON PATENTS, GIVEN OTHER VARIABLES

Year	Given 1955 Sales Only		Given 1955 Sales and Slope Dummies	
	Partial r: π_{ti} on P_i	Regression Coefficient (Profit Dollars per Patent)	Partial r: π_{ti} on P_i	Regression Coefficient (Profit Dollars per Patent)
1955	.089	34,890 (20,600)	.145	71,830 (26,300)
1956	.179	83,790 (24,270)	.224	118,340 (27,500)
1957	.200	88,630 (22,800)	.116	51,640 (23,700)
1958	.358	143,090 (19,640)	.268	147,070 (28,300)
1959	.305	125,640 (20,650)	.039	22,780 (31,300)
1960	.249	101,510 (20,780)	.188	105,220 (29,500)

The growth of sales at more or less constant profit margins, rather than the expansion of profit margins, appears to have been the main reason for the growth of profits associated with inventive output. This is demonstrated in two ways.

First, sales for the year whose profits were being analyzed, rather than base-year 1955 sales, were introduced as the firm-size variable in regression equations analogous to those summarized by Table 1. For all five lagged periods in the analysis omitting industry dummy variables and for four out of five years in the analysis with dummy variables, the effect was a typically substantial fall in regression and correlation coefficients for the P_i variable. This suggests that much of the invention correlated growth in profits was due to increases in firm sales above base-year 1955 levels.

year-to-year sales growth for each period from 1953–54 through 1960–61 and cumulative growth from the base-year 1955 for each year through 1961.[11] In addition, zero-order correlations of $G_{(\Delta t)i}$ with P_i/S_{55i} without dummy variables and the base year sales variable were executed for the same intervals. The results are summarized in Table 2.

TABLE 2

RESULTS OF REGRESSIONS: SALES GROWTH ON PATENTING INTENSITY

Growth Period	Correlation between Percentage Sales Growth and:		β Coefficient from Eq. (2)	Mean Overall Percentage Sales Growth
	P_i/S_{55i}	P_i/S_{55i} Given S_{55i}, Dummies		
1953–54	.039	.011	.001 (.007)	− 4.0
1954–55	−.041	−.086	−.014 (.009)	15.9
1955–56	.175	.183	.020 (.006)	9.1
1956–57	.174	.115	.014 (.006)	5.0
1957–58	−.002	.016	.002 (.006)	− 4.0
1958–59	−.006	−.115	−.017 (.008)	14.2
1959–60	.092	.074	.009 (.006)	2.9
1960–61	−.033	.003	.0004 (.0070)	3.0
1955–57	.248	.205	.037 (.009)	14.8
1955–58	.155	.132	.032 (.013)	11.0
1955–59	.181	.092	.025 (.014)	24.5
1955–60	.192	.105	.034 (.017)	28.1
1955–61	.198	.109	.036 (.018)	31.7

To study the invention-induced sales growth phenomenon further, regressions of the following general form were estimated:

$$G_{(\Delta t)i} = (14 \text{ constants}) + \beta \frac{P_i}{S_{55i}} + \gamma S_{55i} + \epsilon_i, \quad (2)$$

where $G_{(\Delta t)i}$ is the percentage growth in sales of the ith firm during a specified time period Δt. Regressions were run analyzing

We observe initially that the zero-order and partial correlations between P_i/S_{55i} and annual growth are weakly and insignificantly positive for the 1953–54 period and negative for the 1954–55 interval. One implica-

[11] Note that, for this analysis, time coverage extends back to 1953 and up to 1961. The extension was made to gain more insight into the behavior of invention-correlated sales growth over the course of the business cycle. For 1953 and 1961, the number of observations drops to 348 because of further merger complications.

tion of these pre-base-year relationships is that a rapid rate of sales growth does not appear to be an important stimulant to invention.

During the 1955–56 and 1956–57 intervals positive correlations between growth and patenting intensity appeared. They are significant at the .05 level or higher. However, in the recession year 1958 the annual growth correlations fell off badly. In the recovery-year 1959 they did not recover, and indeed the partial correlation with dummy variables taken into account joined the zero-order correlation in turning negative.[12] Not until 1960 did evidence of a positive correlation between year-to-year growth and patenting intensity reappear, only to deteriorate again in the 1960–61 interval.

While one can never rule out chance as a cause, even when statistical significance tests are passed, I am inclined to believe that the observed results (like the similar behavior of the profit regressions) have a more direct economic explanation. Many of the inventions upon which patents were issued in 1959 were undoubtedly entering or just into the Schumpeterian innovative stage in late 1957, when the 1958 recession began.[13] One or both of two mechanisms could have gone into effect then. First, businessmen may have throttled down their innovative plans, deciding to delay the commercial application of many product inventions until business conditions improved. Alternatively, the declining market may have been especially unreceptive to newly introduced products. For some reason these conditions hostile to new-product sales persisted and actually intensified throughout the recovery year 1959—perhaps because innovative plans, once stopped, could not quickly be set into motion again; or because demand in the recovery period was initially directed toward old and familiar goods, with newer products gaining only when more traditional wants were satisfied.[14] The decline in 1960–61, on the other hand, may mark the end of the growth cycle for average 1955 inventions.

It would appear in any event that sales growth was responsible for most of the invention-correlated growth in corporate profits revealed by Table 1. The sharp decline in estimated 1959 inventive profits is similarly attributable to a serious setback in the sales growth of highly inventive firms. Profit margins meanwhile acted as a constant, allowing the dollar volume of profits to vary with the growth of sales.

IV. THE PROFITABILITY OF PATENTED INVENTIONS

Up to this point I have not tried to interpret the c coefficients of equation (1), recorded in Table 1, as unbiased estimates of the profitability of an average patented invention. Considering the adverse effect of the 1958 recession on 1959 profits, the 1958

[12] A broader relationship between patent-correlated sales growth and the business cycle, discovered when this paper was in proof, must be noted. The last column of Table 2 provides a business conditions index, averaging the percentage sales growth of all sampled firms. When this index for over-all year-to-year growth (first eight entries only) is correlated with the β value for the same period (second-to-last column), a rank correlation coefficient of $-.32$ is obtained. But, when the over-all growth index is correlated with the β value lagged one period, the rank correlation jumps to .95. With a two-period lag, the correlation is $-.19$.

It would appear that the sales growth associated with inventions rises one year after a rise in over-all corporate sales and falls one year after a fall in over-all corporate sales. Evidently, the business cycle has a pronounced impact on inventive returns which is explained only imperfectly by the consumption and innovation lag hypotheses advanced in the next paragraph of the text.

A linear regression of one-year-lagged β values on the over-all sales growth index indicates that only when the over-all annual growth rate of corporate sales is at least 4.3 per cent will the next year's invention-correlated sales growth be positive. This is, to say the least, a puzzling result.

[13] The 1958 recession reached a trough in roughly April of that year. Thus the second half of 1958 must also be considered a period of recovery.

[14] Note also that the standard errors of the β coefficients (in parentheses) for years with a negative correlation between sales growth and P_i/S_{55i} are higher than in other years. Apparently, innovative behavior or the response of markets in recession and post-recession years is more variable than in normal years.

figure of $147,000 in after-tax profits per patent seems to be the most suitable candidate as the equation (1) estimator. As indicated previously, it is likely to be biased on the low side due to the imperfect intertemporal correlation of company patenting and to the growth of R & D expenditures. But another approach to the estimation problem warns us that either multicollinearity of patenting with sales or a peculiar distribution of Pareto-type extreme profit values may have imparted a serious bias in the opposite direction.

We have seen that most, if not all, of the invention-correlated growth in profits was due to sales growth. The β coefficient of equation (2) estimates the incremental percentage growth in sales for a specified time interval associated with each incremental patent received per billion dollars of 1955 sales. Consider the β coefficient of .032 given in Table 2 for cumulative growth between 1955 and 1958. Firms in my subsample of 365 received an average of 112 patents per billion dollars of sales, and so the mean firm's growth from 1955 through 1958 presumably was increased by $.032 \times 112 = 3.58$ percentage points as a result of its inventive output. Mean 1955 sales for the sample was $349 million per firm. Growth of 3.58 per cent from this base yields incremental sales in 1958 of $12.5 million. Mean patents per firm were 36.0, giving incremental 1958 sales per patent of $347,000. If we assume that each dollar of incremental sales carried the same after-tax profit margin (4.34 per cent) as the average for all firms, we arrive at an estimated sales growth contribution to profits of roughly $15,000 per patent. This differs by an order of magnitude from the estimate of $147,000 given by equation (1).[15]

When different methods produce such disparate estimates from the same data, we must either pick the superior method on a priori grounds or accept a high level of uncertainty. The equation (2) method has clearly superior statistical properties, avoiding the collinearity problem of equation (1). It may be nearly as good as equation (1) theoretically, given the evidence that profit margins vary little with inventive output. However, the two equations are affected somewhat differently by Pareto-type extreme values, and it is difficult to say which method is preferable on this score without knowing more about the incidence of such values. And both methods may have conservative biases because firms' inventive output varies from year to year and because the R & D costs offsetting lagged inventive returns have been growing. Therefore, we are forced in any event to accept considerable uncertainty. The equation (2) estimate of $15,000 in after-tax profits per patent or $30,000 before taxes is probably a rock-bottom minimum, and the most we can say is that the true average is likely to be higher.[16]

This estimate is at least roughly consistent with the only known independent evidence. In a random sample survey of patents held by U.S. corporations, B. S. Sanders, J. Rossman, and L. J. Harris obtained "best guesses" of cumulative profitability for a subsample of 281 patents. The average estimated profit per patent—apparently before taxes, although the questionnaire was not explicit—was $127,000.[17]

[15] The disparity is not reduced when industry dummy variables are excluded. Table 1 shows an estimated increase in 1958 profits of $143,090 per patent. A regression of $G_{(58-55)i}$ on P_i/S_{55i} estimates that 1955–58 growth increased by 0.032 per patent per billion dollars of 1955 sales. Since this latter estimate is identical to the dummy variable estimate, the sales growth contribution to profits given by the equation (2) method is again $15,000.

[16] It is not possible to compute even a rough return on research investment from these data because no information is available on the additional capital investment required for sales expansion. For what it is worth, however, I estimate from my data and National Science Foundation data on R & D costs per scientist and engineer that the average private R & D cost per 1959 patent was between $80,000 and $120,000.

[17] "The Economic Impact of Patents," *Patent, Trademark, and Copyright Journal*, September, 1958, pp. 340–62. The average cited here was derived by me from their Table 9 and additional information on the most profitable invention, reported in a later article by the same authors. If inventions with profits of $1 million or more are deleted from the sample, the average falls to $31,000.

V. CONCLUSION

The principal findings of this study are as follows: (1) The growth of corporate profits is positively correlated with inventive output, as measured by patenting. (2) The bulk of corporate inventive activity works its favorable effect on profits by facilitating the growth of sales at constant profit margins rather than by widening profit margins. (3) Business recessions appear to have an especially unfavorable impact on the sales and profits of highly inventive corporations.

Through extension of the methodology and data employed here, it should be possible to obtain a more complete picture of the relationships between industrial invention and innovation, sales growth, profits, and the business cycle. It may also be feasible to estimate more precisely the profits attributable to research, development, and patentable invention.

[11]

Inter-industry technology flows in the United States

F.M. SCHERER

Department of Economics, Swarthmore College, Swarthmore, PA 19081, U.S.A.

Final version received March 1982

This paper presents for the U.S. economy a technology flows matrix tracing 1974 industrial R&D expenditures from their industries of origin to industries in which the use of resulting products and processes was anticipated. The distinction between origin and using industries is crucial to understanding the links between R&D and productivity growth. A regression analysis reveals high social rates of return and substantial productivity impacts from R&D attributed to industries of use.

1. Introduction

Economists of as diverse persuasion as Adam Smith, Karl Marx, and Joseph Schumpeter have argued that material standards of living depend critically upon the level of technology. More recently, industrial research and development (R&D) has been identified as a potentially important source of technological change and hence growth in productivity (i.e. output per worker). However, there is surprisingly little solid quantitative evidence about the links between R&D and productivity, whose retarded growth in recent years has become a matter of concern. This article reports some results of an effort to narrow the knowledge gap.

At the heart of the problem is the fact that industrial R&D is directed toward creating and improving both new production processes and new products. A new process in the sense used here is a technical improvement in one's own production methods, while a new product is an improvement sold to others – either to other business enterprises or to consumers. Resources devoted to process R&D normally lead directly to productivity improvements within the enterprise or industry performing the R&D. More output can be produced with a given quantity of labor, capital, and other material input. New products, however, are by definition sold to others. If the new product is a consumer good, it presumably enhances the quality of life directly. If it is a producer good or intermediate sold externally, it serves to improve output/input relationships or the quality of output in the buying industries. With a new turbojet engine product, for example, the R&D is performed in the aircraft engine industry, but the productivity effect often shows up in lower energy consumption or faster, quieter, and more reliable operation of equipment used by the quite distinct airlines industry.

Therein lies the rub. From surveys sponsored by the U.S. National Science Foundation and others, we have extensive statistics concerning expenditures on research and development classified by industries performing the R&D. But there is a dearth of systematic quantitative information on the ultimate beneficiaries of industrial R&D outlays, once they become embodied in new or improved products flowing across industry boundaries. Given this statistical asymmetry, economists investigating the links between R&D and productivity growth have (with only limited exceptions) been forced to assume that the productivity-enhancing effect occurs solely within the R&D-performing industry [1]. This is more wrong than right, since three-fourths of all industrial

* The research underlying this paper was conducted under National Science Foundation Grant PRA-7826526. Use is made of data collected under the Federal Trade Commission's Line of Business program. A review has been conducted to ensure that the figures in this paper do not disclose individual company line of business data.

Research Policy 11 (1982) 227–245
North-Holland Publishing Company

0048-7333/82/0000-0000/$02.75

Table 1
Schmookler's invention flows matrix (hypothetical)

Industry making the invention	Industry using the invention				Total inventions made
	Industry I	Industry II	Industry III	Consumers	
Industry I	3	7	0	5	15
Industry II	2	11	18	9	40
Industry III	5	3	1	1	10
Total inventions used	10	21	19	15	65

R&D is devoted to new or improved products, as distinguished from processes.

A conceptually correct approach to the problem was first proposed by Jacob Schmookler in 1966 [2]. Extending the input–output analysis ideas of Wassily Leontief, Schmookler argued that the structure of industrial technology flows was best characterized in terms of a two-way matrix (table 1), with industries performing R&D and originating inventions comprising the rows and industries (including end consumers) using those inventions comprising the columns. Each numerical element in the matrix represents the flow of technology from a particular originating industry to a particular using industry. Diagonal elements indicate process technology. The sum across a row shows the total amount of technology originating in an industry. Such origin totals (in R&D dollar terms) are available from annual National Science Foundation surveys. The sum down a column gives the total amount of technology used by an industry. In pathbreaking research with invention patents, Schmookler was able to estimate the column sums (but not the individual cell values) for the flow of capital goods technology to a small sample of using industries. Further progress was interrupted by his untimely death in 1967.

Schmookler's research emphasized counts of patents. These have important limitations as cross-sectional measures of industrial efforts to advance technology, in part because the "propensity to patent" varies from one industry to another. The research that will be reported here revealed among other things that the average number of U.S. patents obtained per million dollars of company-financed 1974 research and development ranged from a low of 0.47 in motor vehicles to 3.98 in industrial electrical equipment, with an average across all industries of 1.70 [3].

2. Constructing a technology flows matrix for the U.S. economy

The recent availability of an unusually rich U.S. R&D data source makes it possible to overcome this problem and hence view invention patents as carriers of underlying R&D expenditures rather than as units of invention in their own right. That source is the Federal Trade Commission's Line of Business survey for 1974, which among other things collected data on the R&D outlays of 443 large U.S. industrial corporations disaggregated into some 4,274 separate lines of business pursued by the companies [4]. A company like, say, General Electric was asked under the survey instructions to break its sales and R&D expenditures into such categories as turbojet egines, electric motors, light bulbs, small appliances (such as hair dryers), synthetic resins, etc., rather than reporting them, as is done for National Science Foundation surveys, in one lump assigned to GE's "primary" industry – i.e. industrial electrical equipment. The line of Business survey data allowed one for the first time to achieve a precise match between industrial invention patents and the R&D expenditures supporting activities that gave rise to the inventions.

With this possibility in view, a team of engineering and chemistry students at Northwestern University was put to work examining the printed specifications of the 15,112 U.S. invention patents obtained between June 1976 and March 1977 by the 443 corporations reporting under the FTC's 1974 Line of Business survey. Each individual patent was coded according to industry of origin, industry(ies) in which use of the invention was anticipated, and various other details [5]. The codings by industry of use were based in a rather straightforward way on applicants' descriptions of how their inventions were expected to satisfy the "utility" criterion for patentability. Determining industries of origin was more difficult, since it was necessary to track as closely as possible the industry breakdowns used by companies in reporting their R&D expenditures by line of business. To facilitate the task, the coding team used an extensive accumulation of information on company organization, including data on the products developed or produced at the various geographic locations to which inventors' addresses could be traced. Once the codings were completed and checked, each patent was linked to its originating Line of Business report. All matching problems were checked against additional information in the FTC files, and when appropriate, corrections were made. At the conclusion of the matching procedure, each patent was "tagged" with the average R&D outlay per patent in its appropriate line of business. Thus, the R&D dollars, as well as raw patents, could be flowed out to industries in which the patented inventions' use was anticipated.

2.1. Products versus processes

Consistent with prior surveys, the classification effort revealed that 26.2 percent of the inventions were processes for use within the originating industry. The complete breakdown by use category (including industrial products that also had consumer applications) was as follows:

Internal processes	26.2%
Consumer goods only	7.4%
Industrial materials products	21.6%
With consumer uses too	8.7%
Capital goods products	44.8%
With consumer uses too	7.8%

2.2. Sample coverage

The 443 corporations whose patents were analyzed received 61 percent of all U.S. patents issued to domestic industrial corporations during the sample period. Their 1974 company-financed R&D outlays of $10.6 billion amounted to 72 percent of the comparable all-industry total reported from a National Science Foundation survey [6]. The R&D dollars linked to each patent were inflated by a factor reflecting the sample's fractional coverage of total population sales for each of 263 industry categories. The mean sales coverage ratio was 0.61, i.e. the same as the sample's coverage of industrial corporation patents. However, the inflated sample results do not characterize total industrial technology flow relationships exactly for at least five reasons. First, inventions patented by the characteristically smaller enterprises excluded from the sample might not have use patterns identical to those of sample members. Second, some results of industrial research and development are not patentable, and their inter-industry flows will be overlooked by our methodology. Third, inventions originating in government and university laboratories are not included. The omission of government work (accounting for 2.6 percent of all U.S. patents in 1976) is particularly important for military, space, agricultural, medical, and (only recently) energy applications. University scientists generate few patented inventions, but their basic research enhances the fecundity of industrial laboratories, whose inventive output is our main focus. Fourth, the inventions of "independent inventors" are not covered by the sample. In 1976, individuals with no corporate assignment relationship received 20 percent of all U.S. invention patents. A special small sample revealed that 32 percent of domestic independent inventor patents had consumer goods applications alone (compared to 7 percent for our industrial patent sample). And fifth, the inventions of foreign nationals are also excluded. Foreign corporations accounted for 26 percent of all U.S. patents issued in 1976 and were disproportionately significant in certain industries such as textiles, shoe making, steel, and corn products refining. The difficulty of linking small-firm, university, government, independent, and foreign inventors' inventions to specific origin industries and the impossibility of obtaining corresponding R&D data precluded more exhaustive coverage.

Despite these limitations, the sample systematically covers the preponderance of American industry, which is peculiarly well-suited to such an analysis because of its unusually high degree of technological autarky. Thus, the data are believed to measure tolerably well the flow of technology into the main stream of the U.S. industrial economy. In this respect, it is interesting to note that there are substantial similarities in origin *versus* use and process *versus* product invention patterns between our sample and one analyzed by Pavitt and others covering 2,154 significant technological innovations introduced by British industry between 1945 and 1980 [7].

2.3. Specific-use, general-use, and public goods inventions

Under our coding procedures, a patent could be assigned to as many as three industries of use. Not all inventions are so narrowly constrained in their applications, however. Some, such as new paints, gasoline additives, computers, electrical switchgear, and copying machines, are used to a greater or lesser degree by many or even all industries. Therefore, a dichotomy was drawn (with some further complications ignored here) between "specific" and "general" use inventions. For the 8,852 specific-use inventions, utilization was expected to occur primarily in explicitly identifiable industries, and the R&D dollars linked to those patents were allocated only to the coded industries [8]. R&D underlying the 5,166 general-use inventions, on the other hand, was flowed out in proportion to the origin industry's 1972 sales to as many as 285 potential user industries, as determined from a carrier matrix derived with modifications from the 1972 input–output tables for the United States economy [9].

Numerous other methodological problems had to be solved in estimating the technology flow matrices [5]. One is important enough to merit attention here. It concerns the question of whether inventions are "public" or "private" goods in the sense used by economists. A public good is one whose use by one group (e.g. an industry) does not diminish its availability for use by other groups. Although there are arguments to the contrary, the consensus among economists would probably be that inventions are better characterized as public goods rather than private goods. Nevertheless, technology flow matrices were prepared under both assumptions. Under the public goods assumption, if an invention entailing R&D costs of $1 million was sold in equal amounts to each of three industries, each using industry was presumed to have benefitted from $1 million of embodied R&D. Or with unequal sales, the largest using industry received $1 million of embodied R&D input and smaller industries received benefits scaled down by the ratio of their purchases to the largest user's purchases. A disadvantage of the public goods approach is that using industry R&D embodiments sum to a total greater than origin industry outlays. Because of this problem, I present here only matrices constructed under the alternative "private goods" assumption. That is, if a $1 million invention is used by three industries with equal purchase volumes from the originating industry, each using industry is credited with $333,333 of R&D usage. An exception to this convention was made for consumer goods inventions, which were treated as public goods not diminishing the flow of R&D dollars to industrial using sectors.

3. The technology flows matrix

Oversize table 2 presents the resulting estimates aggregated to the level of 41 origin sectors (the rows) and 53 using sectors (the columns). All entries are in millions of dollars. Blank cells denote R&D flows of less than $50,000. Reading across the row for industry 18, we find that rubber and plastic products makers performed $419.8 million in R&D, of which $13.7 million was embodied in products (such as tractor tires) purchased by the agriculture sector, $203.0 million (from column 18) was used internally within rubber and plastic products (e.g. as process technology), and (skipping to the last column) $119.9 million was embodied in consumer products ranging from passenger car tires to plastic food storage wrap. The diagonal elements of the matrix represent the use of process technology. Entries marked "d" had to be suppressed to comply with the requirement that no underlying R&D data (from the Federal Trade Commission survey) be disclosed for individual companies. Most such deletions involved small amounts of R&D. Over the entire matrix, only 11 entailed R&D flows exceeding $10 million. Reading down column 18, we see that the rubber and plastic products sector purchased products embodying an estimated $21.6 million of R&D from the textile industry (e.g. for tire cord); $169.1 million, mostly for synthetic rubber, from origin sector 13; $9.7 million from the "other machinery" sector; etc.; with total R&D usage (in the last row) summing to $496.5 million.

3.1. Inter-industry balances

In rubber and plastics, the R&D origin and use totals are relatively evenly balanced. This is not true for many other sectors. Printing and publishing businesses used processes and products (including paper, inks, and electronic typesetters) embodying $147.7 million of R&D but performed only $67 million, or 46 percent of used R&D. Other industries with low origin to use ratios were lumber and wood products (0.55), ferrous metals (0.62), textiles (0.71), and apparel and leather (0.75). At the other extreme was the farm machinery industry, originating 10.4 times as much R&D as it used. Other major "exporters" of technology were computers and other office equipment, with an 8.7 origin/use ratio, construction and mining machinery (8.0), and instruments (7.05).

Inspection of table 2 reveals that some industries, such as food and tobacco, apparel, paper products, pharmaceuticals, household appliances, and motor vehicles, devoted the bulk of their product (i.e. off-diagonal) R&D to consumer goods. For all industries together, consumer goods accounted for only $4.1 billion of R&D flows (out of a non-double-counted total of $14.7 billion) when products that were both consumer and producer goods are included, as in table 2, or approximately $1.1 billion when the analysis is limited to products that were solely consumer goods. Some industries such as agricultural chemicals, farm machinery, aircraft, missiles, and other transportation equipment specialize in supplying advanced technology to a very few using sectors (including defense and space operations). Still others such as organic and other chemicals, rubber and plastic products, fabricated metal products (spanning narrowly-defined product lines ranging from bottle caps to nuclear reactors), computers, and instruments are active in providing new technology to a broad array of using industries. These tend to preside over rich science bases that can be tapped, often on a custom-tailored basis, to a host of specific industrial demands

3.2. *Manufacturing versus other sectors*

It has long been known that the vast bulk (i.e., more than 95 percent) of all company-financed industrial R&D is carried out in the manufactur-

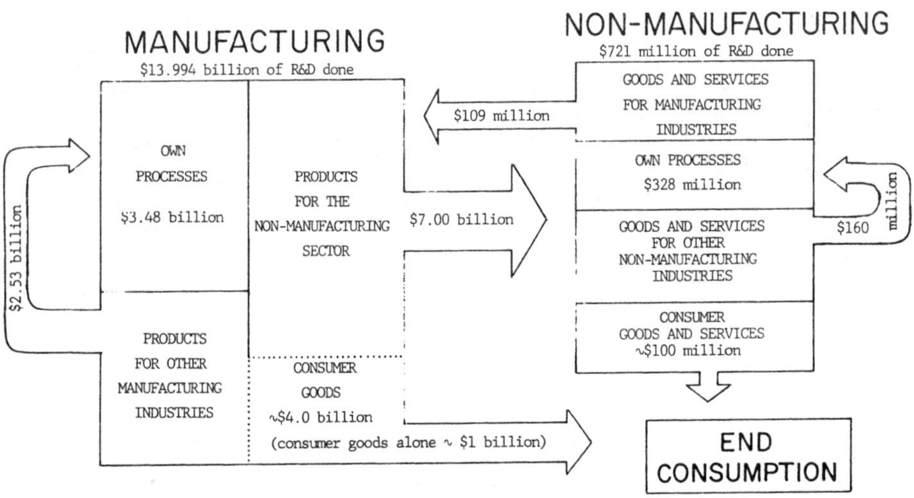

Fig. 1. Company-financed R&D flows between and within manufacturing and non-manufacturing (Consumer goods are treated as public goods, while all other flows are tabulated on a private goods basis. Petroleum extraction is included in non-manufacturing, whereas in table 2, it was included as part of petroleum refining under manufacturing.)

Table 2
Technology flows matrix for company-financed 1974 R&D outlays

		Origin R&D	Agri-culture 1	Mining 2	Food prod-ucts 3	To-bacco prod-ucts 4	Tex-tile prod-ucts 5	Ap-parel prod-ucts 6	Lumber and wood 7	Fur-niture 8	Paper prod-ucts 9	Publish-ing 10	Inor-ganic chem-icals 11	Or-ganic chem-icals 12
1	Agriculture and forestry	128.1	d											
2	Mining, exc. petroleum	60.3		45.2										
3,4	Food and tobacco products	444.9	7.9		257.8	20.4	0.2				d	d		0.1
5	Textile mill products	179.3	0.8		0.2		128.4	8.1	0.3	0.8	0.5	0.2		
6,19	Apparel and leather products	55.5	d	d	d		d	16.5	d	d	d	d		d
7	Lumber and wood products	72.6	0.4		0.2		0.2		64.2	0.6	0.1			
8	Furniture	51.1	0.1	0.1	0.3		0.1	0.1	0.3	7.8	0.2	0.2		
9	Paper mill products	202.3	6.0	0.1	25.8	0.1	1.2	0.3	0.4	0.4	86.4	3.1	0.1	0.1
10	Printing & publishing	67.4	0.1		0.4		0.1				0.1	32.7		
11	Industrial inorganic chemicals	159.2	1.2	0.1	d	0.8	1.7	0.1	0.1		4.6	0.3	90.8	3.3
12	Industrial organic chemicals	297.2	3.0	0.2	2.7	d	11.8	0.4	1.3	0.1	5.9	d	1.1	163.3
13	Synthetic resins, fibers, rubber	601.6	1.6	0.5	2.8	d	32.0	11.8	1.1	0.8	13.9	0.5	d	0.8
14	Pharmaceuticals	557.3	32.0		0.2						d			
15	Agricultural chemicals	186.7	142.8		d						d			0.1
16	Paints, toiletries, explosives, and other chemical products	485.7	3.9	4.2	11.1	0.2	26.3	0.2	7.8	4.5	20.5	13.4	0.7	1.4
17	Petroleum extraction and refining	380.3	4.0	0.7	1.1	0.1	0.3	0.4	0.9	0.1	1.0	0.3	0.2	0.3
18	Rubber & plastic products	419.8	13.7	2.8	18.1	0.7	3.0	1.1	d	3.5	5.0	2.2	0.7	0.8
20	Stone, clay, & glass products	265.0	0.9	0.1	d	0.1	d	0.1	2.4	1.0	d	0.3	0.1	0.2
21	Ferrous metals	189.3	d	0.1	0.1				0.1	0.4				d
22	Nonferrous metals	156.9	d	0.1	d		0.1	0.1	0.3	0.2	0.1	0.4	0.2	0.4
23	Fabricated metal products	552.7	5.3	3.0	28.0	0.2	0.7	0.7	4.5	2.7	7.4	0.9	1.0	2.7
24	Engines and turbines	282.2	10.4	3.2	2.2	0.3	0.6	0.6	1.1	0.3	2.5	0.9	0.4	1.0
25	Farm machinery	199.3	165.4											

cont. overleaf

26	Construction, mining, and materials handling equipment	351.2	2.0	65.9	4.6	0.3	1.2	0.3	7.3	0.3	1.3	0.6	0.3	0.8
27	Metalworking machinery	121.5	5.4	0.2	0.4		0.1		2.8	0.5	d	0.1		
28	Other machinery	691.0	17.6	8.0	56.9	1.5	18.9	1.9	7.8	1.1	20.9	40.1	5.0	10.1
29	Computers and office equipment	1153.0	2.9	2.4	15.9	0.6	9.4	3.4	1.6	1.7	10.1	13.4	1.8	5.0
30	Industrial electrical equipment	205.9	0.3	1.6	1.3		1.1	0.2	0.5	0.3	2.5	0.3	0.7	1.6
31	Household appliances	102.7	0.1		0.1		0.4	2.2	0.6					
32	Lamps, batteries, ignition, X-ray, and other electrical equipment	233.3	8.1	0.6	2.5	0.1	0.5	0.3	1.3	0.4	0.7	0.7	0.2	0.4
33	Radio and communication equipment	1227.7	5.1	1.6	5.3	0.7	1.5	1.5	1.5	0.7	1.7	2.4	0.5	0.9
34	Electronic components	594.9	0.5	0.2	0.7		0.6	0.2	0.2	0.1	0.4	0.7	0.1	0.3
35	Motor vehicles and equipment	1518.0	78.0	10.6	25.4	1.1	5.0	3.6	17.1	3.9	5.4	10.0	0.9	2.2
36	Aircraft	659.4	0.3		0.2				0.1		0.2	0.1		0.1
37	Missiles, spacecraft, and ordnance	122.7			d									d
38	Other transportation equipment	140.1		d										d
39	Measuring and medical instruments, photo equipment, and time-pieces	1036.4	6.3	2.1	8.1	0.8	3.3	1.5	2.0	0.9	7.2	20.7	2.4	5.3
40	Miscellaneous manufactures	211.6	d	0.1	0.6	0.1	0.9	1.3	0.6	0.3	0.2	0.5	0.1	0.1
41,42	Trade and finance and real estate	39.7												
43	Transportation and public utilities	47.2												
44,45	Construction and services, including R&D services	266.0	0.7	d	0.7	0.1	0.2	0.2	0.2	0.1	0.3	0.3	d	5.5
	Total R&D dollars used		561.8	157.3	493.4	29.8	250.8	57.2	131.1	33.7	206.0	147.7	108.8	207.6

Table 2 (continued)

		Origin R&D	Synthetic resins etc. 13	Pharmaceuticals 14	Agri. chemicals 15	Other chem. products 16	Petroleum 17	Rubber and plastics 18	Leather products 19	Stone, clay, and glass 20	Ferrous metals 21	Nonferrous metals 22	Fab. metal products 23	Engines and turbines 24
1	Agriculture and forestry	128.1												
2	Mining, exc. petroleum	60.3	d			d		d						
3,4	Food and tobacco products	444.9	0.1	d		d	0.1	0.1		0.1	0.1		d	
5	Textile mill products	179.3					0.1	21.6	0.9	0.1		0.1	d	
6,19	Apparel and leather products	55.5				d	d	d		d	d	d	d	
7	Lumber and wood products	72.6					0.1	0.1		0.3	0.2	0.2	0.3	
8	Furniture	51.1				0.1	0.1	0.1		0.1	0.2	0.1	0.2	
9	Paper mill products	202.3	1.0	0.7	0.1	1.0	1.5	1.2	0.2	0.9	0.1	0.1	0.6	
10	Printing and publishing	67.4	0.1	0.1		0.1	0.1	0.1		0.1	0.1		0.1	
11	Industrial inorganic chemicals	159.2	1.8	0.2	0.3	6.1	9.3	0.4		0.9	2.5	d	1.6	d
12	Industrial organic chemicals	297.2	24.1	1.2	2.3	19.0	7.1	13.8	d	2.7	1.6	2.2	0.8	
13	Synthetic resins, fibers, rubber	601.6	70.8	d		23.9	2.3	169.1	1.7	6.4	0.9	10.0	2.7	
14	Pharmaceuticals	557.3		71.0		d								
15	Agricultural chemicals	186.7		d	34.2	d		d						
16	Paints, toiletries, explosives, and other chemical products	485.7	9.6	1.0		85.4	13.8	16.2	2.5	5.4	3.5	3.8	31.8	0.1
17	Petroleum extraction and refining	380.3	0.3	0.1	0.1	0.8	312.7	0.7	0.1	0.6	0.6	0.4	1.5	0.1
18	Rubber and plastic products	419.8	1.3	1.3	0.1	4.3	3.8	203.0	1.9	2.5	1.0	5.1	4.6	0.2
20	Stone, clay, and glass products	265.0	0.1	d		1.3	d	d		155.3	11.3	0.3	2.7	1.1
21	Ferrous metals	189.3		d	d		d	0.1		0.1	164.2	0.3	9.7	d
22	Nonferrous metals	156.9		0.1		0.2	d	0.1	0.1	0.3	2.8	102.9	7.0	0.8
23	Fabricated metal products	552.7	1.5	0.7	1.4	d	7.4	1.1	0.3	1.3	12.3	2.3	127.7	3.0
24	Engines and turbines	282.2	0.5	0.2	0.3	0.5	6.5	0.6	0.1	1.1	2.0	1.1	1.9	38.9
25	Farm machinery	199.3											0.1	
26	Construction, mining, and materials handling equipment	351.2	0.7	0.2	0.2	0.3	17.7	0.4	0.2	5.8	3.4	1.3	1.9	1.0

cont. overleaf

27	Metalworking machinery	121.5	0.1			d	0.1	1.4		0.9	14.5	7.5	15.1	1.0
28	Other machinery	691.0	5.4	2.1	1.6	3.4	16.2	9.7	d	13.6	35.4	9.1	16.0	1.1
29	Computers and office equipment	1153.0	4.8	6.2	1.3	8.6	20.6	8.1	1.6	8.5	19.0	5.3	10.2	5.5
30	Industrial electrical equipment	205.9	0.6		0.5	0.3	4.0	0.5		d	8.2	4.9	5.1	0.7
31	Household appliances	102.7					0.1						0.1	
32	Lamps, batteries, ignition, X-ray, and other electrical equipment	233.3	0.3	0.2	0.1	0.3	1.2	2.0	0.1	0.8	4.6	0.5	1.3	0.4
33	Radio and communication equipment	1227.7	0.6	0.6	0.2	1.0	7.6	1.6	0.3	1.9	2.4	1.0	3.4	0.5
34	Electronic components	594.9	0.2	0.2	0.1	0.3	0.9	0.4	0.1	0.6	0.7	0.2	0.6	0.2
35	Motor vehicles and equipment	1518.0	0.9	1.7	0.6	2.3	16.7	2.6	1.2	11.4	4.2	2.6	14.2	0.8
36	Aircraft	659.4					0.5	0.1		0.1	0.1	0.1	0.2	
37	Missiles, spacecraft, and ordnance	122.7		d			d							
38	Other transportation equipment	140.1			d		d				0.1			
39	Measuring and medical instruments, photo equipment, and timepieces	1036.4	2.8	4.2	1.3	4.3	13.5	4.0	0.4	5.1	6.6	3.1	5.7	0.7
40	Miscellaneous manufactures	211.6	0.1	0.1		0.2	1.4	0.3	0.1	0.2	0.4	0.1	0.7	
41,42	Trade and finance and real estate	39.7												
43	Transportation and public utilities	47.2												
44,45	Construction and services, including R&D services	266.0	d	0.1	d		24.4	0.3		0.3	5.0	0.7	0.4	
	Total R&D dollars used		332.1	95.3	45.7	180.3	496.5	470.0	16.6	232.0	307.8	166.1	270.3	56.9

Table 2 (continued)

		Origin R&D	Farm ma-chinery 25	Con-struc-tion equip. 26	Metal-work-ing mach'y 27	Other ma-chinery 28	Office equip-ment 29	Indust. elec-trical 30	Appli-ances 31	Other elec-trical 32	Com-muni-cations equip. 33	Elec-tronic com-ponents 34	Motor vehi-cles 35	Air-craft 36
1	Agriculture and forestry	128.1												
2	Mining, exc. petroleum	60.3												
3,4	Food and tobacco products	444.9											d	
5	Textile mill products	179.3				0.2			0.1	0.1			0.6	
6,19	Apparel and leather products	55.5		d		d				d	d		d	
7	Lumber and wood products	72.6				0.1					0.1		0.1	
8	Furniture	51.1		0.1		0.2	0.1		0.1	0.1	0.5	0.1	0.5	0.1
9	Paper mill products	202.3				0.3	0.1	0.1	0.2	d	0.2	0.1	0.3	0.1
10	Printing and publishing	67.4				0.1					0.1		0.1	0.1
11	Industrial inorganic chemicals	159.2	0.1	0.1	0.1	0.2	d			d	0.2	1.1	2.7	0.3
12	Industrial organic chemicals	297.2				0.4	d		0.1	0.6	d	2.0	0.3	0.1
13	Synthetic resins, fibers, rubber	601.6				0.5	0.1	0.8	1.9	1.9	1.1	1.8	12.0	1.2
14	Pharmaceuticals	557.3												
15	Agricultural chemicals	186.7												
16	Paints, toiletries, explosives, and other chemical products	485.7	0.6	0.8	0.3	2.0	2.4	2.0	5.7	0.9	d	4.7	19.3	1.3
17	Petroleum extraction and refining	380.3	d	0.1	0.2	0.7	0.1	0.2		0.1	0.1	0.1	0.5	0.3
18	Rubber and plastic products	419.8	0.4	0.9	0.3	1.6	1.1	1.0	1.5	2.6	1.0	1.2	7.3	0.8
20	Stone, clay, and glass products	265.0	0.1	0.4	0.7	d	0.1	d	0.3	0.7	1.0	d	5.6	2.1
21	Ferrous metals	189.3	0.3	0.6	0.3	1.2	d	1.7	0.4	0.3	0.3	0.1	2.8	0.1
22	Nonferrous metals	156.9	0.2	0.3	d	1.3	0.2	1.3	0.3	2.3	0.6	1.2	1.4	1.3
23	Fabricated metal products	552.7	1.0	2.0	1.5	7.2	0.8	1.6	1.4	1.6	1.8	0.9	11.0	16.3
24	Engines and turbines	282.2	0.3	1.7	0.3	3.9	0.2	0.9	0.2	0.3	0.5	0.3	1.8	0.9
25	Farm machinery	199.3	d										0.1	

cont. overleaf

26	Construction, mining, and materials handling equipment	351.2	0.5	16.0	0.4	2.0	0.7	0.4	0.4	0.5	1.0	0.9	5.7	0.2
27	Metalworking machinery	121.5	1.3	2.2	d	7.4	0.8	1.8	0.7	2.0	1.3	1.1	12.2	3.2
28	Other machinery	691.0	0.8	1.6	2.7	64.9	1.5	1.3	8.0	2.5	1.8	3.3	16.2	1.5
29	Computers and office equipment	1153.0	3.5	7.3	2.9	12.9	110.5	4.2	1.4	3.3	16.5	6.6	17.1	20.5
30	Industrial electrical equipment	205.9	0.6	2.9	2.2	15.6	2.6	34.4	3.8	1.7	0.7	0.8	1.4	0.4
31	Household appliances	102.7							6.8	d	0.1		0.1	
32	Lamps, batteries, ignition, X-ray, and other electrical equipment	233.3	1.6	1.4	0.3	1.3	0.6	0.2	0.4	27.8	0.6	0.3	12.2	0.4
33	Radio and communication equipment	1227.7	0.4	0.9	0.7	2.2	1.2	1.1	0.5	1.2	106.3	2.0	6.2	17.6
34	Electronic components	594.9	0.1	0.3	0.1	0.8	3.9	0.6	0.4	0.6	6.5	386.4	3.1	4.6
35	Motor vehicles and equipment	1518.0	2.4	2.9	1.2	5.1	1.5	1.0	0.8	1.1	2.0	1.5	158.8	0.8
36	Aircraft	659.4				0.1							0.4	160.5
37	Missiles, spacecraft, and ordnance	122.7												d
38	Other transportation equipment	140.1											d	
39	Measuring and medical instruments, photo equipment, and timepieces	1036.4	0.5	1.0	0.9	5.5	2.3	3.0	5.9	2.1	12.4	11.4	6.8	9.7
40	Miscellaneous manufactures	211.6	0.1	0.1	0.1	0.3	0.1		0.1	0.1	0.1	0.1	0.5	0.1
41,42	Trade and finance and real estate	39.7												
43	Transportation and public utilities	47.2												
44,45	Construction and services, including R&D services	266.0		0.1	0.1	0.3	0.1			d	d	d	0.5	0.1
	Total R&D dollars used		19.2	43.8	22.4	141.1	131.9	59.0	41.7	72.2	185.6	446.4	308.1	245.0

Table 2 (continued)

		Origin R&D	Missiles and ord-nance 37	Other trans. equip. 38	Instru-ments 39	Misc. manu-fac-tures 40	Trade 41	Finance, insurance, and real estate 42	Ground trans-porta-tion 43A	Air trans-porta-tion 43B	Water trans-porta-tion 43C	Tele-com-munica-tions 43D	Elec-tric, gas, and sanitary utilities 43E	Con-struc-tion 44
1	Agriculture and forestry	128.1												
2	Mining, exc. petroleum	60.3											d	
3,4	Food and tobacco products	444.9					16.0	0.4	d	0.3		d	d	0.3
5	Textile mill products	179.3		d	0.2	0.3	0.8	1.0	0.2	0.3	0.1	0.1	0.1	0.8
6,19	Apparel and leather products	55.5			d	d	d	d	d	d		d	d	d
7	Lumber and wood products	72.6	0.1	0.2		0.3	0.6	0.4	0.2			0.1	0.4	0,3
8	Furniture	51.1		0.2	0.1	0.1	7.0	2.0	0.3	d		0.2	0.2	0.6
9	Paper mill products	202.3			0.9	0.6	14.7	1.0	0.5	0.3		0.2	0.3	1.4
10	Printing and publishing	67.4			0.1		2.5	1.1	0.2	0.1		0.2	0.2	0.4
11	Industrial inorganic chemicals	159.2	0.1	0.4	d	0.1	1.7	0.9	0.3	0.1		0.2	d	1.0
12	Industrial organic chemicals	297.2		0.1	d	0.3	1.6	0.7	0.3	0.1	d	0.2	0.8	0.6
13	Synthetic resins, fibers, rubber	601.6	0.2	0.7	3.7	4.8	1.2	1.0	0.5	0.2	0.1	0.3	0.4	1.1
14	Pharmaceuticals	557.3												
15	Agricultural chemicals	186.7					0.2	0.1	0.2				0.1	0.1
16	Paints, toiletries, explosives, and other chemical products	485.7	0.3	3.6	1.3	3.4	9.5	2.7	2.5	0.3	0.5	2.1	7.5	16.3
17	Petroleum extraction and refining	380.3	0.1	0.1	0.2	0.3	13.3	2.9	5.1	2.9	1.0	0.3	5.7	6.1
18	Rubber and plastic products	419.8	0.3	0.7	1.9	2.9	22.2	9.5	12.4	0.9	d	1.3	5.8	9.7
20	Stone, clay, and glass products	265.0	0.2	0.3	d	0.7	8.3	2.0	0.9	0.3	d	1.6	5.3	9.0
21	Ferrous metals	189.3	0.2	0.5	0.2	0.3	d		d		d		d	1.6
22	Nonferrous metals	156.9	0.5	0.3	0.5	0.4	0.8	0.5	0.7			d	1.8	0.4
23	Fabricated metal products	552.7	0.7	2.9	1.7	1.0	11.7	9.4	8.1	3.3	0.4	5.1	151.2	17.1
24	Engines and turbines	282.2	0.3	8.5	0.6	0.3	12.0	8.9	7.8	1.0	6.4	1.7	75.8	10.2
25	Farm machinery	199.3		0.1			d	5.9					0.1	2.4

cont. overleaf

26	Construction, mining, and materials handling equipment	351.2	0.1	2.4	0.4	0.3	10.2	1.6	3.9	2.3	0.2	0.4	1.2	154.4
27	Metalworking machinery	121.5	0.8	0.7	3.9	1.1	2.3	0.1	0.3				d	6.6
28	Other machinery	691.0	0.9	2.7	2.4	2.1	64.0	11.5	8.4	3.5	1.2	2.3	33.0	20.5
29	Computers and office equipment	1153.0	3.8	3.5	16.8	3.7	97.5	176.4	20.9	4.3	1.5	25.8	26.1	2.3
30	Industrial electrical equipment	205.9	0.2	2.4	1.6	0.8	2.1	1.8	2.5	0.3	0.4	0.5	52.0	1.8
31	Household appliances	102.7		0.2			4.1	2.8	0.2			0.1	0.1	0.2
32	Lamps, batteries, ignition, X-ray, and other electrical equipment	233.3	0.1	1.1	2.4	0.5	13.0	6.1	11.4	19.7	0.6	8.8	11.5	5.7
33	Radio and communication equipment	1227.7	3.6	2.2	1.7	0.9	31.8	20.9	11.7	75.6	12.2	417.7	11.0	16.1
34	Electronic components	594.9	1.5	0.5	2.2	0.8	4.7	5.7	2.2	3.9	0.2	43.7	2.3	1.7
35	Motor vehicles and equipment	1518.0	0.8	1.1	7.0	3.8	288.8	99.0	185.3	5.9	2.5	9.5	43.9	125.1
36	Aircraft	659.4	0.9	0.2			0.7	0.4	0.5	360.4		0.1	4.1	0.5
37	Missiles, spacecraft, and ordnance	122.7	22.0							0.3		d	d	d
38	Other transportation equipment	140.1		d			d	d	59.9	d	6.4	d		
39	Measuring and medical instruments, photo equipment, and timepieces	1036.4	2.5	2.1	88.8	1.2	33.1	26.8	14.7	35.4	2.7	9.3	35.0	13.9
40	Miscellaneous manufactures	211.6		1.2	0.4	77.5	3.9	2.3	0.7	d	d	d	0.6	1.6
41,42	Trade and finance and real estate	39.7					(d	)						
43	Transportation and public utilities	47.2									d	d		
44,45	Construction and services, including R&D services	266.0		0.2	0.2	0.2	4.3	2.5	0.9	d	d	0.6	4.7	2.2
	Total R&D dollars used		40.8	44.8	147.0	108.8	727.9	409.7	364.9	524.1	84.1	542.8	493.9	432.9

Table 2 (continued)

		Origin R&D	Medical services 45A	All other services 45B	General government 46	Defense and space 47	Final consumption 48	Row code
1	Agriculture and forestry	128.1					d	1
2	Mining, exc. petroleum	60.3					0.0	2
3,4	Food and tobacco products	444.9	0.9	2.5	0.4	d	143.1	3,4
5	Textile mill products	179.3	0.7	d	0.4	0.2	18.2	5
6,19	Apparel and leather products	55.5	d	d	d	d	37.7	6,19
7	Lumber and wood products	72.6	0.2	1.5	0.4	0.2	2.8	7
8	Furniture	51.1	d	7.1	1.3	1.1	22.4	8
9	Paper mill products	202.3	2.4	4.0	1.6	0.4	73.3	9
10	Printing and publishing	67.4	0.2	d	0.8	1.3	13.9	10
11	Industrial inorganic chemicals	159.2	0.8	1.6	1.7	1.9	8.8	11
12	Industrial organic chemicals	297.2	2.9	2.3	1.5	9.9	14.1	12
13	Sythetic resins, fibers, rubber	601.6	2.0	4.6	1.5	2.9	10.0	13
14	Pharmaceuticals	557.3	321.4	d	0.8	0.1	462.0	14
15	Agricultural chemicals	186.7	d	1.0	0.5	0.1	80.2	15
16	Paints, toiletries, explosives, and other chemical products	485.7	4.7	23.7	9.1	3.9	161.9	16
17	Petroleum extraction and refining	380.3	1.0	5.5	3.1	3.2	53.2	17
18	Rubber and plastic products	419.8	4.8	18.1	7.2	3.9	111.9	18
20	Stone, clay, and glass products	265.0	1.7	3.7	4.6	2.1	36.8	20
21	Ferrous metals	189.3		d		0.1	0.0	21
22	Nonferrous metals	156.9	1.2	0.7	0.5	3.0	9.6	22
23	Fabricated metal products	552.7	3.2	19.8	8.4	29.7	106.6	23
24	Engines and turbines	282.2	2.4	21.1	15.6	15.3	59.6	24
25	Farm machinery	199.3	0.3	4.6	0.5	0.1	33.1	25

cont. overleaf

26	Construction, mining, and materials handling equipment	351.2	4.7	3.5	8.1	10.5	4.0	26
27	Metalworking machinery	121.5		6.6	1.6	0.9	12.2	27
28	Other machinery	691.0	7.8	64.3	12.1	26.8	111.6	28
29	Computers and office equipment	1153.0	19.8	224.2	75.6	75.2	45.0	29
30	Industrial electrical equipment	205.9	13.6	7.9	5.2	4.7	25.4	30
31	Household appliances	102.7	0.2	10.1	0.4	d	94.1	31
32	Lamps, batteries, ignition, X-ray, and other electrical equipment	233.3	26.0	13.5	7.2	20.9	120.0	32
33	Radio and communication equipment	1227.7	8.8	38.1	39.4	220.6	223.2	33
34	Electronic components	594.9	2.6	28.2	11.7	50.0	46.2	34
35	Motor vehicles and equipment	1518.0	16.9	204.7	79.8	27.7	1345.9	35
36	Aircraft	659.4	0.2	0.4	6.5	120.1	d	36
37	Missiles, spacecraft, and ordnance	122.7	d	d	d	93.2	d	37
38	Other transportation equipment	140.1	d	d	d	d	51.7	38
39	Measuring and medical instruments, photo equipment, and timepieces	1036.4	207.0	142.0	35.6	60.2	342.5	39
40	Miscellaneous manufactures	211.6	1.7	29.0	4.3	1.8	86.9	40
41,42	Trade and finance and real estate	39.7					0.0	41,42
43	Transportation and public utilities	47.2					0.0	43
44,45	Construction and services, including R&D services	266.0	d	72.6	d	d	14.0	44,45
	Total R&D dollars used		684.6	1000.3	378.1	842.2	4111.4	

ing sector, which in 1974 originated 23.7 percent of U.S. gross national product. Figure 1 exploits the additive property of technology flow estimates made under a "private goods" assumption to show the gross flow relationships between and within manufacturing and non-manufacturing. (The non-manufacturing sector includes government as a user of R&D, but neither R&D done in government laboratories nor R&D performed by industry or universities and financed by the government is counted in the present analysis. It should be noted too that the estimates of R&D flows originating within non-manufacturing are subject to much larger sampling errors than those for manufacturing.) From the figure, it becomes clear how dependent the non-manufacturing economy is upon manufacturing for new technology. Manufacturing enterprises "exported" to non-manufacturing half of the $14 billion in R&D they originated. The amount of industrial (i.e. non-governmental) R&D originated, financed, *and* used within non-manufacturing was only 7 percent of the amount imported from manufacturing. Within manufacturing, the balance of flows is quite different. Of the $6.1 billion of R&D used, 57 percent consisted of process R&D – that is, R&D performed in the specific narrowly-defined industry of use. Even more striking are the results of an analysis focusing only on capital goods inventions with specific uses in at most three narrowly-defined manufacturing industries. Three-fourths of all such using industry-specific inventions originated within the using industry. Evidently, if one has comparative advantage at doing industrial R&D, as most larger manufacturing enterprises have, there is a tendency toward considerable self-sufficiency in developing one's own specialized production technology. When that comparative advantage is absent, as in most non-manufacturing industries, one tends to rely heavily upon others for the technology one uses.

3.3. Government-financed R&D

The R&D flows analyzed thus far have been those financed by the originating business enterprises. Industrial corporations also perform a substantial amount of R&D under contract, usually with the Federal government reimbursing costs. Our analysis of the technology flows associated with this R&D is less complete and precise because of limitations in the facilitating patent data. The companies in our sample conducted $333 million of contract R&D in narrowly-defined industries in which no government contract-related patents were recorded. This had two probable explanations: a low propensity to patent contract-related inventions (on which the patent holder's right to exclude is constrained by government license provisions), and incomplete company compliance with the requirement that patents obtained under Federal contracts include notice of the contract connection. Through resort to contracting agency records, it was possible to identify a total of 325 sample company inventions originating under government contracts, but it is believed that another 75 patents eluded the search. The contract-related patents were linked to estimate the average amount of contract R&D per patent in their industry of origin, and these R&D dollars were flowed out to using sectors on the basis of the inventions' coded anticipated uses.

Table 3 summarizes the results. Sample companies obtained patents linked, after inflation by sampling ratios that probably underestimate the concentration of Federal contract work, to contract R&D outlays of $6.77 billion. All but 16 percent of this total originated in five industry groups for which there were sufficiently many companies active to report spending details without violating confidentiality rules. Not surprisingly, the most prolific performers of Federally-financed R&D were the missiles and ordnance, communication equipment (i.e. radar and guidance system), and aircraft industries. Most of the technology developed thereby, at least as discerned through our analysis of patent specifications, was specific to defense and space applications – e.g. jet engine ducting applicable only to fighter-type aircraft, radars that operated at military-blocked frequencies or with special jam-proof features, semiconductors able to function in unusually severe environments, etc. The airline industry was the largest beneficiary of specifically identifiable spillovers, especially from the development of lighter, more reliable turbojet engines. Consumer product uses embodying $158 million of Federal contract R&D were also identified, in every instance from inventions that also had industrial uses. Federally-supported R&D in "all other" industries yielded inventions with a higher proportion of industrial uses outside defense and space than R&D in the

Table 3
Truncated federally-supported industrial R&D flows matrix

	Industry of Origin	Federal R&D originated	Internal process R&D	43B Air transport	47 Defense and space	All other industrial uses
24	Engines and turbines	34	n [a]	n	11	23
29	Office equipment	336	n	n	214	122
33	Radio and communication equipment	1631	133	16	53	
34	Electronic components	202	100	n	99	3
36	Aircraft, including engines	1342	213	190	927	12
37	Guided missiles, spacecraft and ordnance	2162	448	0	1639	74
	All other industries	1063	42	9	503	509
	Total R&D originated	6770				
	Total R&D used		936	214	4822	796

[a] n: not separately reported, but included in "all other industrial uses".

traditional aerospace industries. If the results of this incomplete Federal contract R&D analysis are combined with those for privately-financed R&D, we find that, of total 1974 U.S. R&D outlays amounting to \$21.5 billion (\$14.7 billion company-financed and \$6.8 billion Federal), \$5.66 billion, or 26.4 percent, were associated with product inventions flowing to the defense and space fields.

4. R&D and productivity growth

A principal motive for tracing R&D outlays from industries of origin to industries of use is to understand better how R&D affects productivity growth. The influence of R&D on productivity growth operates with a lag. Our R&D data are for 1974, and so one should look to the late 1970s for the first signs of an effect. Data limitations and the need to control for business cycle influences directed the main analysis toward changes in output per worker hour over the business cycle peak-to-peak period 1973–78. Three sets of productivity data for this period, with quite different measurement approaches and industry coverage, were analyzed. We focus here on the results for the most comprehensive sample, with 87 industry group observations covering virtually all of the manufacturing sector plus agriculture, crude oil and gas, railroads, air transport, communications, and the combined electric-gas-sanitary services sector. To obtain the productivity indices for this sample, the current-dollar value of an industry's sales was divided by an average price index for the industry, usually derived from the U.S. Producer Price Index series. This estimate of "real output" was then divided by industry hours of work (including salaried personnel) to obtain an index of output per worker hour. Needless to say, productivity growth comparisons based upon such indices can be no more reliable than the comprehensiveness of the output value deflators (price indices) used and the deflators' effectiveness in tracking product quality changes.

The principal methodology was ordinary least-squares regression analysis, where ΔLP is the annual percentage change in an industry's real output per worker hour from 1973 to 1978, ΔK is the annual percentage change in the industry's gross constant-dollar plant and equipment/labor ratio over the same period, PRODRD is 1974 company-financed product R&D classified by industry of origin and divided through by 1974 origin industry sales, and USERD is R&D (product plus process) flowed through to industry of use and divided by using industry sales. The USERD variable is defined under the "private good" assumption; variables constructed under a "public good" assumption had less explanatory power.

With all 87 industries included and with t-ratios given in subscripted parentheses, the fitted regression equation is:

$$\Delta LP = -0.14 + \underset{(3.30)}{0.347}\,\Delta K + \underset{(2.01)}{0.289}\ \text{PRODRD} + \underset{(1.89)}{0.742}\ \text{USERD}; \qquad R^2 = 0.193. \tag{1}$$

Both R&D variables have positive, statistically significant coefficients, as does the capital/labor change variable.

The strength of the PRODRD variable, with a t-ratio marginally higher than the USERD variable, is surprising, since the logic underlying the technology flows matrix suggests that most productivity benefits would be realized by R&D-using, not product R&D-originating industries. A strong product R&D-originating effect is plausible only to the extent that product innovators exercise strong monopoly power in their pricing or when there are systematic errors in measuring productivity. To test for the latter, the sample was divided (before any analyses were run) into two subsets: one for which the real output indices used in computing productivity growth were based upon reasonably comprehensive output value deflators, and another for which the price deflator coverage was sufficiently sparse that there was reason to doubt the reliability of the computed real output and hence productivity indices. For the 51 industry groups with "well measured" output indices, the regression equation was as follows:

$$\Delta LP = -0.16 + \underset{(2.90)}{0.400}\ \Delta K - \underset{(0.54)}{0.182}\ PRODRD + \underset{(2.53)}{1.039}\ USERD; \quad R^2 = 0.241. \tag{2}$$

Here, as expected, used R&D plays a powerful role while PRODRD is statistically insignificant (with an implausible negative coefficient sign). For the 36 industry groups with "poorly-measured" output indices, the regression equation is:

$$\Delta LP = 0.08 + \underset{(1.82)}{0.310}\ \Delta K + \underset{(2.10)}{0.431}\ PRODRD + \underset{(0.10)}{0.096}\ USERD; \quad R^2 = 0.197. \tag{3}$$

An opposite pattern emerges; used R&D is insignificant while PRODRD is significant. That the results with the poorly-measured data are atrributable to measurement error is suggested by the fact that when similar regressions are computed for a completely different sample using unusually high quality productivity indices based upon physical (not value-deflated) output data, their coefficient significance pattern is the same as that of equation (2) rather than equation (3).

The R&D coefficient values in these equations can be interpreted, subject to certain caveats, as steady-state rates of return on investment in research and development. Except in equation (3), the USERD values indicate quite substantial social returns, on the order of 75 to 100 percent, for the processes and products flowing from R&D. Moreover, the coefficient magnitudes are such that, if used R&D were increased by two standard deviations, annual productivity growth would rise by 1.1 to 1.5 percentage points. Thus, important economic impacts are implied. Interpretation of the PRODRD coefficients is complicated by the possibility of serious measurement errors in the raw productivity data. For what they are worth, the coefficient estimates suggest private returns to product R&D (captured by the originator) in the range of 0 to 43 percent.

Additional analyses discussed more fully in [10] add further insights. Perhaps most important, in contrast to the results obtained by both myself and other scholars using highly aggregated data, there was no evidence with disaggregated industry data of a decline during the 1970s in the ability of R&D to drive productivity growth. There were also less conclusive indications that capital goods R&D has a more potent impact on productivity growth than materials R&D, and that at least some of the capital goods effect is best characterized as "embodied" in new capital investment. That is, the new technology and the capital equipment embodying it are joint and inseparable inputs into industrial production processes. Finally, dollar-denominated R&D usage variables had more explanatory power than raw patent counts, revealing the importance of weighting patents by their underlying R&D expenditures if one is to use patent data as an index of innovative output.

5. Conclusion

The new data presented in this article provide an unusually rich picture of the complex interdependences among industries in the process of technological change. Thus, the statistics arc of considerable interest in their own right. But more importantly, such data are vital to achieving a proper understanding of the links between technological innovation and economic progress, given the high fraction of industry R&D that is product-oriented and the evidence that productivity growth, when well-measured, comes from *used* R&D rather than product R&D at its point of origin.

Recognition of the importance of these linkages is widespread, and in response to the perceived need, there has been classic simultaneous invention. One approach is the work on significant British innovations by Pavitt and others at the University of Sussex [7]. Also, quite independently of my research with patents, the Canadian Patent Office has made impressive strides toward classifying by industries of origin and use a sample of Canadian patents from the 1973–77 period and *all* such patents since 1978 [11]. The availability of such data should permit substantial progress toward understanding how new technology affects productivity growth in an "open" economy heavily dependent upon foreign inventions. For the United States, further progress requires an extension of my data, which were unable to cover all sources and uses of new technology and which provide a snapshot for only a single year. The obvious direction for improvement lies in classifying by origin industry (linked to FTC R&D data) and using industries all patented inventions, domestic and foreign, for additional years. This could be done either by specially-trained Patent Office examiners, as in Canada, or by having inventors and their attorneys – the individuals in the best position to know – provide appropriate origin and use information. One hesitates to recommend still another government data collection program in a world of budgetary cutbacks and paperwork rebellions. Yet such a step, which would increase the costs of patent processing only trivially, would provide significant new leverage on the productivity growth question.

References

[1] The exceptions are Nestor E. Terleckyj, *Effects of R&D on the Productivity Growth of Industries: An Exploratory Study* (National Planning Association, 1974); and two other analyses using the Terleckyj indices, Edwin Mansfield, Basic Research and Productivity Increase in Manufacturing, *American Economic Review* 70 (December 1980) 867; and John W. Kendrick and Elliot S. Grossman, *Productivity in the United States: Trends and Cycles* (Johns Hopkins University Press, Baltimore MD., 1980) pp. 100–111. See also Edwin Mansfield et al., Social and Private Rates of Return from Industrial Innovations, *Quarterly Journal of Economics* 91 (May 1977) 221.

[2] Jacob Schmookler, *Invention and Economic Growth* (Harvard University Press, Cambridge, MA., 1966), Chapter VIII.

[3] F.M. Scherer, Research and Development Expenditures and Patenting, forthcoming in the American Patent Law Association *Quarterly Journal*.

[4] U.S. Federal Trade Commission, *Statistical Report: Annual Line of Business Report; 1974* (Washington, DC, September 1981).

[5] For further methodological details, see F.M. Scherer; Using Linked Patent and R&D Data To Measure Inter-Industry Technology Flows, paper presented at the National Bureau of Economic Research conference on R&D, Patents, and Productivity, October 1981.

[6] U.S. National Science Board, *Science Indicators: 1978* (Washington, DC, 1979), p. 201.

[7] Keith Pavitt, Some Characteristics of Innovative Activities in British Industry, manuscript, Science Policy Research Unit, University of Sussex, 7 January 1982, pp. 11–12.

[8] It is for such specific-use inventions that one is likely to find the kind of originator-user locus indeterminacies and coordination emphasized by von Hippel. That is, an invention might be made as an internal process by the ultimate user, as a product by a specialized supplier to that user, or through a coordinated effort by both parties. See Eric von Hippel, The Dominant Role of Users in the Scientific Instrument Process, *Research Policy* 5 (July 1976) 212; and Appropriability of Innovation Benefit as a Predictor of the Functional Locus of Innovation, working paper, Sloan School of Management, Massachusetts Institute of Technology, June 1979. Although no systematic evidence is available from our study, it was clear from patent specifications that some "product" inventions resulted from a cooperative effort with users.

[9] U.S. Department of Commerce, Bureau of Economic Analysis, *The Detailed Input-Output Structure of the U.S. Economy: 1972*, vol. I (Washington, DC, 1979); and P.E. Coughlin et al., New Structures and Equipment by Using Industries, 1972, *Survey of Current Business* (July 1980), 45–54.

[10] F.M. Scherer, Inter-Industry Technology Flows and Productivity Growth, forthcoming in the *Review of Economics and Statistics*.

[11] E.D. Ellis, Canadian Patent Data Base: The Philosophy, Construction and Uses of the Canadian Patent Data Base PATDAT, *World Patent Information* 3 (1981), 13–18.

THE JOURNAL OF INDUSTRIAL ECONOMICS
Volume XXX March 1982 No. 3

DEMAND-PULL AND TECHNOLOGICAL INVENTION: SCHMOOKLER REVISTED

F. M. Scherer*

During the 1950s and 1960s, quantitative research on the economics of invention was spurred by awareness that technical change was a primary source of the rapid economic growth achieved by industrialized nations. A leader in this effort was Jacob Schmookler, whose analysis of patent statistics [2] [4] provided the first demonstration that inventive activity is responsive to the pull of demand. The motivation now is different but no less compelling: a decade of retarded productivity growth makes it imperative to understand the forces governing the rate of technical advance. This paper retests Schmookler's demand-pull hypothesis with a new and more comprehensive data set.

I. THE SCHMOOKLER HYPOTHESES

Schmookler's main contention, contrary to the prevailing emphasis on changes in scientific and technological knowledge, was that demand played a leading role in determining both the direction and magnitude of inventive activity. His basic underlying premises were two: (1) That the ability to make inventions is widespread, flexible, and responsive to profit-making opportunities; and (2) That the larger an actual or potential market is, the more inventive activity will be directed toward it, partly because the profitability of invention rises with market size, all else equal, and partly because chance encounters between inventive talent and a problem needing solution are more frequent, the more productive activity there is devoted to meeting some demand. In his most comprehensive work [4, Chapter VIII], Schmookler took into account the undeniable importance of differences in knowledge by observing that firms in certain industries (e.g., chemicals, electronics, and machinery) had comparative advantage at making inventions by virtue of the access to rich and growing scientific and technological knowledge bases in their fields of specialization.[1] However, these knowledge

* This research was supported under National Science Foundation grant PRA-7826526. The author is indebted to Mary Gianos, Chun Yue Lai, Brett Spencer, Pin Tai, and William Cork for classifying the patents analyzed here; to Mr. Lai for computer programming; and to Edward Klotz, David Walker, Marilyn Wagner, and Catherine Conrad for research assistance.
Use is made here of aggregated data collected under the Federal Trade Commission's Line of Business program. A determination has been made that the results do not disclose individual company data. The conclusions drawn are the author's and not necessarily those of the Commission.

[1] For a critique of Schmookler emphasizing the role of technological opportunity, see [3].

bases were highly adaptable, he argued, and so the applications to which they were put depended upon relative profitability, which in turn depended upon demand. Invention then flowed through a kind of input-output matrix from knowledge-rich originating industries to high-demand using industries.

Although there were other (i.e., time series) tests supporting his simpler demand-pull hypotheses, the main confirmation of Schmookler's theory of the relative roles of demand-pull and technological opportunity came from analyses of patenting for various industry cross-sections. His prime focus was on capital goods inventions, measured by the number of U.S. patents linked by Patent Office sub-class codes to particular industrial end uses. The more investment there was in a using industry, Schmookler claimed, the more patented capital goods inventions directed toward that industry's needs one should observe. The source of those inventions could be the using industry, other industries that specialize in supplying capital goods, or independent (i.e., non-corporate) inventors. Schmookler's patent counting methodology made no distinction among sources, but that was viewed as unimportant; what mattered was the pull of using industry demand. His cross-sectional sample covered up to 23 using industries, including petroleum refining, synthetic fibers, glass, sugar, tobacco, railroads, diverse textile and apparel industries, and several lumber and paper products industries. For a variety of time periods extending from 1936 to 1950, Schmookler found that the greater using industry capital investment was, the larger the number of relevant capital goods invention patents from all sources was. The associations were strong, with r^2 values ranging between 0.85 and 0.92, and in logarithmic form, the estimated regression coefficients were insignificantly different from 1.0, implying essential linearity of the demand-pull relationship.

To isolate the role of technological opportunity, Schmookler turned to a different sample of large corporations' patents classified by *industry of origin*. Since roughly three-fourths of business enterprises' patented inventions comprise new or improved products sold to others and only a fourth are internally-used processes, a classification of patents by industry of origin is quite different from one by industry of use. The question arises, is there a difference in the statistical relationship between patents classified by industry of origin and some index of demand, e.g., total inventing corporation sales, as compared to a classification by industry of use and a demand index (e.g., for capital goods inventions, user investment)?[2] Breaking his corporate sample into 14 broad industry classes, Schmookler found that the relationship between patenting and origin industry sales was essentially linear, as it had been for patenting and using industry investment. However, the association was much weaker, with r^2 values on the order of 0.43—half the levels obtained for the industry of use analyses. This difference Schmookler attributed to the

[2] From the perspective of a technology-originating industry, sales measures the purchases of its products. For industries *using* capital goods, investment measures the purchases of those goods.

intervening role of technological opportunity differences: firms in industries with rich knowledge bases originated many more inventions per million dollars of sales than less favorably disposed firms, causing the overall sales—patent origination correlation to deteriorate. Yet when differences in opportunity were aggregated out by flowing inventions through to *using* industries, regardless of origin, strong correlations demonstrated the pervasive force of demand-pull.

In essence, then, Schmookler's case for the relative importance of demand-pull rested upon three empirical results: (1) the linearity of the relationship between demand, as measured by new capital investment, and the flow of capital goods inventions to investing industries; (2) the strong correlations between those two indices; and (3) the much weaker correlations between counts of patents assigned by industry of *origin* and indices of demand or industry size.

Among the various possible limitations of these tests by Schmookler, two deserve emphasis. First, the sample of industries chosen for his investment—capital goods patent analysis was small and perhaps not representative. "Older" industries such as apparel, lumber, and paper-making were over-represented; and the number of patents included in the analysis totalled only 6 to 8 percent of all U.S. patents issued to individuals and domestic corporations during the late 1940s and early 1950s. Second, the sample of patents classified by industry of origin was quite different from the use-oriented sample, covering a much larger fraction of all patents, but excluding unaffiliated inventor and small-firm patents. The sample of origin industries was also much broader and defined at higher levels of aggregation. A new sample makes it possible to retest the Schmookler hypotheses while substantially avoiding these problems.

II. THE NEW DATA

As part of a broader effort to estimate the technology flows matrix first proposed by Schmookler, a sample of 443 large U.S. corporations was drawn. For reasons only peripherally related to the research reported here, these were the corporations covered by the Federal Trade Commission's 1974 Line of Business survey. An attempt was made to identify all the U.S. invention patents originating from U.S. resident inventors and issued to the 443 enterprises between June 1976 and March 1977. Some 15,112 patents, or 61 percent of all patents issued to U.S. industrial corporations during that period, were identified in this way. They comprised the basic working sample.

The printed specification of each individual patent was examined by a team of four Northwestern University students (an organic chemist, a chemical engineer, an electrical engineer, and a "utility infielder") to determine the nature of the invention (e.g., internal process vs. capital good product vs. material product), the industry(ies) in which the invention originated, and the industry(ies) in which use of the invention was anticiated.[3] The industry code system upon which these assignments were made tracked closely the FTC Line of Business codes and included a total of 245 manufacturing categories, usually defined at the four- or three-digit S.I.C. Level. Assignments according to industry of use were based upon information patent applicants provided to show the "utility" of their inventions. Roughly 7.4 percent of the sample inventions had consumer goods uses only, and another 34 percent had uses too broad to associate with any identifiable subset of using industries.[4] These were excluded from the sample analyzed here. Only the 8,875 inventions coded as having one, two, or three specifically identifiable industries of use remain. Classifications according to industry of origin were based upon information concerning the nature of the inventions, the patent-receiving corporations' product structures, and the specific activities undertaken at laboratories and plants near which the relevant inventors resided. With each specific-use patent coded as to both industry of origin and industry(ies) of use, both sides of Schmookler's two-way classification were available for the same set of inventions. Thus, the dual-sample problem of his research was avoided.

Our sample, however, covers only a subset of all industrial corporations. To analyze the forces affecting industrial invention, it was necessary to inflate the sampled patents so that their totals corresponded to the relevant inventive output of the U.S. industrial corporation universe. This was done by estimating for each origin industry i the fraction w_i of that industry's sales accounted for by companies included in our sample and then inflating each patent from industry i by $1/w_i$. Among the 245 industries, these sales coverage ratios w_i ranged from 0.06 to 0.99, with a value added weighted average of 0.61. The implicit assumptions underlying this inflation approach are that patent coverage for an origin industry is proportional to sales coverage and that the patents of non-sampled corporations are similar, e.g., in use patterns, to those of sampled companies. No independent verification of the second assumption is possible. Concerning the first assumption, it is noteworthy that the weighted average sales coverage ratio of 61 percent is identical to the sampled corporations' 61 percent share of all domestic industrial corporation patents. Thus, the proportionality assumption is correct on average, even though there may be positive or negative inflation errors for individual industries.

The sampling error characteristics of our approach differ significantly from those of Schmookler's. His use-oriented sample included *all* patents, from large enterprises and small, foreign and domestic, as well as from independent inventors, in the use classes selected for analysis. His sampling of patents linked according to patent classification codes to using industries is exhaustive,

[3] For further methodological details, see [8].

[4] On this problem, see the letter Schmookler wrote to Robert Baldwin, quoted in [4, p. 91].

but he sampled only a small subset of using industries. Our sample of using manufacturing industries is exhaustive, but we include only a subset of the inventions flowing to those industries, correcting in part for this non-exhaustiveness through origin industry sampling ratio inflation. No correction is made in our sample for the exclusion of inventions made overseas (15 percent of all U.S. patents in 1940 and 35 percent in 1976) or by independent inventors (20 percent of all 1976 U.S. patents). Both Schmookler's analysis and ours, needless to say, are subject to possible errors in the classification of individual patents' using industries.

The industrial use patents in our sample were divided into two mutually exclusive groups—materials inventions and capital goods inventions. The latter included originating industry products that became production equipment for some other (using) industry plus process inventions for internal use by the origin industry, but excluding process patents on compositions of matter, since such patents are often used to protect *products* on which patent protection might otherwise be unavailable. The capital goods part of this dichotomy was designed to replicate as closely as possible the spirit of Schmookler's capital goods invention selection criterion.[5] The capital goods invention sample was further subdivided for some analyses into two components—processes for internal use by the originating firm and capital goods product inventions for sale across industry lines.

The sum of patented inventions used in the jth industry, each individual patent multiplied by the ith originating industry sampling ratio $1/w_i$, will henceforth be designated PM_j for industrial material inventions and PC_j for capital goods inventions. When an invention had more than one industry of use, it was counted fully for each industry to which a prospective use was traced.[6]

Following Schmookler, the preferred measure of demand-pull for capital goods inventions is new capital investment in using industries. Data constraints required that the analysis be limited to manufacturing industries of use only, for which consistent Census statistics are available. Schmookler found that the strongest correlations emerged when patent applications were lagged two years after the year of investment [4, p. 147]. Since it took 19 months on average for a patent application to mature into an issued patent during the mid-1970s,[7] this implies a maximum effect on our patents, whose midpoint issuance date was October 1976, from 1973 capital investment.

[5] However, Schmookler's classification was not completely faithful to his capital goods use criterion. See [5, pp. 90, 266, and 278]. For instance, his textile dyeing patent classes include azo dye compositions, which we would count as industrial materials. And more importantly, his synthetic fibers industry classes include cellulose derivative compounds that we would classify as material *products* of the fibers industry, not capital goods *processes*. These cases appear to be exceptional.

[6] Thus, we are implicitly treating inventions as public goods—an assumption whose logic is analyzed in [8]. Sixty-nine patents obtained by corporations with no patents in the sample period came from a broader three-year sample and were accorded a weight of $\frac{10}{36}$.

[7] For a fuller discussion, see [8].

Investment data for 1972 through 1974 were therefore collected. The investment variable for the jth using industry in year t will be designated I_{tj}. For industrial materials inventions, the preferred measure of demand-pull is using industry material purchases M_{tj}, although data on using industry value added V_{tj} were also gathered.[8]

III. RESULTS

Since 31 manufacturing industries (out of 245) had no assigned capital goods inventions, and 72 were recorded as being users of no industry-specific materials inventions, regressions of patenting on demand-pull indices were run first in untransformed (i.e., additive) variables for all industries,[9] and then (following Schmookler) in logarithms for a restricted sample with non-zero patent use.

In every case, the strongest association was between the relevant patenting variable and 1974 capital investment, material purchases, or value added. In most but not all cases, 1973 lagged variables performed slightly better than 1972 variables. When the lagged variables for all three years were introduced together, the characteristic pattern was for the 1974 variable to be positive and statistically significant and the 1972 and 1973 variables to be insignificant, occasionally with perverse negative signs. When a triangular lag structure with weights of 0.6 for 1974, 0.3 for 1973, and 0.1 for 1972 was imposed, r^2 values were 0.004 to 0.013 lower than with 1974 data alone as the independent variable. The persistently greater explanatory power of 1974 demand-pull indices is a bit surprising, since it implies no lag between the emergence of demand-pull influences and the time when invention occurs (which precedes the filing of a patent application by nine months on average). These results could have occurred by chance, or they may reflect a tendency for corporate inventors to anticipate favorable demand conditions even before they materialize fully. In either event, the results are not appreciably different for any of the three lags or combinations thereof, and the exposition can be simplified by focusing only on the results for 1974 demand-pull variables, which are summarized in Table I.

Several points stand out. First, in all cases, the relationship between demand-pull indices and associated patent flows is positive and statistically significant, with t-ratios ranging from 5.54 (for equation 1.4) to 17.5 (equation 1.1). Second, the association between capital goods inventions and investment is much stronger than the association between industrial materials inventions

[8] The data were drawn from the U.S. Census of Manufactures for 1972 and from the Annual Surveys of Manufactures for 1973 and 1974.

[9] That the regressions have numerous zero-valued dependent variable observations suggests the use of Tobit analysis. However, the nonlinearities implied thereby are inconsistent with both the Schmookler findings and the relationships observed here. As a check on the validity of the full sample OLS regressions, zero-valued observations were deleted. For the analogue of equation 1.1, the regression coefficient was 0.227 and the r^2 was .537. For the analogues of regressions 1.3 and 1.5, the regression coefficients were .00247 and .00627 respectively (differing from their counterparts by less than one standard error), and the r^2 values were .119 and .161.

TABLE I
SUMMARY OF BIVARIATE REGRESSION RESULTS

Equation	Dependent Variable	Constant	Regression Coefficient and Independent Variable*	r^2	N
1.1	PC_j	−0.8	$+0.228I_{74j}$ (.013)	.544	245
1.2	$\log PC_j$	−0.1	$+0.686 \log I_{74j}$ (.060)	.382	214
1.3	PM_j	7.2	$+0.00278M_{74j}$ (.00043)	.149	245
1.4	$\log PM_j$	−0.5	$+0.443 \log M_{74j}$ (.080)	.153	173
1.5	PM_j	1.4	$+0.00673V_{74j}$ (.00087)	.197	245
1.6	$\log PM_j$	−0.9	$+0.588 \log V_{74j}$ (.092)	.194	173
WITH CAPITAL GOODS INVENTIONS SUBDIVIDED					
1.7	PCI_j	−0.7	$+0.1719I_{74j}$ (.0126)	.435	245
1.8	PCE_j	−0.2	$+0.0562I_{74j}$ (.0031)	.578	245
WITH AUTO AND STEEL INDUSTRY OBSERVATIONS DELETED					
1.9	PC_j	−7.7	$+0.298I_{74j}$ (.014)	.651	242
1.10	$\log PC_j$	−0.2	$+0.697 \log I_{74j}$ (.062)	.374	211
1.11	PM_j	6.3	$+0.00334M_{74j}$ (.00049)	.161	242

* Investment, materials purchases, and value added are all scaled in millions of dollars. Standard errors of the regression coefficients are given in parentheses.

and either material purchases or value added (the latter of which unexpectedly has somewhat greater explanatory power). Professor Schmookler evidently chose the best-suited class of inventions for analyzing the role of demand-pull. Third, the r^2 values are at best much lower than those obtained by Schmookler. It is uncertain whether this is so because of sampling or classification errors under our methodology, the shorter time period covered by our data (with less cancelling out of random measurement errors), extraordinary sensitivity to demand-pull influences of the industry and patent class sample chosen by Schmookler,[10] or structural changes over the quarter century separating our samples. Given the broad coverage of our sample, it seems unlikely that the differences can be ascribed solely to origin industry sampling ratio errors. Fourth, logarithmic regressions 1.2, 1.5, and

[10] That Schmookler's results are not invariant across samples is shown by the considerable increases in r^2 and logarithmic regression coefficient values as his sample was changed somewhat between his 1963 article with Griliches [2] and his 1966 book [4, p. 145]. The principal changes entailed eliminating the shoemaking industry, with high patenting but poorly measured investment, and adding railroads (which in most years had the highest investment and patenting of sampled industries).

1.6 reveal that the linear demand-pull relationships observed by Schmookler do not hold up for our sample. Rather, with elasticities in the range of 0.44 to 0.69, appreciable diminishing marginal returns appear to prevail.

To replicate Schmookler's test of the relative roles of demand-pull and technological opportunity, the 1974 sales of sample companies, broken down by individual lines of business, were aggregated to an almost identical industry code structure and level as that assumed in Table I, and those aggregated origin industry sales S_{74i} were taken as the right-hand side variable in regressions explaining patents P_i classified and aggregated to *industries of origin*. The matching between patents, firms, and industries in this instance is nearly exact, stemming from a link effort exploiting extraordinarily rich data.[11] Owing to the precision of the link, one might expect less "noise" to infiltrate the relationship between origin industry sales and patents than the relationship between diversely inflated using industry patents and demand-pull indices. R^2 values should be correspondingly higher, all else equal.

The simple regression of patents on origin industry sales (in millions of dollars) was as follows:[12]

$$(1) \qquad P_i = 40.2 + \underset{(.0015)}{0.0135} S_{74i}; \quad r^2 = .243, \mathcal{N} = 241,$$

with the regression coefficient's standard error given in parentheses. When logarithms are taken (requiring the omission of 17 industries originating no patents), the regression is:

$$(2) \qquad \log P_i = -1.45 + \underset{(.083)}{0.904} \log S_{74i}; \quad r^2 = .351, \mathcal{N} = 223.$$

In both instances, the r^2 are somewhat to substantially higher than those for the industrial materials use regressions. In untransformed regression (1), r^2 is less than half its capital goods industry-of-use counterpart, but in logarithmic regression (2), it is nearly as high as, and insignificantly different from, the value in equation 1.2. Here, in contrast to the Table I regressions, taking logarithms suppresses the noise associated with a few very large but low-patenting industries (notably, automobile assembly). The support for Schmookler is again equivocal. The flow of appropriately matched inventions appears to be more closely correlated with using industry demand (measured by investment or materials purchases) than with originating industries' sales only for capital goods inventions, and then only for one regression specification, and not for the logarithmic form emphasized in Schmookler's 1966 work.

[11] See [8].

[12] When the same regression is run with observations weighted by the reciprocal of the industry coverage ratios, making the weights given industry observations correspond more closely to those of Table I, the slope estimate is 0.0131 and the r^2 is slightly lower at 0.187.

The capital goods inventions analyzed in Table I consist of two rather different subsets: inventions that were internal processes to their originators, and those that were products to the originators but processes to their users. It would not be unreasonable to suppose that demand-pull influences are transmitted differently between these two cases, e.g., intra-firm markets (for process inventions) might work more efficiently than inter-industry markets. To explore this possibility, the invention sample was bifurcated into internal process inventions PCI_j (where using industry j = origin industry i) and externally sold products PCE_j (where $j \neq i$).

An insight provided by this procedure merits a brief digression. Both our data and National Science Foundation surveys reveal that 96 to 98 percent of all industrial inventive activity and corresponding patenting occur in the manufacturing sector. The bifurcation disclosed that manufacturers are their own most active suppliers of specialized capital goods inventions. Fully 75 percent of the 7,935 (inflated) capital goods patents with specific manufacturing industries of use were internal processes. Of 214 manufacturing industries with non-zero capital goods invention use, 78 percent drew a larger fraction of inventions from inside than outside. However, the companies in our sample also supplied 5,699 (inflated) capital goods inventions to nonmanufacturing industries—e.g., agriculture, mining, telecommunications, and national defense operations. Eighty-eight percent of those inventions crossed industry bounds, being products to the originating industry and processes to the using industry.

Focusing on manufacturing uses alone, equations 1.7 and 1.8 in Table I reveal somewhat higher r^2 values for external use inventions than internal processes—i.e., 0.578 for the former as compared to 0.435 for the latter. Evidently, external markets are at least as responsive in transmitting demand-pull stimuli as internal markets.[13] And they may even function with less slippage.

One further elaboration of the basic results is warranted. A plot of equation 1.1 and its diversely lagged variants showed that three industries—auto assembly, auto parts, and steel—were ill-behaved outliers. They had particularly high levels of investment, but drew far fewer capital goods inventions than one would have predicted from the general linear relationship estimated for all industries. When the three were deleted from the sample, the untransformed regression r^2 (equation 1.9) rose by 0.107 to 0.651. No similar increase occurred for logarithmic analogue equation 1.10, apparently because the logarithmic regressions were better able to accommodate the

[13] The correlation coefficient differences are significant in a z-test at the 0.03 level.
A referee commented that regression 1.7 might be viewed as the stronger of the two because its regression coefficient is substantially higher. On this, two observations are required. First, the regression 1.7 coefficient is higher simply because there are roughly three times as many internal as external inventions. Second, the Schmookler hypotheses are concerned primarily with the *tightness* of the relationships between demand indices and invention, as reflected in a measure such as r^2, and not with the level of particular regression coefficient values.

nonlinearities associated with these outliers,[14] and perhaps also because there were other industries which, although much smaller, deviated proportionately from all-sample patterns at least as much as autos and steel. When capital goods inventions were broken down into internal and external processes, deletion of the three industries led to 0.07 and 0.08 increases in r^2 relative to full-sample untransformed regressions 1.7 and 1.8. It would appear that both external and internal demand-pull transmission mechanisms were unexpectedly weak. One cannot help wondering whether this failure to elicit process inventions might be related to the auto and steel industries' well-known import competition problems.[15]

IV. FURTHER ANALYSIS

We depart now from the direct Schmookler tradition to explore several further hypotheses. It is conceivable that demand-pull influences manifest themselves in ways other than, or in addition to, the level of using industry investment, purchases, or value added. One plausible hypothesis is that profit possibilities are signalled in part by *changes* in using industry output. Also, labor-saving capital goods invention may be stimulated by unusually rapid increases in using industry wages. To test the first hypothesis, a variable ΔQ_j measuring percentage changes in real using industry output between the Census years 1967 and 1972 was computed.[16] To test the second, the percentage change ΔW_j in using industry production worker wage payments per manhour between 1968 and 1973 was estimated from Census data. The multiple regression incorporating these two variables along with 1974 investment to explain using industry capital goods inventions was as follows:

$$(3) \qquad PC_{wj} = \underset{}{22.9} + \underset{(.014)}{0.229 I_{74j}} + \underset{(.090)}{0.271 \Delta Q_j} - \underset{(.468)}{0.828 \Delta W_j}; \qquad R^2 = .567, \; N = 245.$$

[14] Note the increase in the logarithmic regression coefficient toward unity when the three outliers were deleted.

[15] Compare Abernathy [1], who argues that the auto industry has been deficient in product but not in process innovation. For autos, at least two extenuating explanations might exist. First, the technology of auto production may be such that productivity is most readily enhanced by general-purpose machine tools, computers, and the like, whose uses are less apt to be associated with specific industries and which therefore would not be included in our subsample. Second, although the number of capital goods inventions with specific auto industry uses is small, the average 1974 research and development outlay leading to an auto industry patent, \$3.55 million, was much higher than the average of \$588,000 for all sample companies. But this low "propensity to patent" is almost surely the result of an R&D orientation that stresses styling and model testing rather than the creation of new mechanical features. It is noteworthy that the auto parts industry, with equally low patent pull relative to investment, spent only \$230,000 on R&D per patent received. For steel, the dearth of process inventions probably reflects a lack of imaginative internal research and development plus the atrophying of specialized steel industry process equipment suppliers' R&D when investment for U.S. steel-making capacity expansion remained stagnant between 1955 and 1964.

[16] The source is [9].

Relative to otherwise comparable equation 1.1 in Table I, the increase in R^2 is 0.023, which is statistically significant in an incremental F-ratio test, with $F = 6.26$. The output change variable is positive as predicted and statistically significant. The wage change variable has a paradoxical negative sign and a t-ratio of 1.77, i.e., not quite significant. Similar results emerged when logarithms were taken of all but the output change variable (which had some negative values) and when the automobile, auto parts, and steel industry observations were deleted. For industrial material inventions, the most closely corresponding regression is:

$$(4) \qquad PM_{wj} = \underset{}{2.9} + \underset{(.00043)}{0.00272M_{74j}} + \underset{(.054)}{0.168\Delta Q_j} + \underset{(.276)}{0.024\Delta W_j};$$

$$R^2 = .182, \; N = 245.$$

The change in R^2 compared to bivariate equation 1.3 is 0.033, with $F = 4.86$. The output change variable is again positive and significant, but the level of demand variable continues to have much greater explanatory power. The wage variable falls far short of statistical significance. Evidently, past increases in using industry output provide modest additional stimulus to invention above and beyond current *levels* of demand. No such stimulating role is detected for rapid rates of wage increase in using industries.

Finally, we consider an alternative way of assessing the role of technological opportunity in the industries that originate inventions. Extending the concepts employed in two earlier articles [6] [7], industries were classified into seven groups according to the perceived richness of their knowledge bases: organic chemicals, other chemicals, electronic systems and devices, other electrical equipment, the metallurgical trades, industries with "traditional" technologies (such as sugar refining, textile weaving, and cement making), and a base case consisting mostly of industries with mechanical technologies. These judgments were made *ex ante*, that is, before any analyses were run and indeed before sales and other linked financial data were assembled. The technology class dummy variables were introduced into analogues of origin industry patenting regressions (1) and (2). In the additive (i.e., nonlogarithmic) version, the appropriate specification is for the dummy variables to modify the base case (general and mechanical) regression slopes, and so the standard errors reported here test the hypothesis that the modified slopes differ from the base case slope:

$$(5) \qquad P_i = 30.1 + \underset{(.0014)}{0.0116S_{\text{base}}} - \underset{(.0039)}{0.0154S_{\text{traditional}}} + \underset{(.0044)}{0.0721S_{\text{organic}}}$$

$$- \underset{(.0017)}{0.0018S_{\text{other chemicals}}} + \underset{(.0046)}{0.0837S_{\text{electronics}}} + \underset{(.0164)}{0.0452S_{\text{electrical}}}$$

$$- \underset{(.0026)}{0.0071S_{\text{metallurgical}}}; \quad R^2 = .809, \; N = 241.$$

The most important insight is that this simple method of taking into account differences in technological opportunity has raised the proportion of patenting variance explained from 0.243 in equation (1), with origin industry sales alone as the explanatory (i.e., homogeneous demand-pull) variable, to 0.809.[17] Clearly, differences in opportunity play a large and easily systematized role. Not surprisingly, the slope dummy coefficients for organic chemicals and electronics are positive and highly significant, with t-ratios of 16.47 and 18.27, while those for the traditional and metallurgical technology industries are negative ($t = 3.92$ and 2.74), indicating less patenting per million dollars of sales than base case industries. When the analogue of logarithmic equation (2) is estimated, with the technology class variables introduced as intercept dummies,[18] the result is as follows:

$$(6) \qquad \log P_i = [7 \text{ constant terms}] + 0.882 \log S_{74i};$$
$$R^2 = .653, \; N = 223.$$

Although not as dramatic as the change between equations (1) and (5), the proportion of variance explained has more than doubled relative to equation (2) as a consequence of taking into account opportunity differences. These augmented industry of origin regressions show a good deal more explanatory power than the Schmooklerian regressions analyzing only the linkage between using industry demand and the flow of inventions from any industrial source favored by opportunity. The latter relationship is apparently attenuated by considerable noise.

V. CONCLUSION

The analysis here does some damage to Professor Schmookler's findings on the role of demand-pull. His detailed results replicate imperfectly when all manufacturing industry, rather than a small subset, is investigated and when industrial materials inventions are the focus. Nevertheless, at least for capital goods inventions, the main thrust of his theory survives. Although weaker than those obtained by Schmookler, the correlations between capital goods patenting and using industry investment are impressive. They persist not only for internal process inventions, but also for capital goods product inventions sold across industry lines. Markets work, both internally and externally, in transmitting demand-pull stimuli. Both the pull of demand and differences in technological opportunity, which determine the specific industries in which

[17] Very similar results are achieved when industry research and development spending rather than patenting is taken as the dependent variable.

[18] That is, where $P = \alpha S^{\beta}$, they modify the slope term α, which is analogous to the slopes in equation (5), but when logarithms are taken, is estimated as an intercept constant. If $\beta = 1$, regression (6) would, except in its error properties, be identical to a regression (5) with the intercept term constrained to zero.

inventive activity is concentrated, must be taken into account for an adequate conception of how technological change occurs.

NORTHWESTERN UNIVERSITY, EVANSTON ACCEPTED JULY 1981

REFERENCES

[1] Abernathy, William J., *The Productivity Dilemma: Roadblock to Innovation in the Automobile Industry* (Johns Hopkins University Press, Baltimore, 1978).

[2] Griliches, Zvi and Jacob Schmookler, "Inventing and Maximizing," *American Economic Review*, 53 (September 1963), pp. 725–29.

[3] Rosenberg, Nathan, "Science, Invention and Economic Growth," *Economic Journal*, 100 (March 1974), pp. 90–100.

[4] Schmookler, Jacob, *Invention and Economic Growth* (Harvard University Press, Cambridge, 1966).

[5] ———, *Patents, Invention, and Economic Change: Data and Selected Essays* (Zvi Griliches and Leonid Hurwicz, ed.) (Harvard University Press, Cambridge, 1972).

[6] Scherer, F. M., "Firm Size, Market Structure, Opportunity, and the Output of Patented Inventions," *American Economic Review*, 55 (December 1965), pp. 1097–1125.

[7] ———, "Market Structure and the Employment of Scientists and Engineers," *American Economic Review*, 57 (June 1967), pp. 524–531.

[8] ———, "Using Linked Patent and R&D Data To Measure Inter-Industry Technology Flows," paper presented at a National Bureau of Economic Research conference on R&D, Patents, and Productivity (October 1981).

[9] U.S. Bureau of the Census, *1972 Census of Manufactures*, vol. IV, "Indexes of Production," (Washington, 1977).

International Journal of Industrial Organization 1 (1983) 107–128. North-Holland

THE PROPENSITY TO PATENT*

F.M. SCHERER

Swarthmore College, Swarthmore, PA 19081, USA

This paper analyzes the relationship between 1974 R & D expenditures and invention patenting by 4,274 narrowly-defined lines of business in 443 U.S. industrial corporations. The probability of non-zero patenting rose systematically with company-financed R & D outlays, as did the number of patents received. Within industries, patents tended most frequently to rise in proportion to R & D; among the exceptions, the tendency was toward diminishing returns. Variables measuring overseas sales, federal R & D support, diversification, scope of invention use, and invention type had only modest explanatory power. There was no significant evidence of disproportionate patent accumulation in the more highly concentrated industries.

1. Introduction

Invention patents are of interest to economists for three main reasons. They help stimulate investment in innovation; they contribute to monopoly power; and they are a rich source of qualitative and quantitative information on technological change. This paper focuses on the third of these characteristics and, to a much lesser degree, on the second. Patents have both advantages and disadvantages for measuring the quantity of technological innovation [Kuznets (1962), Sanders (1962b), Schmookler (1966)]. On one hand, their temporal and technological coverage is nearly comprehensive, unlike case studies and counts of 'major innovations', and they imply some minimum standard of technical novelty, unlike raw statistics on research and development spending or employment. On the other hand, the quantity and quality of industrial patenting may depend upon chance, how readily a technology lends itself to patent protection, and business decision-makers' varying perceptions of how much advantage they will derive from patent

*This research was supported under National Science Foundation grant PRA-7826526. The author is indebted to Mary Gianos, Chun Yue Lai, Brett Spencer, Pin Tai, and William Cork for classifying the patents studied here; to Mr. Lai and Joe Cholka for computer programming; to Edward Klotz and Jim Hagerman for statistical assistance; and to Stephen Cosslett, Marc Nerlove, Paul Joskow, and anonymous referees for helpful comments.

Use is made here of data collected under the Federal Trade Commission's Line of Business program. FTC staff have reviewed the results and determined that individual company data are not disclosed. The conclusions drawn are the author's and not necessarily those of the Commission.

rights [Taylor and Silberston (1973, pp. 201–208, 245, 290–291)]. Not much of a systematic nature is known about these phenomena, which can be characterized as differences in the propensity to patent.

Specifically, the propensity to patent is measured here in terms of the number of patents industrial corporations obtain per million dollars of company-financed research and development (R&D) expenditures.[1] We investigate the closeness of the dependence between patenting and R&D expenditure, the extent of inter-industry and inter-firm differences in patent yield, and the degree to which differences in patent yield depend upon the type of technology, company characteristics, and market structure.

2. The data

The analysis is made possible by the linking of two uniquely rich data sets. One is the Federal Trade Commission's Line of Business survey for 1974, which required 443 corporations to report domestic sales, costs (including R&D outlays), and other financial variables broken down into some 276 standardized industry categories — 262 for manufacturing and 14 for non-manufacturing. See U.S. Federal Trade Commission (1981). The second data set covers 15,112 U.S. patents originating from U.S. resident inventors and issued to the same 443 corporations during the ten-month period from June 1976 through March 1977.[2] The lag of 28 months from the midpoint of the financial data period (June 1974) and the patent issuance period reflects two considerations. First, the lag between conceiving an invention and filing a U.S. patent application has been found to average nine months. See Sanders (1962b, p. 71, and 1962a).[3] Second, the average lag from patent application to issuance in the mid-1970s was 19 months. On this, there is an appreciable variance, with a few sample patents having been applied for during World War II and suspended until 1976 owing to national security. However, 92 percent of the sample patents had applications dated in the years 1974–1976 and 59 percent were applied for in 1975 alone.[4]

Each sample patent's printed specification was individually inspected to

[1]Discussants' comments reveal that this definition is not uniformly accepted. It is consistent, however, with the author's prior work [Scherer (1965b, pp. 1101–1102)] and the important study by Taylor and Silbertson (1973, p. 63).

[2]The total number of U.S. patents issued during this period was 83.27 percent of the number issued during the surrounding 12-month period. To annualize the sample, all patent counts are therefore multiplied by 1/0.8327=1.2009. Inquiries at the U.S. Patent Office revealed no significant clustering of patent issues by technological field or industry sufficient to make a ten-month sample unrepresentative.

[3]Similarly, on West Germany, see Grefermann et al. (1974, pp. 34–37). According to a referee, a 1982 paper by D. Bosworth and A. Westway shows a one-year lag between R&D and patent filing in the United Kingdom.

[4]To avoid this variance problem, one might attempt to match with R&D outlays patents classified by application date rather than issue date. However, this was infeasible.

extract such information as the industry of origin, industry(ies) in which the innovation's use was anticipated, economic characteristics of the invention (e.g., internal process vs. capital good product vs. consumer product), and technological characteristics. Of greatest immediate interest is the industry of origin assignment. The 443 sample corporations broke down their financial data (including R&D outlays) into a total of 4,274 individual domestic lines of business (LBs), or an average of 9.65 per company, spanning 249 usable categories (consolidated owing to disclosure and matching problems from the original list of 276). Each patent was manually linked to one or (for joint inventions) more company LBs, so that an unprecedentedly detailed matching of patents with R&D expenditures was achieved.[5]

The 443 corporations together expended $10.6 billion on company-financed research and development in 1974, or 73 percent of the figure for all U.S. corporations reported by the National Science Foundation from its annual survey. They obtained 59 percent of all invention patents issued to U.S. corporations during the sample time period and roughly 61 percent of patents issued to industrial corporations (i.e., excluding universities, non-profit research institutes, patent management firms, public utility enterprises, retailing companies, and the like). Thus, their share of patents was considerably less than their share of R&D, mirroring the previously demonstrated tendency for small U.S. corporations (uniformly excluded from the Line of Business sample) to obtain more patents per unit of R&D expenditure than large corporations. See Schmookler (1966, pp. 32–35) and Scherer (1980, p. 418).

Table 1 shows the distribution of annualized sample company patents, company-financed R&D expenditures, and patents per million dollars of R&D aggregated into 22 relatively broad industry groups. The mean yield of patents per million dollars of R&D is 1.70, corresponding to an average R&D outlay per patent of $588,000. There is considerable variation in patent yields — i.e., in the propensity to patent — across broad industry groups. The lowest yields are found in motor vehicles and other transportation equipment, with the aircraft industry next lowest and office equipment third. These are industries in which much of the R&D effort is devoted to the building and testing of prototypes — activities relatively unlikely to generate many patentable inventions. Also, in the automobile assembly segment of motor vehicles, much of the R&D expenditure has historically been devoted to styling innovation, whose patentable invention contribution is undoubtedly low.[6] The most patents per million dollars of R&D emerged in industrial and residential electrical equipment; stone, clay,

[5]For methodological details, see Scherer (1981).

[6]In automobile assembly, 0.28 patents were obtained per million dollars of R&D outlay. In automobile parts, on the other hand, which is more oriented toward developing new technology, the yield was 4.35 patents.

Table 1

R&D and patenting in broad industrial groups.

Industry group	Company R&D outlays (millions)	Total patents[a]	Patents per million dollars R&D
Food and tobacco products	$272	311	1.14
Textiles and apparel	65	147	2.62
Wood products and furniture	37	50	1.35
Paper products	150	292	1.95
Chemicals, excluding drugs	1,260	3.492	2.77
Drugs	449	868	1.93
Petroleum extraction and refining	360	1,046	2.91
Rubber and plastics products	287	637	2.22
Stone, clay, and glass products	149	477	3.20
Primary metals	239	424	1.77
Fabricated metal products	246	450	1.83
Machinery, except office	852	1,657	1.94
Computers and office equipment	1,054	1,045	0.99
Industrial electrical equipment (SIC 361, 362, 364, 367)	210	836	3.98
Household appliances	78	232	2.97
Communications equipment and electronic devices	1,136	2,384	2.10
Motor vehicles and other transportation equipment (SIC 371, 373, 374, 375, 379)	1,791	809	0.45
Aircraft and aircraft engines	653	501	0.77
Guided missiles and ordnance	103	173	1.68
Scientific and professional instruments	652	1,351	2.07
Miscellaneous manufacturing	279	514	1.84
Service industries (including R&D services)	184	208	1.13
All industries combined	$10,641	18,088	1.70

[a]Sample patents have been multiplied by 1.2009 to put them on an annualized basis comparable to the R&D expenditure data.

and glass products; and household appliances, where the high propensity to patent may have more to do with firms' incentives to seek patent protection than with the intrinsic ease of coming up with patentable inventions. On these differences, more will be said later.

3. Patenting and the level of R&D

We proceed now to analyze the association between patenting and the level of R&D expenditure.

One strong result is that patenting, although not barred to part-time corporate inventors,[7] is closely associated with formally organized R&D

[7]Compare Schmookler (1957), who concluded that full-time inventors accounted for less than half of the inventions studied in a 1953 mail survey.

activity. Among the 4,274 individual company lines of business, 1,851 had non-zero patenting activity. R&D expenditures were reported by 3,003 LBs. Of the 1,271 lines with no R&D, 1,111 also received no patents. Thus, 160 lines received patents but reported no R&D.

Griliches and associates have suggested that patenting by companies can be modelled as a Poisson process in which patented inventions appear only rarely for firms spending little or nothing on R&D, but where the probability of patenting rises with R&D expenditures. See Bound et al. (1981) and Hausman et al. (1981). Fig. 1 provides insight into one facet of this hypothesis. Sample LBs were arrayed in ascending order of company-financed R&D expenditure, and the relative frequency of lines with non-zero patenting was tabulated. The plotted points and fitted smooth curve reveal a steady rise in the probability of patenting as R&D increases. Only 12.6 percent of the LBs reporting no R&D received patents. The probability of patenting increases sharply when R&D outlays exceed $300,000 per year, approaching unity as outlays exceed $50 million. Attempts to discriminate among alternative stochastic processes generating the observed plot of points met with ambiguous success. Of three distributions tested, a logistic curve fitted the data least well. A two-parameter Weibull function, which results from a cumulative Poisson process dependent upon R&D outlays, had the best fit at low levels of R&D (where the Poisson hypothesis is most plausible),[8] but was inferior to a log normal distribution at intermediate and high levels of R&D.

For the analysis of patenting *levels*, the presence of many zero observations suggests the use of nonlinear methods such as Poisson or Tobit regressions. Poisson methods were inadequately developed at the time the analysis here was carried out; they are being extended by Hausman et al. (1981). The dependence of patenting on R&D was investigated using bivariate (and, with explanatory variables motivated later, multivariate) Tobit analysis, but with unsatisfactory results. The Tobit estimates at low R&D levels were more severely biased than ordinary least squares regressions, apparently because of sensitivity to a non-normal error term distribution. Therefore, ordinary least squares was used in the analysis that follows despite some risk of specification error. Sensitivity tests showed no substantial difference in hypothesis test results with the elimination of zero-valued patent observations or the substitution of Tobit for OLS methods.

The simple OLS regression of patenting P on R&D (RD) (in millions of dollars) for all 4,274 company lines of business was as follows:

$$P = \underset{[t=15.3]}{2.83} + \underset{[t=62]}{0.562\,RD}, \qquad r^2 = 0.487, \quad \sigma_{P \cdot RD} = 12.02. \tag{1}$$

[8]The fitted Weibull function was $\Pr[Pat > 0] = 1 - e^{-0.0104 RD^{0.647}}$, where RD is measured in thousands of dollars. I am grateful to Jim Hagerman for his help on this analysis.

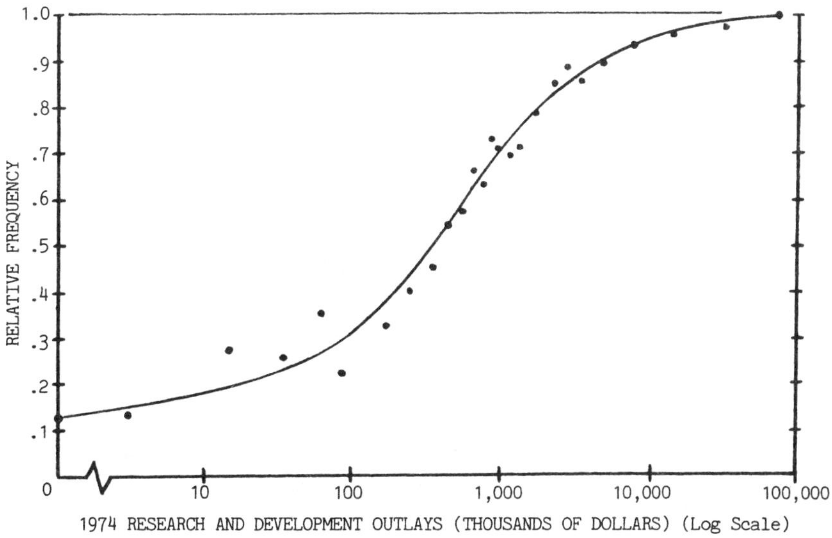

Fig. 1. Probability of non-zero patenting in a line of business as a function of R & D outlays.

Here, as in later regressions, *t*-ratios are reported in brackets. Nearly half of the variance in patenting levels is 'explained' by differences in the level of R&D outlays.[9]

What accounts for the 'unexplained' variance was investigated in three main steps: (1) through an analysis of differences in technology classes; (2) by isolating the effects of inter-industry differences; and (3) by introducing additional systematic explanatory variables. The first two analyses are undertaken in this section, the third in section 4.

3.1. Technology class and industry differences

An attempt was made to account for differences in the propensity to patent at a high level of generality by classifying each industry into one of eight technology classes: the 'traditional' industries such as staple foods, weaving, and Portland cement; organic chemicals; other chemicals; electrical equipment; electronic equipment and components; the metallurgical industries; nonmanufacturing fields, and a base case consisting mostly of industries oriented around mechanical technologies. These classifications, which are a modest extension of earlier work by the author [Scherer (1965b,

[9]In all regressions reported here, companies' basic and central laboratory research outlays were spread over other R&D outlays in proportion to the latter. The simple r^2 between patenting and R&D without such allocations was 0.492. Inspection revealed that many patents resulted from such central research, although an exact split between basic and applied R&D and their associated patenting was impractical owing to substantial differences in the ways companies chose to report their data.

1967)], were completed before the FTC R&D data were inspected. The technology classes were coded as zero-one dummy variables, and each individual company LB was assigned the appropriate technology class dummy.

When both intercept and slope dummy variables (the latter interacting multiplicatively with the R&D expenditure values) were introduced, R^2 rose from 0.487 in bivariate regression (1) to 0.613, with an incremental F-ratio of [(0.126/14)/(0.387/4258)]=99. Plainly, there are substantial differences among technology groups in the dependence of patenting on R&D. Slope values ranged from 0.4 patents per million R&D dollars in the mechanical technology group to 1.83 in electrical equipment.

It was evident, however, that there were also appreciable differences in patenting behavior within the technology groups — e.g., as noted earlier, between automobile assembly and automobile parts. Therefore, each FTC code industry with at least five patent-receiving constitutent company lines of business was permitted to have its own best-fitting linear slope and intercept parameters. The five-or-more criterion was satisfied for 124 of the 249 usable industry categories.[10] Industries with fewer than five patent-receiving LBs (accounting in total for 7 percent of sample patents) were consigned to a common residual group. Through this approach, 250 separate regression parameters were in effect estimated. The median slope value among five-or-more subset industries was 1.31 patents per million dollars of R&D. For the pooled industries, the single estimated slope was 1.66. Only four industries had negative slope terms, all insignificantly different from zero. The R^2 resulting from this regression allowing inter-industry differences to manifest themselves was 0.853. Relative to the regression with technology class slope and intercept dummy variables alone, the 0.240 increase in explained variance implies a highly significant F-ratio of 27.8. The near doubling of R^2 as one moves from bivariate regression (1) to estimating 125 separate linear regressions suggests that inter-industry differences in the propensity to patent are of roughly the same quantitative importance as differences in the level of patenting associated with varying R&D levels. The high R^2 value implies that rather little variation in patenting is left to explain once inter-industry differences and individual LB R&D levels within industries are accounted for.

3.2. *Nonlinearities and the 'Schumpeterian' hypothesis*

It remains open whether the relationship between R&D and patenting within industries is in fact linear as assumed thus far. The question is of considerable interest in part because of the light it helps shed on debate over

[10]The five-or-more criterion was adopted to maintain at least two degrees of freedom in quadratic regressions reported later.

the so-called 'Schumpeterian' hypothesis, i.e., that large business units are in some sense more progressive technologically than smaller firms. This in turn, as Kohn and Scott (1982), modifying the model of Fisher and Temin (1973), have shown, hinges on three distinguishable questions: whether large firms support more R&D in relation to their size than small firms; whether the R&D of large firms is more 'productive' in yielding inventions and innovations; and whether the R&D output of large companies enjoys wider application than its small-firm counterpart. The analysis in this section and portions of the next is relevant mainly to the R&D 'productivity' question. Specifically, do large-scale R&D programs yield disproportionately more or fewer patented inventions than smaller programs, other variables such as industry characteristics held constant? The R&D support question is addressed using the same data in a separate paper [Scherer (1982)].

The theoretical discourse over the Schumpeterian hypothesis has focused on elasticities — e.g., the elasticity of inventive output with respect to R&D input; and most empirical studies have derived elasticities by estimating levels regressions in the logarithms. This empirical approach has, as I cautioned in another article (1965a), several serious difficulties. For one, as we have seen, R&D inputs and outputs often have zero values, and the arbitrary convention one adopts to get around the 'logarithm of zero' problem can have a profound effect on the results, changing estimated relationships from elastic to inelastic or vice versa. Second, as we have also seen, companies can obtain patents without doing R&D, so the assumption of a zero *P*–R&D intercept and hence a homogeneous input–output relationship may be violated. Third, estimated output–input regressions sometimes exhibit points of inflection and other violations of the homogeneity assumption. Fourth, logarithmic transformations tend to place maximum weight on the more numerous smaller firms, whereas the Schumpeterian hypothesis is normally taken to emphasize the behavior of the (relatively few) very large enterprises. There is no simple escape from these difficulties, but it is clear that logarithmic regressions are unacceptable and that some other nonlinear method must be used.

The principal approach adopted for our new data was to add quadratic terms to the 124 linear industry-by-industry regressions of patenting on R&D. Compared to the analogous disaggregated linear regression analysis, there was a small but statistically significant increase in the amount of variance explained. R^2 rose by 0.036 to 0.889, with an incremental F-ratio (124/3900 d.f.) of 10.16. Negative quadratic coefficients (reflecting diminishing marginal patent output as the scale of R&D increased) outnumbered positive coefficients by 73 to 51. Forty-five of the individual industry quadratic coefficients passed incremental F-ratio tests at the 5% level; 26 of them were negative and 19 positive. Among the industries with significant negative coefficients, the curvature was sufficiently pronounced that 21 exhibited local maxima at R&D levels lower than the expenditures of the largest R&D performer. In no case, however, did the maximum occur at an R&D level less than twice the mean level of R&D spending per industry member.

If intercept coefficient values were zero, regressions with negative quadratic terms would unambiguously imply an elasticity of patent output with respect to R&D input of less than 1; positive quadratic terms would imply an elasticity greater than 1. However, the zero-intercept condition is not always satisfied. Of the 124 quadratic regressions, 10 had intercepts significantly different from zero at the 5% level in a two-tailed test. Only one of these was negative. Changes from an initially elastic regime at low R&D levels to an inelastic regime can occur when the intercept is negative and the quadratic term is also negative. For the one case of this type, the transition from elastic to inelastic occurred at 11 percent of the largest LB's outlay. When both intercept and quadratic terms are positive, the patent output elasticity can change from inelastic at low size levels to elastic. Three industries had significant positive intercept and quadratic terms. In all three, the transition from inelastic to elastic occurred at an R&D level not exceeding 61 percent of the leading LB's outlay. Taking into account these complications, we can classify our 124 industries as follows on the elasticity of patent output with respect to R&D input:

	No. of industries	*Percent*
No significant departure from constant returns	74	59.7
Increasing returns ($E > 1$)		
Throughout	16	15.3
At higher R&D levels	3	
Diminishing returns ($E < 1$)		
Throughout	30	25.0
At higher R&D levels	1	

Thus, the preponderant pattern is toward essentially constant returns in the patent output — R&D input relationship. To the extent that there are deviations, they tend to be more on the side of diminishing rather than increasing returns.

It is worth recalling that the characteristically small corporations excluded from the Line of Business sample obtained a share of patents disproportionate to their share of U.S. industrial R&D. This also runs contrary to an increasing returns hypothesis. However, this evidence on patent output — R&D input relationships could have either of two rather different interpretations: that the largest firms in an industry generate fewer patentable inventions per dollar of R&D than their smaller counterparts, or that they choose to patent fewer inventions of a possibly proportionate or even disproportionate inventive output. Subsequent analysis will add limited insights on this critical point. Meanwhile, one further strand of outside evidence is immediately relevant. Earlier surveys have shown that the largest corporations commercially utilize a smaller fraction of their patented inventions than do relatively small corporations, which implies possible qualitative inferiority of the largest firms' 'typical' inventions. See Sanders (1964) and (on West Germany) Grefermann et al. (1974, app. table 57). Taken together, the evidence at least compels skepticism toward the neo-Schumpeterian argument that giant-scale enterprises are uniquely effective engines of technological invention.

4. Determinants of propensity to patent differences

We probe more deeply now into the reasons why some industries and firms obtain more patents per unit of R&D outlay than others. Obviously, there can be numerous ad hoc reasons such as differential creativity of an organization's R&D scientists and engineers or differing beliefs about the value of patent protection. Here we test hypotheses concerning more systematic and quantifiable determinants of the propensity to patent.

One involves the role of government R&D support. In addition to $10.6 billion of company-financed R&D, sample companies conducted in 1974 $5.97 billion of contract R&D, most of it associated with defense and space applications. It might be thought that such outside R&D support would enhance the patent-generating power of companies' internal R&D programs. However, it is at least as plausible that enterprises heavily engaged in federal R&D contracting have low propensities to patent because the government insists upon either title or a royalty-free license to any inventions made under its contracts[11] and because it can legally infringe upon privately-developed patents for national defense purposes, subject to the payment of 'reasonable' (and typically modest) royalties. We test these alternate hypotheses by introducing a variable FED, which measures the ratio of contract R&D outlays to total (company-financed plus contract) outlays.

Government contract work tends also, although not uniquely so, to be 'systems' oriented — that is, entailing the development of complex aggregations on which it is difficult to get strong patent protection, and for which there is extensive prototype testing that adds greatly to R&D costs without yielding many patentable inventions. Sample patents were classified as to whether their underlying inventions were best described as systems or subsystems, much simpler devices, circuits, compositions of matter, processes, or methods. Lines of business with a preponderance of systems or subsystems

[11] Of the 15,112 patents in our sample, only 325 were identified as resulting from federal contracts. Incomplete Patent Office and contracting agency records make this an underestimate; roughly 75 military contract-originated inventions are believed to have eluded the search.

inventions proved to generate fewer patents per million dollars of R&D than the sample as a whole: 1.41 as compared to 1.70. We shall use in subsequent tests a variable SYST measuring the proportion of an LB's patent portfolio consisting of systems and subsystems inventions. Similar variables were constructed for other invention types and will be described subsequently.

Another hypothesis is prompted by a peculiarity in the FTC R&D data set. Companies were required to provide line of business breakdowns only for their domestic U.S. unregulated activities. Foreign operations were exempted. Some companies have R&D establishments in the United States that support manufacturing operations abroad. They were permitted to prorate domestic R&D costs between domestic sales (covered by LB reports) and foreign sales (excluded). Their domestic patents, however, were counted in full under our methodology.[12] One would expect the number of patents received per million R&D dollars to be greater, therefore, the larger foreign sales are relative to domestic unregulated sector sales. The variable *PDOM* used to test this hypothesis measures domestic LB sector sales as a fraction of total company sales. Since it approximates one minus the foreign sales ratio, we expect it to have a negative sign.

Under a hypothesis first advanced by Nelson (1959) and extended by others, company diversification might affect inventive activity favorably in a number of ways. Most relevant to our current focus, more diversified enterprises might have higher propensities to patent either because cross-fertilization among diverse industry technologies enhances the fecundity of R&D or because a diversified company is better able to commercialize off-beat inventions and hence is more likely to seek patent protection for them. To measure sample companies' diversification, we compute the Herfindahl numbers equivalent index over sales in a company's unregulated domestic lines of business. That is, where S_i is a company's 1974 sales in the ith of its n lines of business, $DIVERS = 1/\sum_{i=1}^{n} S_i^2$. The five most diversified companies in the sample had diversification indices averaging 19.8. The median company's numbers-equivalent diversification index was approximately 3.0.

The structure of the markets in which sellers operate might also affect the propensity to patent. There are two plausible linkages. For one, firms in concentrated industries, and especially the leading sellers, may attempt to retard entry into their lines by erecting a fence of accumulated patents. This implies a high propensity to patent for the largest firms in concentrated markets. A more benign conjecture with similar quantitative implications is that firms in concentrated industries invest more in securing patent protection because they can expect to appropriate a higher share of the rewards from exclusively exploiting the inventions they develop. Given the

[12]Patents derived from overseas R&D activity did not appear in our sample because only inventions made by U.S. resident inventors were included. Approximately 950 inventions assigned to sample companies by foreign resident inventors were excluded.

possibility of licensing and the greater threat of rapid imitation in less concentrated industries, however, it is easy to visualize the appropriability incentive working in the opposite direction — i.e., toward higher propensities to seek patent protection in less concentrated industries. The relevant test variable will be a weighted average index CR4 (in ratio form) of 1972 four-firm seller concentration in meaningfully defined components of the FTC line of business categories, where the components are measured at either the four- or five-digit S.I.C. level, depending upon the degree of product substitutability.[13] No correction is made for geographic markets, since both the exclusionary and appropriability hypotheses are of doubtful relevance to isolated local markets.

A final set of hypotheses was motivated by conceptual difficulties encountered in developing a matrix tracing the flow of technology from industries of origin to industries of use. See Scherer (1981). Some inventions serve a narrow set of demands, e.g., only a single using industry. Others (such as new computers, truck components, and general-purpose adhesives) tap the demand of many industries. In virtually all models of inventive activity that assume profit maximization, the size or scope of the potential market plays a key role. Fig. 2 portrays the implications of varying market scope. The horizontal axis measures 'the extent of invention', which may entail any of several phenomena such as the percentage cost reduction achieved through a new process,[14] the reliability of the resulting product, the number of special features or variants a product has, etc. There is a cost function relationship $C(\Delta)$ between required R&D outlays (vertical axis) and the extent of invention Δ. Normally, one expects this function eventually to exhibit diminishing returns, manifested in increasing marginal costs, although when the extent of invention variable measures the number of product features or variants, there may be a protracted stage of essentially constant marginal cost. The quasi-rents $Q(\Delta, N)$ realized from innovating also depend upon the extent of invention as well as upon the intrinsic scope of the market, denoted here in terms of N. Intuition suggests the existence of diminishing marginal returns in the relationship between quasi-rent realizations and the extent of invention; i.e., $\partial Q/\partial \Delta > 0$ and $\partial^2 Q/\partial \Delta^2 < 0$. Here too, however, exceptions can be imagined, e.g., when market potential increases greatly once a certain quality threshold is exceeded. Increases in the number of using industries N should shift the Q function upward and increase its slope; that is, $\partial^2 Q/\partial \Delta\, \partial N > 0$. Given this, an increase in intrinsic market scope moves the profit-maximizing choice to a larger value of Δ and higher R&D outlay level, as illustrated by the alternative Δ optima OA and OB in fig. 2. If the

[13]The determination of which industry definitions best approximated market realities was initiated by Leonard W. Weiss, with some subsequent modifications by the author. I am indebted to Professor Weiss for his contribution.

[14]See Nordhaus (1969, ch. 5), from which this formulation is adapted.

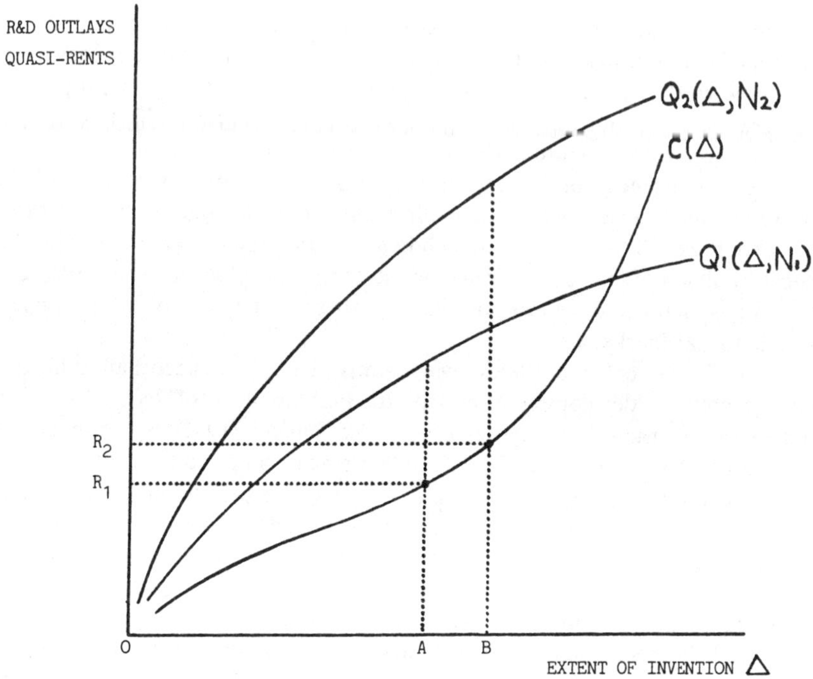

Fig. 2. Profit-maximizing extent of invention with increasing market scope.

cost relationship exhibits continuously diminishing returns, this necessarily implies an increase in the average R&D cost per unit of invention Δ. If patents are proportional to Δ, and a fortiori if patenting expands less than proportionately with Δ, this in turn implies a rising average cost per patent as the scope of the market increases. However, this result is not necessary. With a stage of increasing or constant returns in the R&D cost function *and* diminishing returns in the quasi-rent relationship, the change in optima as N increases could occur within a range of constant or even falling average cost.

Tests of the 'rising average cost with wider scope' hypothesis are permitted by the fact that our sample patents are classified by industry(ies) of use as well as industry of origin. The classification scheme includes five industrial use possibilities: one specific industry of use, two users, three users, general use distributed in proportion to the origin industry's input–output table sales, and ubiquitous use (e.g., small trucks or copying machines). For the inventions classified as having specific industries of use, the average number of using industries was 1.4. For general-use inventions, the median Herfindahl numbers equivalent index (measuring in effect the number of industries, assumed to be of equal size, in which an invention's use was expected to occur) was 8.8; the mean value was 13.4. For ubiquitous use inventions, the numbers equivalent was even higher at 23.9. We hypothesize that the lowest yield of patented inventions per million dollars of R&D (or the highest R&D cost per invention) will be in LBs with the highest proportion of ubiquitous-use inventions (*UBIQ*) and the next lowest yield in LBs with a high fraction of general use inventions (*GEN*). Specific industry of use inventions comprise the base case unless otherwise indicated.

4.1. Regression model specification

Two of the explanatory variables proposed here, *PDOM* and *DIVERS*, are company-specific. The *CONC* variable is industry-specific; i.e., a set of industry dummy variables would characterize it completely. Most of the variables lie somewhere in between; e.g., some industries such as aircraft, guided missiles, and communications equipment have much higher values of *FED* and *SYST* than others, but there are also appreciable intra-industry differences. Complicating the problem is the existence of collinearity between industry and measured explanatory characteristics. Thus, the auto industry is very large, has unusually high concentration, and also exhibits an unusually low patent yield per million R&D dollars. As such, its company lines of business exerted a disproportionate impact on some analyses, but it is questionable whether this reflects systematic causal relationships as distinguished from simple coincidence. Such collinearity problems led to instability in some coefficient estimates. To guard against spurious inferences, two different analytic approaches were pursued, and these were supplemented by special tests when appropriate.

One main approach controlled for inter-industry differences only to the extent that they were associated with the seven technology class dummy variables discussed in section 3. The level of patenting *P* for a line of business was regressed on the level of R&D expenditure *RD* plus technology class slope dummy variables, as in section 3,[15] and the new explanatory variables were introduced as multiplicative interaction terms with the R&D variable, e.g., for *SYST*, as $SYST \cdot RD$.[16] There were several grounds for adopting this specification. For one, the distribution across LBs of patenting per million R&D dollars was skewed, and an analysis using *P*/*RD* in raw form would have been dominated by the extreme values associated with lines of business doing little R&D but having finite patenting. The 'levels with interaction variable' approach, in contrast, weights LBs in proportion to

[15]Intercept dummy variables are suppressed owing to their low explanatory power and the lack of a priori theory as to why intercept values should differ among technology classes. The hypothesis tests were unaltered by the inclusion of intercept dummies.

[16]Except in its error term characteristics and weighting, the model estimated here, i.e., $P = a + b_1 RD + \cdots + c_1 X_1 RD + \cdots + \varepsilon$, is nearly equivalent to the ratio model: $P/RD = a/RD + b_1 + \cdots + c_1 X_1 + \cdots + e$, where X_1 is a characteristic whose effect on the propensity to patent is to be tested.

their R&D, thereby emphasizing quantitatively important relationships. Also, the interaction approach implies higher estimated changes in LB patenting associated with given changes in some characteristic for large lines of business than for small. This appeared more appropriate than using non-interacting variables, which implies equal effects across all LB sizes.

An alternative, quite different approach was to eliminate industry effects to the maximum possible extent by letting, as before, each industry with five or more patent-receiving LBs have its own best-fitting R&D coefficient. With 250 parameters to be estimated for the R&D variables alone, this posed difficult computational problems. They were circumvented by exploiting the property that the correlation between the residuals of an $(N-1)$-order regression and an Nth variable is the same as the partial correlation of the Nth variable with the dependent variable in an N-order regression, holding the other $N-1$ independent variables constant. Thus, a separate linear regression of P on RD was computed for each of 124 industries (plus one pooled regression for the industries with four or fewer patent-receiving LBs). The residuals from these 125 regressions were calculated, and they were taken as dependent variable in regressions on the additional explanatory variables identified in this section. With all or (given the pooled industry regression) most of the inter-industry variance eliminated in the first stage, these 'residual' regressions place maximum emphasis on inter-firm effects.

Individual patent characteristics data (averaged within LBs) were available only for 1,819 LBs with patents.[17] Therefore, analyses of those characteristics had to be performed using only the 1,819 observation data subset.

4.2. Hypothesis test results

Table 2 summarizes the principal regression results. It includes four regressions — two each in the patent levels—R&D interaction form and the industry regression residuals form, each for the full 4,274 LB sample (less six LBs with zero sales) and the truncated 1,819 sample including only LBs with patents.

The results vary little between the full and truncated samples. However, there are major differences between interaction regressions (2) and (3), which control only for eight technology class differences, and regressions (4) and (5), which control more fully for individual industry differences. Many variables with coefficient estimates significantly different from zero in regressions (2) and (3) turn insignificant and/or change signs when industry effects are taken into account. Only the *PDOM* and *DIVERS* variables plus a logarithmic

[17]Thirty-two other LBs counted in fig. 1 as having non-zero patenting were dropped because they received zero weight in an R&D-weighted allocation procedure for patents with multiple industries of origin.

Table 2

Multiple regression results.[a]

Variables	Regression (2) Levels with interaction, $N=4{,}268$	Regression (3) Levels with interaction, $N=1{,}819$	Regression (4) Residuals, $N=4{,}268$	Regression (5) Residuals, $N=1{,}819$
R&D slope and intercept terms	9 parameters estimated but not reported		250 parameters estimated to obtain residuals	
FED	−0.739 [14.32]	−0.459 [3.38]	+1.68 [2.84]	+2.80 [2.45]
PDOM	−0.823 [19.29]	−0.351 [2.34]	−2.32 [3.84]	−4.46 [3.34]
DIVERS	+0.041 [9.44]	n.e.	+0.055 [2.56]	+0.165 [3.19]
CR4	−2.03 [25.67]	−3.35 [16.82]	−2.17 [0.41]	−1.80 [1.44]
SYST	n.e.	−0.632 [3.38]	n.e.	−0.681 [1.02]
COMPO	n.e.	−0.939 [3.73]	n.e.	−0.694 [0.81]
CONS	n.e.	+1.083 [5.42]	n.e.	−0.285 [0.30]
GEN	n.e.	+0.584 [4.39]	n.e.	−0.740 [0.98]
UBIQ	n.e.	+0.316 [0.90]	n.e.	−0.372 [0.33]
Log_{10} *SALES*	+2.20 [10.26]	n.e.	+1.07 [6.81]	+1.66 [4.31]
R^2	0.684	0.672	0.019	0.026
Increase in R^2 compared to regression with *RD* only	0.197	0.079	—	—
Incremental *F*-ratio	221.3	54.44	16.73	4.46

[a]n.e. = not estimated.

SALES variable introduced to control for nonlinearities perform consistently and significantly.

The *PDOM* variable's significant negative coefficient is in accord with expectations: companies whose operations are preponderantly domestic obtain fewer patents per million dollars of R&D than companies with substantial overseas operations, to which some domestic R&D expenses may have been allocated. For the interaction regressions, the elasticity of patenting with respect to the fraction of sales in domestic LBs lies in the range of −0.42 to −0.14. The coefficients of regressions (4) and (5) can be interpreted as indicating that moving from having no overseas or regulated sector sales ($PDOM = 1.0$) to $PDOM = 0$ is accompanied by an increase of from 2.3 to 4.5 patents per million dollars of R&D. The coefficient for regression (5) is higher than that for regression (4) because regression (4) includes 2,449 LBs with no patenting.

Diversification appears to be unambiguously stimulative to patenting. The estimated coefficients suggest that a unit increase in the numbers equivalent diversification index, which would be a 16 percent increase over the mean level of diversification, is accompanied by a 3 to 9 percent increase in the average patent yield per million dollars of R&D.

The consistently positive coefficients on log *SALES* might seem inconsistent with our finding in section 3 of a tendency toward diminishing returns in the patents–R&D relationship. The puzzle is easily resolved, however. When the level of patenting increases less than proportionately with R&D, fitting a linear regression of patents on R&D will leave positive residuals for LBs of intermediate size and negative residuals for LBs of small and large scales. Since there are very few LBs in the largest-scale tail of the size distribution and since taking logarithms compresses measured size differences between those LBs and units of intermediate size, the log *SALES* coefficients estimated here are dominated by nonlinearities in the small to medium size range, where the progression is from negative to positive residuals as scale increases. Consistent with this explanation, when regressions analogous to those of table 2 were run for R&D on *SALES*, either by technology class as in regression (2) or by taking the residuals from linear industry regressions, the log *SALES* coefficients were significantly negative, whereas quadratic industry-by-industry R&D on *SALES* regressions showed a tendency toward increasing returns. See Scherer (1982).

The variable *FED*, measuring (mostly) federal contract R&D outlays as a fraction of total R&D, is significant in all cases, but its coefficient changes sign from the technology class to the industry-by-industry regressions. Evidently, industries with a strong federal R&D orientation seek fewer patents, all else equal. But within particular industries, LBs with relatively more federal funding obtain more patents, possibly (but speculatively) because the invention-generating potential of their company-financed R&D is supplemented.

The most powerful explanatory variable in regressions (2) and (3) is the four-firm concentration ratio, but it falls to insignificance when one controls for individual industry effects. Since concentration is industry-specific, this leaves unresolved the question of whether concentration's strong role in the technology class regressions is behaviorally meaningful or spurious. That there is a substantial spurious or idiosyncratic component is suggested by the fact that when auto and truck industry LB observations are deleted from the sample, the *t*-ratios for *CR4* drop sharply into the neighborhood of 4.0. In an attempt to achieve further insight, two additional tests were conducted. First, variations in the slope coefficients from the 124 individual industry patents on R&D regressions reflect systematic inter-industry propensity to patent differences. For the 119 manufacturing industries on which such regressions were computed, the estimated slope coefficients were regressed on *CR4* plus intercept dummy variables for the traditional, organic chemical, other chemical, electrical and electronic technology classes. The coefficient for *CR4* was negative, as in the table 2 regressions, but with a *t*-ratio of only 1.62, which is significant in a two-tailed test at only the 0.12 level. To the extent that significance can be ascribed to the persisting negative sign, the implications contradict the hypothesis that patenting per million R&D dollars is systematically *higher* in highly concentrated industries, whose members are best able to capture monopoly benefits from blocking entry through patent proliferation. Second, it seems reasonable to suppose that the prospect of gaining monopoly rewards through patent proliferation mainly affects the largest sellers in concentrated industries. If so, one would expect to find a positive correlation between *CR4* and the extent of convex-downward nonlinearity in the individual industry *P*–R&D regressions. This hypothesis was tested by regressing on *CR4* plus five technology class dummies the normalized quadratic β term (i.e., with variables deflated by their standard errors) from the 119 manufacturing industry quadratic regressions. The resulting *t*-ratio for *CR4* was +0.10, which falls far short of statistical significance. All-in-all, the analyses provide no support for the hypothesis that firms in highly concentrated industries systematically accumulate patents disproportionately to their R&D efforts in an attempt to forestall entry. This does not, of course, rule out the possibility of unsystematic entry-deterring behavior in isolated cases.

Two variables, *SYST* (measuring the fraction of an LB's patents pertaining to systems or subsystems) and *COMPO* (measuring the fraction covering chemical compositions of matter) were introduced into the restricted sample regressions to test the effect of invention type. Both were significant in the interaction regression but fade to insignificance in residuals regression (5), indicating that the effects do not persist within industries. The *SYST* variable, implying as predicted a relatively low yield of systems inventions per million R&D dollars, is of special interest because of its possible correlation with firm size. The elasticity of patenting with respect to it in regression (2) is modest (−0.16), as is its incremental contribution to R^2 (less than 0.005). Further insight into the size question is obtained by regressing *SYST* on seven technology class dummies plus log *SALES* (in millions of dollars), as follows:

$$SYST = 0.45 + 0.028 \log_{10} SALES - 0.26\ TRADITIONAL$$
$$[t=15]\quad [1.98] \qquad\qquad [13]$$

$$-0.47\ ORGANICS - 0.41\ CHEMICALS$$
$$[15.7] \qquad\qquad [13.7]$$

$$+0.13\ ELECTRICAL + 0.03\ ELECTRONICS$$
$$[3.3] \qquad\qquad [1.0]$$

$$-0.27\ METALLURGICAL - 0.0079\ NONMFG;$$
$$[6.8] \qquad\qquad [0.16]$$

$$R^2 = 0.215, \qquad N = 1{,}819. \tag{6}$$

The proportion of systems inventions does increase as expected with LB size. But the relationship is weak. The t-ratio is only 1.98, and increasing LB sales from \$100 million to \$1 billion leads to only an increase from 51.0 to 53.8 percent in the fraction of base case inventions that are systems or subsystems. Not surprisingly, the highest incidence of systems and subsystems inventions was in electrical equipment, electronics, and the mechanical technology base case. The chemicals and traditional industries have particularly few systems inventions.

To test the scope of use hypotheses, the general input–output use variable *GEN* and ubiquitous industrial use variable *UBIQ* were introduced along with a variable *CONS* measuring the fraction of inventions whose use was solely by end consumers. Contrary to expectations, *GEN* has a positive and significant coefficient while *UBIQ*, which should have been most strongly negative, is insignificant in technology class interaction regression (3). Both the *GEN* and *CONS* effects fade to insignificance when individual industry effects are taken into account [regression (5)]. To explore further the complex relationships that underlay these results and others revealing considerable *GEN* and *UBIQ* coefficient instability, the average amount of R&D per patent by LB was computed for 15,063 individual patents, and a two-way analysis of variance was executed, with eight technology classes as one basis of classification and *CONS*, *GEN*, *UBIQ*, and the three separate specific industry-of-use categories as the other.[18] Ignoring technology class effects, *GEN* inventions had the highest average cost, with the following hierarchy of

[18]To avoid extreme value problems, 49 inventions whose average R&D outlay exceeded \$5 million are excluded. The results are similar when those inventions are included.

values:

GEN	$722,000
2 Uses	587,000
CONS	530,000
UBIQ	493,000
3 Uses	485,000
1 Use	474,000

Although the ordering hardly suggests any consistent tendency for inventions of broader use to be more intensively developed, the category differences are highly significant statistically, with an F-ratio of 60.15. As one would expect, technology class differences had even higher F-ratios. Two-way tests revealed appreciable interactions, with an F-ratio on the cross-effects of 26.06. *UBIQ* inventions, hypothesized to entail the most R&D per patent, fulfilled expectations in electronics but were *least* costly in four other technology classes. The *GEN* inventions had the highest R&D in the mechanical industries and the second highest in four other classes, but they had the second-lowest average R&D backing in the traditional technologies. Single industry-of-use inventions had the lowest average R&D cost only in 'other chemicals'. They were second lowest in organic chemicals and the mechanical industries and second highest in electronics. Clearly, support for the 'rising average cost with wider scope of use' hypothesis is at best weak and erratic. This is disappointing from the standpoint of finding new supported hypotheses. However, the lack of systematic relationships between scope of use and the amount of R&D undertaken greatly simplifies the construction of technology flows matrices based upon linked patent and R&D data. See Scherer (1981).

More generally, the explanatory power of the special variables introduced to test for propensity to patent differences is modest. Most of the systematic variance in individual line of business patenting appears to be associated not with measurable company, product, or use characteristics, but with the level of R&D effort, differences in technology, and differences in opportunity and incentive peculiar to individual industries.

5. Conclusion

Although puzzles remain, our analysis of matched patent and R&D data sheds considerable light on the determinants of business units' patenting. Industrial patenting is closely associated with the conduct of research and development. The more R&D a business unit does, the more likely it is to obtain patents and the more patents it secures. Within individual industries, patenting most commonly rises roughly proportionately with R&D effort. For the minority of industries with significantly nonlinear input–output relationships, the tendency is more often toward diminishing than increasing returns, contrary to what would be found if the largest firms had especially productive R&D operations. Company diversification appears mildly conducive to patenting. There is no persuasive evidence of systematic patent proliferation in the more highly concentrated manufacturing industries. The propensity to patent is higher in industries (but not necessarily across firms within industries) lacking substantial federal government R&D support. Systems invention patents, which have a slightly higher incidence in larger business units, tend to be backed by larger R&D expenditures. The links between the scope of an invention's industrial uses and the intensity of underlying R&D effort are complex and probably without much of a systematic component.

References

Bosworth, D. and A. Westaway, 1982, The influence of demand and supply side pressures on the quantity and quality of inventive activity, Occasional research paper no. 64, Jan. (Loughborough University of Technology, Department of Economics).

Bound, John, C. Cummins, Z. Griliches, B.H. Hall and A. Jaffe, 1981, Who does R&D and who patents? Paper presented at a National Bureau of Economic Research conference on R&D, patents, and productivity, Oct.

Fisher, F.M. and P. Temin, 1973, Returns to scale in research and development: What does the Schumpeterian hypothesis imply?, Journal of Political Economy 81, 56–70.

Grefermann, K., K.H. Oppenlaender, E. Peffgen et al., 1974, Patentwesen und technischer Fortschritt (Schwartz, Goettingen).

Hausman, J., B. Hall and Z. Griliches, 1981, Econometric models for count data with an application to the patents–R&D relationship, National Bureau of Economic Research technical working paper no. 17 (NBER, Cambridge, MA).

Kohn, Meir and J. Scott, 1982, Scale economies in research and development: The Schumpeterian hypothesis, Journal of Industrial Economics 30, 239–249.

Kuznets, S., 1962, Inventive activity: Problems of definition and measurement, in: The rate and direction of inventive activity, National Bureau of Economic Research conference volume (Princeton University Press, Princeton, NJ) 19–43.

Nelson, R.R., 1959, The simple economics of basic scientific research, Journal of Political Economy 67, 297–306.

Nordhaus, W.D., 1969, Invention, growth, and welfare (M.I.T. Press, Cambridge, MA).

Sanders, B., 1962a, Speedy entry of patented inventions into commercial use, Patent, Trademark, and Copyright Journal 6, 87–116.

Sanders, B., 1962b, Some difficulties in measuring inventive activity, in: The rate and direction of inventive activity, National Bureau of Economic Research conference volume (Princeton University Press, Princeton, NJ) 53–77.

Sanders, B., 1964, Patterns of commercial exploitation of patented inventions by large and small corporations, Patent, Trademark, and Copyright Journal 8, 51–93.

Scherer, F.M., 1965a, Size of firm, oligopoly, and research: A comment, Canadian Journal of Economics and Political Science 31, 256–266.

Scherer, F.M., 1965b, Firm size, market structure, opportunity, and the output of patented inventions, American Economic Review 55, 1097–1125.

Scherer, F.M., 1980, Industrial market structure and economic performance (Houghton-Mifflin, Boston, MA).

Scherer, F.M., 1981, Using linked patent and R&D data to measure inter-industry technology flows, Paper presented at a National Bureau of Economic Research conference on R&D, patents, and productivity, Oct.

Scherer, F.M., 1982, Technological change and the modern corporation, paper presented at a Columbia University Law School conference on the impact of the modern corporation.

Schmookler, J., 1957, Inventors past and present, Review of Economics and Statistics 39, 321–333.

Schmookler, J., 1966, Invention and economic growth (Harvard University Press, Cambridge, MA).

Taylor, C.T. and Z.A. Silberston, 1973, The economic impact of the patent system (Cambridge University Press, Cambridge).

U.S. Federal Trade Commission, 1981, Statistical report: Annual line of business report: 1974 (Washington, DC).

Economic Systems Research, Vol. 15, No. 3, September 2003

Routledge
Taylor & Francis Group

Technology Flows Matrix Estimation Revisited

FREDERIC M. SCHERER

ABSTRACT *This paper revisits the methodological problems of estimating matrices showing how technological advances—measured by industry research and development outlays—flow from industries of origin to using industries. An early effort relied upon the analysis of 15 112 US patents. Several alternative methods are explored to address methodological questions concerning the choice of carrier matrices, the handling of diagonal elements, and the treatment of capital goods flows. Technology flow matrices estimated using diverse combinations of assumptions are tested for goodness-of-fit relative to the original patent-based matrix and for their ability to 'predict' productivity growth in Solowian regression equations. Although some anomalies emerge, the best results are obtained using combined first-order transactions and capital flows matrices with diagonal elements adjusted to reflect the ratio of internal process to all R&D spending. However, flow data compiled using the Leontief inverse matrix add explanatory power in productivity growth regressions.*

KEYWORDS: *Productivity growth; technology; research and development*

1. Introduction

During the early 1980s, I estimated a highly disaggregated matrix of technology flows from US industries that performed research and development (R&D) to industries expected to use the R&D outcomes. The results, extended to analyse how technology flows affected productivity growth in the 1960s and 1970s, are reported in Scherer (1982a, 1982b, 1984). In this paper, I return to the scene of the crime two decades later to see whether the desired matrix of technology flows could have been obtained using publicly available information, or information that could be gleaned as a by-product of existing surveys, without a costly effort extracting micro-data from a large sample of individual invention patents.

2. Significance of the Problem

It is well accepted among economists that the huge gains in consumers' material prosperity achieved in industrialized nations during the past two centuries are attributable in significant measure to technological change—see, for example, Schumpeter (1942), Solow (1957), Denison (1979), and Mokyr (1990). Determin-

Frederic M. Scherer, Emeritus, Harvard University, John F. Kennedy School of Government, Cambridge, Massachusetts 02138, USA. E-mail: mike_scherer@harvard.edu. *Editor's note*: This invited paper is based on the keynote address delivered at the 14th International Conference on Input-Output Techniques, Montreal, Canada, 14 October 2002. A shortened version, which was translated into French, is included in Cloutier *et al.* (2003).

ISSN 0953-5314 print; 1469-5758 online/03/030327-32 DOI: 10.1080/0953531032000111790

ing the precise contribution to those gains of new technology, as distinguished from augmented human capital, more intense collaboration of physical capital with labour inputs, shifts in demand from the goods and services of low-productivity to high-productivity industries, favourable governmental institutions and infrastructure, and the like, is more difficult. To solve the puzzle, one must understand how each of these factors is linked to productivity growth changes, usually measured over the span of a few years or decades. The received consensus is that technological change defined narrowly has been responsible for a substantial but minority fraction of observed productivity growth—see, for example, Griliches (1995).

On that inference there are of course dissenting views. Dale Jorgenson (1990) has tended in the past, even if less so recently, to assign relatively more weight to physical capital accumulation and less to technical change in the Solowian residual sense.[1] More recently, William J. Baumol (2002) has argued that the received consensus underestimates the role of technological change because, without the scientific and technological advances that have occurred since the Industrial Revolution, it would have been difficult or even impossible to reach beyond immediate human subsistence needs, undertaking the education with which human capital has been augmented and accumulating complementary physical capital. On this broader interpretation, toward which I incline, my paper will have little to say. Rather, I focus on the problem of measuring more precisely the relatively short-run links between industrial R&D, as one source of technological change, and the growth of productivity.

This has typically been done by regressing estimates of industrial productivity growth on diverse indices of industrial technological advance, usually proxied by some estimate of research and development performed. The basic difficulty with this approach has been known at least since the publication of a brief paper by Gustafson (1962), who showed that the vast majority of industrial R&D, estimated from my own research (Scherer 1982a) to be on the order of 75%, was aimed at developing new and improved products sold to other firms and to end consumers. The technological advances from such product R&D would normally be embodied in the goods and services sold by the R&D performers and from which purchasers derive benefits, including enhanced productivity. Only about a quarter of industrial R&D was process-oriented, that is, aimed at improving the performing firms' internal production processes and hence arguably raising the performers' labour or total factor productivity. To illustrate, most of the R&D performed by the Pratt & Whitney Division of United Technologies leads to improved turbojet engines that increase the reliability, fuel economy, and range of the civilian and military aircraft in which they are embodied. The new drugs developed by Merck are sold to health maintenance organizations and end consumers, reducing the frequency and length of hospital stays and improving consumers' health and their productivity in work environments. Quite generally, significant benefits from product R&D are derived by those who purchase the goods and services in which the results of the R&D are embodied. For such product R&D, again, the majority of all industrial R&D, relating the productivity growth of industry *i* to the R&D performed in industry *i*, as all too many economists have done, could lead to seriously erroneous insights.

To move beyond this facile generalization requires an analysis at two levels of subtlety. At the first level, one focuses on what happens in an exact economic analysis of the changes wrought by product R&D. In Figure 1, we assume that a firm's R&D efforts lead to a new product for which the demand curve (taking into account the existence and prices of inferior substitute products) is represented by

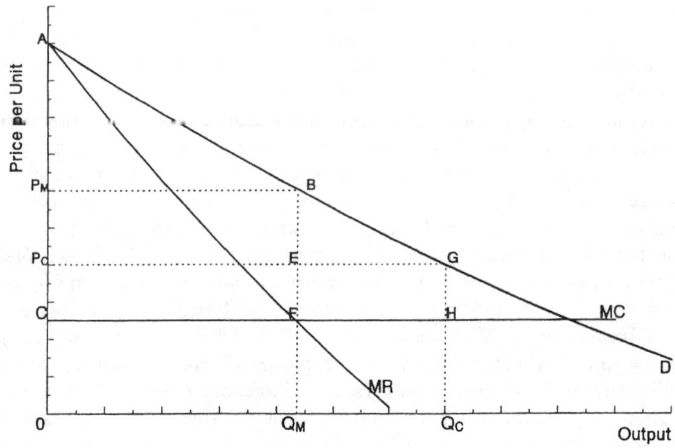

Figure 1. Theoretical Distribution of Benefits from Product Innovation.

AD. If the firm has a monopoly in the new and superior product, it will equate marginal revenue with marginal cost (affected by process R&D, and assumed constant at OC per unit) and set price OP_M, realizing profits of P_MBFC. If, previously, the firm was in competitive equilibrium with revenues barely covering input costs, the profit represents an increase of revenues over input costs correctly attributable to the originating firm's benefit. But the firm's customers also gain a surplus measured by near-triangle ABP_M. Thus, in a correct accounting, part of the social surplus from the R&D is captured by the firm, part by its customers.[2] If, however, several firms come up with similar new products, they may compete on a price basis and force the subject firm's price down to OP_C. Now the lion's share of the benefits from the R&D is realized by consumers and only the smaller quantity P_CGHC is appropriated by firms performing the R&D. The more price competition there is, the smaller is the originating firm's share of the social benefits from its innovation.[3] Ignoring second-order general equilibrium effects on other monopolistically competitive firms' demand curves, which are sometimes substantial, this is the division a theoretically correct analysis of the benefits from industrial R&D would reveal.

However, the data with which economists must work in the real world of productivity analysis often fall short of theoretical ideals. To measure productivity growth, we attempt to assess output changes in real—i.e. constant purchasing power—terms. Normally, output changes are measured by comparing the value of a firm's (or more likely, industry's) sales (or value added) at an initial point in time with the value at a terminal point in time. However, to perform the comparison correctly, nominal values must be deflated by price indices reflecting price level changes *and* changes in the product mix for a given industry. Product R&D leads to improved products which displace inferior products from market baskets. Obtaining price indices that correctly account for the change in product quality is difficult. Most analyses have concluded that the price indices compiled by government agencies, such as the Bureau of Labor Statistics, tend to underestimate the value-enhancing effects of product quality increases, and hence when used as deflators, to underestimate the gain in real output value from an initial period to a post-innovation period—see, for example, Griliches (1979). The more they underestimate the real value gain, all else equal, the lower is the productivity growth attributed to the industry selling improved products, and the lower is the imputed input cost to industries using the products, whose total factor productivity gains may be overestimated as a consequence.

The computer industry was, for many years, singled out as one in which official price deflators egregiously underestimated the rate at which technological improvements reduced the cost of computing operations—estimated to be falling at roughly 28% per year during the 1960s and 1970s (see, for example, Flamm, 1987, pp. 27–28). To correct the problem, the US Bureau of Economic Analysis and then the Bureau of Labor Statistics began using essentially hedonic (i.e. function cost-based) price indices that implied a much more rapid rate of implicit computer services price decline and hence a much more rapid rate of real output value and productivity increase for the computer industry—e.g. 26.8% per year for the 1973–88 productivity growth data compiled by the National Bureau of Economic Research (see Young, 1989 and Scherer, 1993, p. 10).

The implications of these measurement conventions are illustrated in Figure 2. Suppose computer users base their purchasing decisions on the real cost, adjusted for general purchasing power changes, of computing services per gigaflop (billion floating point operations). In base period 1, the price is OP_1 and the quantity consumed, given the assumed demand relationship, OQ_1. Now suppose the price per gigaflop falls by period 2 to OP_2, leading to an increase in the quantity of computer services demanded to OQ_2. Ignoring changes in the economy-wide price level, one would conclude from contemporary Census reports that the value of computer industry output has increased slightly from P_1AQ_1O to P_2BQ_2O. But with hedonic methods, the price index in period 2 in terms of period 1 prices is P_2/P_1 and so the deflated real output of the industry increases to P_1HQ_2O. This is a very substantial increase. Indeed, the implied consumers' surplus gain P_1HBP_2 exceeds by the near-triangle AHB the actual increase P_1ABP_2. Given these assump-

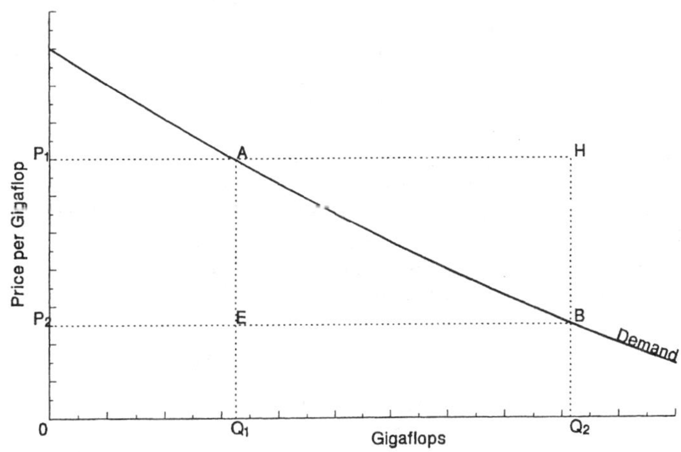

Figure 2. Prices and Quantities with Hedonic Price Changes.

tions, hedonic or function cost-based price indices tend to overestimate the real value gains from improved products, whereas traditional price indices tend to underestimate them.

For purposes of tracing where and how much productivity growth occurs, the use of hedonic price indices tends to fix the locus of gains as the industry from which the improved products originate and indeed to overestimate those gains relative to the actual economic benefits. For industries purchasing the improved computers, which from Figure 1 can be assumed to derive substantial net benefits, the use of function cost-based price indices tends to exaggerate the value of the capital goods purchased and hence to reduce or perhaps eliminate the measured value of total factor productivity growth (even if not labour productivity growth without a capital intensity adjustment). Quite generally, the more price deflators underestimate the value of product improvements, the less productivity growth one is likely to attribute to industries originating the improvements. The more (as a result of hedonic methods) they overestimate the value of the improvements, the more the productivity growth one is likely to attribute to industries originating the improvements. The shift to hedonic price deflators for computers has tended to show the total factor productivity gains from rapid technological progress to be concentrated in the computer producing sector even though, because of competitive pricing, virtually all sectors of the economy have benefited substantially from that progress.[4]

These considerations have an important bearing on attempts to estimate econometrically the impact of R&D on productivity growth. If, as is often the case, price deflators do not fully account for the benefits of product improvement, it is necessary to trace the flow of product R&D out to using industries in order to estimate its full contribution. On the other hand, when hedonic price deflators are employed, most if not all of the impact will be found within the industry originating the product R&D. My experiments with alternative deflator assumptions for computers within a much larger sample of industries (Scherer, 1993) support this generalization.

The focus here on benefits captured by either the innovator or purchasers of innovative products does not exclude the possibility of technology flowing through the economy in other ways. Zvi Griliches (1979) distinguishes between 'rent spillovers', which encompass the technology flows analysed here, and 'knowledge spillovers', which occur without embodiment in goods exchanged through market transactions. To the extent that 'upstream' producers provide disembodied know-how to their customers along with the sale of hardware or software—and there is reason to believe that such transfers are widespread[5]—the two will indeed be correlated. But disembodied knowledge may also flow through the economy in ways unrelated to market transactions, e.g. through the third parties' examination of patent specifications and articles in technical journals. These flows are best analysed, not with the techniques employed here but through the tracking of citations data (see, for example, Jaffe & Trajtenberg, 2002).

3. The Original Technology Flows Measurement Effort

Persuaded, rightly or wrongly, that existing measurement methods required an analysis of industries using innovative products to assess how R&D affected productivity growth, I embarked in the late 1970s upon a project seeking to trace the flows of embodied technology from originating to using industries. The conceptual basis for the effort was laid in Jacob Schmookler's (1966) pioneering book. Nestor Terleckyj (1974) had estimated a small-scale predecessor technology flows matrix and used it to evaluate the contributions of industrial R&D to productivity growth.

The event precipitating my research was the impending publication (US Federal Trade Commission, 1981) of data on industrial research and development expenditures for 1974 much more richly disaggregated (to 263 sectors) and with much less cross-industry contamination than any that had been available previously. Collection of the so-called Line of Business data, it should be noted, proceeded with valuable support from an affidavit submitted by Wassily Leontief in litigation that eventually reached the US Supreme Court. Given these new data plus consternation over the decline of U.S. productivity growth rates, foreshadowed by an unprecedented drop in constant-dollar industrial R&D spending, it appeared worthwhile to develop a detailed matrix tracing technology flows from industries performing R&D to industries using the fruits of that R&D.

The link from R&D spending by individual firms in narrowly-defined (three or four-digit) industries was effected by analysing 15 112 invention patents obtained by 443 typically large US corporations filing Line of Business reports with the Federal Trade Commission. A team of four Northwestern University students—an electrical engineer, a biology specialist, a chemical engineer, and a mechanically gifted farm boy—devoted roughly three months each to extracting from each patent a battery of information, including the line of business in which the underlying R&D was done and the industrial fields, identified in patent specifications to justify the 'utility' of claimed inventions, in which use of the inventions was likely. Each coding was reviewed by the author and in questionable cases rechecked, sometimes through direct contacts with companies. The coded patents were then linked to individual lines of business on which companies reported confidentially to the Federal Trade Commission. For each of 4274 individual lines of business, an average R&D cost per assigned patent was computed. For each patent, its R&D cost, adjusted upward to reflect origin industry sampling ratios, was then flowed out to one or more of 286 using industries, including personal consumption, to estimate the technology flows matrix. For 66% of the patents, from one to three specific industries of use could be identified.[6] Their underlying R&D outlay averages were allocated among multiple industries according to the using industries' relative purchase volume, as determined from the 1972 input–output transactions matrix. The remaining third were categorized as inventions of general industrial use. The R&D costs of those inventions were allocated out to using industries in proportion to 1972 input–output transactions flows or (for inventions of ubiquitous use) economy wide value added shares, with various modifications clarified later. Further methodological details are reported in Scherer (1984).

Figure 3 provides an aggregated schematic view of the resulting technology flows matrix. In 1974, 95% of all company-financed industrial R&D in the United States was performed within manufacturing industries. (Since then the contribution of non-manufacturing industries, and especially the software and biotech industries, classified in services, has expanded to a reported 36% in 1999.) Roughly half of the technology originating in the manufacturing sector during 1974 flowed out to non-manufacturing industries, arguably driving productivity growth in those industries. Only about 7% of the R&D performed was directed solely toward creating new and improved consumer goods.

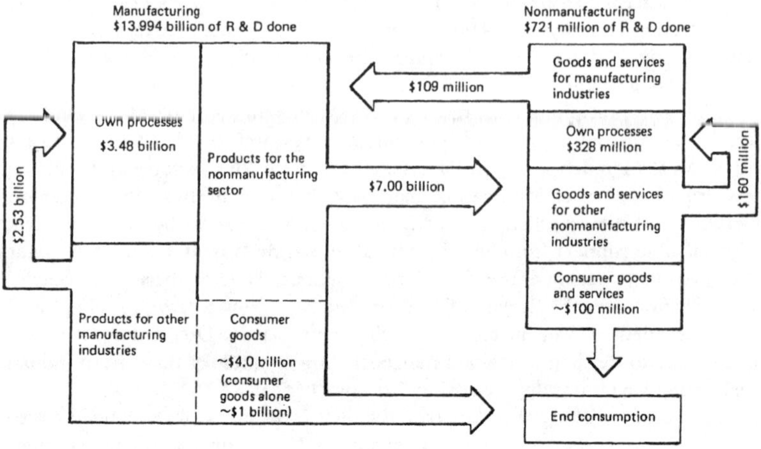

Figure 3. Industrial Technology Flows in the US Economy: 1974.

4. Re-estimation Using Less Labour-intensive Methods

When the technology flows matrix estimation methodology was articulated at a National Bureau of Economic Research conference in 1982, discussant Edwin Mansfield asked inter alia (Scherer 1984, p. 462), 'I wonder whether it would be possible for Professor Scherer to compare his findings with what would have resulted if he had simply used an input–output matrix to allocate R&D expenditures?' Certainly, if similar results could be obtained in the way Mansfield suggested, much less effort would have been required to do the job, and the effort could be replicated economically at regular intervals. At the time, I was exhausted both psychologically and financially by the work that had been accomplished and had plunged into a quite different project, so I did not follow up on Mansfield's suggestion. This, I gradually came to realize, was a serious mistake. I was induced to return to the question by a Bureau of Economic Analysis query as to whether it would be feasible to estimate technology flows matrices using input–output data for an extension to the Bureau's satellite accounts program. With help from Bureau staff, the requisite machine-readable transactions data for 1972 were retrieved and a re-estimation effort could be attempted.[7] The remainder of this paper reports the results, compares them to estimates obtained using my original more labour-intensive approach, and speculates on future opportunities.

4.1. Which Input–Output Matrix?

Several conceptual questions had to be solved. Among them, perhaps the most fundamental is whether first-order input–output transaction matrices should be used to carry R&D from originating to using industries, or whether the total requirements (Leontief inverse) matrix, calculating inputs used both directly and indirectly to produce a given vector of outputs, was a better candidate. In his pioneering effort, Nestor Terleckyj used the first-order transactions matrix. More recently, an ambitious OECD (1996) effort opted for a modified total requirements matrix approach. My own approach was eclectic. For two-thirds of the patents, little or no resort to input–output data was required because the patents had been linked directly to using industries. For the other third, arguments in favour of each approach were recognized. The first-order 'make and use' matrix was taken as a starting point, but for 22 technology-originating industries, transactions were carried farther from the first-order using industry to the industry purchasing the output of that industry or in one case (synthetic fibres) to the third-order user.[8] This choice can be criticized as arbitrary. Its principal defence is that it was based upon a detailed understanding, from reviewing 15 112 patents, of how the technology originating in diverse industries affected the activities of downstream industries.

The case for taking into account second and subsequent order flows is best summarized by considering the computer industry. Our basic objective is to place technological innovations, wherever they originate, in the industrial sector where they are likely to yield measured productivity improvements. (A different objective might lead to alternative choices.) As we saw earlier, improvements in computer technology have enormously reduced the costs of processing data. Many of those improvements have come from innovations in storage devices, which were, in 1972, and are again under NAICS, included as part of the four-digit computer industry and, following Moore's Law, from innovations increasing the capacity and speed of microprocessors and memory chips. Since only the integrated circuits cross industry lines, we focus on them. The first-order transaction is from the semiconductor industry to the computer industry. But who derives the productivity benefits? Under the logic of Figure 1, the ultimate beneficiary, since cost reductions per semiconductor function tend to be passed on by computer makers to computer buyers, is the computer buyer, not the computer industry. However, by the logic of Figure 2, if hedonic price indices are used to deflate nominal computer industry output, the productivity effects are likely to show up in the computer industry (unless hedonic price indices are also used in deflating semiconductor industry output.) Thus, depending upon how the productivity data are compiled, an argument can be made for either a first-order flow (from semiconductors to computers) or for a second-order flow (from semiconductors to computer buyers). My choice in compiling my original technology flows matrix, recognizing that computer price deflators at the time understated true cost reductions per data processing operation, was to implement a second-order flow for semiconductors.

Ignoring the price deflator problem, consider an improved synthetic resin originating in old SIC industry 2821, used to make engineered plastic parts in automobile parts plants. If the utility of the innovation comes from faster or less waste-prone moulding in the automobile parts plant, the productivity gains are likely to show up in the automobile parts industry. If, however, the benefit of the innovation is lighter weight or superior durability relative to parts previously produced, the productivity benefits are likely to be realized by industries and end consumers who buy the vehicles incorporating the parts.

Such ambiguities abound when one tries to trace the locus at which productivity gains are realized. Thus, the case for using first-order transaction as opposed to total requirements matrices is intrinsically equivocal. One way to resolve it is to make ad hoc choices depending upon the perceived dominant character of usage patterns, as I tried to do in constructing my original matrices. However, the goal of my revisited effort was to find simpler solutions, which meant an either-or carrier matrix choice. The first-order transactions matrix was initially given preference, in part because it was closer to my original approach.[9] A still better solution is to pursue both global alternatives and see which one yields results conforming more closely to my original matrix (in which, again, two-thirds of the allocations were based upon patent data) and which approach explains productivity growth more successfully. This dual assault on the problem is pursued here.

4.2. The Diagonal Problem

When published input–output matrices are used as the 'carrier' to trace R&D from the industry where it is performed to the industry(ies) using it, there is inevitably a problem of mismatch between what the diagonal values measure and what one wishes them to measure. What *should* be on the diagonal of an appropriate carrier matrix is the fraction of R&D performed by industry i and used by industry i, which consists preponderantly of process R&D. Industries vary widely in their orientation toward process, as contrasted to product R&D—in my original technology flows sample, from 0% to 100% process, with a simple average value across industries of 29.9% process R&D and a median of 19%. For industries with a strong internal process orientation, the input–output transactions' diagonal values are almost always much too low. For industries (typically the more research-intensive ones) with low process orientation, the diagonals are often too 'fat', especially when there are extensive inter-plant shipments within an industry as defined. For 154 industries with positive process R&D values, the simple correlation between process R&D as a percentage of total R&D, measured using R&D-weighted patents, and diagonal transaction matrix elements as a percentage of total intermediate industry shipments (excluding final demand vectors) was -0.03. Plainly, the use of input–output data to estimate the amount of own process R&D fails badly.

In my original application of input–output data to allocate general-purpose inventions (again, roughly one third of all inventions) to using industries, I tried to alleviate the intra-industry shipments problem by reducing diagonal elements to values not exceeding a fraction measuring the row industry's share of industry output. However, if input–output matrices were used in the future as the principal R&D allocator in constructing technology flows matrices, the results might be improved greatly through the incorporation of individual industry process R&D ratio estimates. Such estimates were collected in early US R&D censuses and could be added to future questionnaires with little additional burden on respondents. Given this possibility, I constructed alternative carrier matrices in which the diagonal elements, expressed as fractions of total intermediate industry output (excluding personal consumption, gross private fixed investment, inventory changes, and other final demand items), were replaced by individual industry process R&D fractions derived from the data on 15 112 patents, with other elements renormalized so that all included row elements summed to unity. The technology flows carrier matrices derived in this way will be called 'process-adjusted' matrices; those without diagonal adjustments will be called 'naive' matrices.

4.3. The Capital Goods Problem

Under standard input–output conventions, the capital goods output of an industry is allocated in the transactions matrix to a gross domestic fixed investment category analogous to the personal consumption category, and not to individual using industries. If this convention were accepted at face value, flows of technology embodied in capital goods delivered to individual using industries would be lost or mismeasured. This would be most unfortunate, especially given the evidence from our analysis of individual patents that 44.8% of all the patents were associated with capital goods products sold to other industries, not including the 26.2% of patents covering process inventions, many of which would affect internally modified capital equipment.[10] In contrast, only 21.6% of the patents pertained to industrial materials. Thus, somewhat more than two-thirds of the technology flowing from an origin industry to other using industries and potentially affecting the productivity of the using industry was embodied in capital goods. To exclude capital goods transaction flows from the carrier matrix could lead to serious errors. Flows of spare parts and other non-capital items treated as inter-industry transactions originating in capital goods producing industries could be poor proxies for the flow of capital goods to users.

The US Bureau of Economic Analysis (and apparently the input–output compilation agencies of other national governments) publishes separate matrices tracing capital goods flows from producing or origin industries to using industries. These are much more highly aggregated than the transactions matrices—in the rows, because many industries sell no capital goods, but more importantly, in the columns, to only 80 using industries in the 1972 version.

In my original technology flows matrix estimation, I disaggregated the capital flow transactions in proportion to more narrowly-defined using industry new capital investment (or for some non-manufacturing industries, value added) as a fraction of capital flows for the more aggregated industry category. The resulting capital flow estimates were added to the transactions estimates to arrive at the basic carrier matrix allocating R&D from origin to using industries.

A similar procedure was followed in re-estimating carrier matrices to determine technology flows from input–output data alone, without recourse to the detailed patent use codings. The disaggregations were from 77 using industries out to 211 more narrowly defined industries. For 70 of the 192 technology-originating industries, capital flows were at least 5% of the sum of transactions plus capital flows. The mean capital flow value was 19.4% of combined transactions plus capital flows.

As with the transactions matrix lacking capital flow values, the combined transactions plus capital flows carrier matrix was computed in two ways—one without correction of diagonal elements, and one with diagonals proportional to internal process R&D as a fraction of total originating industry R&D (and with other row elements re-normalized to ensure unit row sums for all using (i.e. column) industries, excluding personal consumption and other final demand items). Capital goods supplying industries were, on average, much less process invention-intensive than industries with negligible capital goods flows. For capital goods suppliers, R&D devoted to process inventions was 8.6% of total R&D; for the latter (i.e. industries with at most trivial capital flows), process R&D averaged 42.1% of total R&D.

Ambiguity over whether one should base technology flows on the first-order transactions matrix or the total requirements matrix largely vanishes when capital flows are the focus. Except when the capital goods developer is the company that will utilize the new technology, in which case the R&D should be characterized as process R&D, the productivity benefits of new capital goods are normally realized by the first downstream purchasing industry.[11]

5. Re-estimated versus Original Values

Two criteria are applied to judge the superiority of alternative technology flows matrix estimation methods. For one, the technology flows matrix created during the early 1980s by classifying 15 112 patents can be viewed as a standard against which alternatives should be evaluated. To be sure, the original sample covered only the activities of 443 corporations conducting approximately 73% of all US company-financed research and development in 1974. Excluded firms, which were for the most part smaller, could have had different usage patterns than those of included companies. And even for the included companies, one would expect usage patterns to undergo some statistical variation over time. Nevertheless, the classifications were made with extreme care, and because (with an exception to be noted later) alternative benchmarks do not exist, no better standard for assessing the revised matrices' accuracy is known. Second, the output of alternative matrix estimation methods can be used to predict productivity growth to see which contender yields the most satisfactory predictions. Both approaches are pursued here.

For purposes of predicting productivity growth in technology-using industries, the most relevant variables are the sums of the technology flows matrix columns—that is, the sum of the various amounts of technology an industry imports from other industries, along with the diagonal element measuring process technology originated by the industry in question. These were available for, at most, 205 industries, excluding inter alia personal consumption and gross private fixed investment, but including various government activities. Many of the non-manufacturing industries were highly aggregated—e.g. finance, insurance, and real estate services—and tended to dwarf the more highly disaggregated observations for manufacturing. Therefore, separate prediction error computations are also reported for manufacturing only.

The relatively disaggregated data were also aggregated back to a matrix with 41 technology-originating rows and 50 using industry sectors, including personal consumption, replicating as closely as possible the 41 × 53 matrix published (with various deletions owing to data element confidentiality) in Scherer (1982a, 1984). The new matrix resulting from these aggregations is presented here as Appendix Table A1. That matrix was constructed implementing the assumptions considered most suitable on a priori grounds—i.e. with the first-order transactions and capital flows carrier matrices combined, and with corrections on the diagonal for the incidence of process technology.

Personal consumption column sums were excluded from most of the tests that follow because they posed special conceptual problems. Only 7.4% of the 15 112 patents pertained solely to consumer goods. Another 8.7% had joint consumer goods and producing sector applications. In the original matrix estimated two decades ago, R&D outlays linked to inventions identified as consumer goods only were allocated on the basis of the patent classifications to that use column. However, when there was joint use, personal consumption usage was treated as a public good ancillary to the industrial uses, and so carrier matrix row weights summed to unity for the industrial uses, with double-counting of consumer goods usage. For the cruder input–output matrix-based approach here, normalizing row sum shares to unity excluding most final demand items—the assumption most consistent with my original multi-use convention—assigns too much weight and hence R&D cost to using industries other than personal consumption for the 7.4% of inventions (with slightly lower R&D per patent) actually used only in personal consumption.[12] Accepting this error was deemed the least of various alternative evils.

5.1. Goodness-of-Fit Analysis

Table 1 summarizes the tests conducted. Used to assess goodness of fit is the simple Pearsonian correlation coefficient r and four summary measures—mean, median, and values computed at the first and fourth quartile distribution boundaries—of the percentage deviations between newly estimated and original column sum technology usage, in millions of 1974 dollars.

Among the computations using first-order transactions, the naive transactions matrix performs least well as an R&D carrier by nearly all measures. When technology flows are based on either the transactions matrix or the sum of transactions plus capital flows, unambiguously better fits result when diagonal elements are corrected for the observed incidence of own-process usage.

The principal surprise relative to a priori expectations is the superior performance, at least for percentage errors, of the estimation using only the transactions matrix rather than the theoretically preferred combination of transactions and capital flows matrices. However, the preferred combination performs better in

Table 1. Goodness-of-fit measures: 1982 estimates as basis

			Percentage errors*			
	N	Corr.	Mean	Median	1st Quart.	3rd Quart.
Naive Transactions						
All industries	205	0.701	78.60	18.41	−37.31	116.82
Manufacturing	181	0.528	90.04	22.01	−30.20	126.25
Process-Adjusted						
All industries	205	0.716	33.85	8.78	−15.23	63.50
Manufacturing	181	0.878	37.53	12.24	−11.23	63.50
Transactions plus Capital Flows Combined						
All industries	205	0.859	62.73	25.30	−20.37	111.22
Manufacturing	181	0.715	68.14	36.22	−14.48	122.18
Combined, Process-Adjusted						
All industries	205	0.899	41.52	19.64	−3.91	68.78
Manufacturing	181	0.921	43.62	24.12	−2.30	68.86
Naive Leontief Inverse						
All industries	203	0.113	669.60	229.44	6.85	718.68
Manufacturing	180	0.218	754.45	280.00	55.37	859.44
Process-Adjusted Leontief Inverse						
All industries	203	0.246	574.26	162.84	27.86	629.91
Manufacturing	180	0.381	645.85	242.43	46.83	703.45
Aggregated Appendix Matrix (Process-Adjusted, with Capital Flows)						
All industries	49	0.843	25.25	10.77	−1.49	37.98

*100(New Estimate—Original Estimate)/Original Estimate.

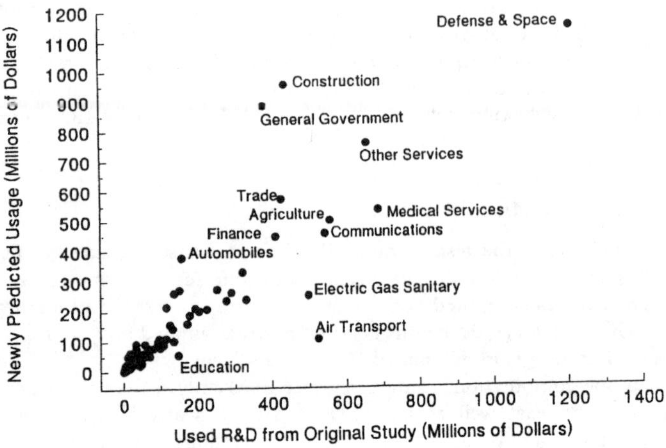

Figure 4. Scatter Diagram of Combined and Process-Corrected User Data.

terms of simple correlation coefficients. Evidently, the latter had smaller prediction errors for relatively extreme values, which tend disproportionately to influence correlation coefficients, while using industries with intermediate used R&D values in the original study had somewhat larger prediction errors when capital flows were added to transaction values in computing the relevant carrier matrices.

Figure 4 is a scatter diagram arraying the observations for all 205 using industries according to R&D usage (in millions of dollars) predicted with full use of the data on 15 112 patents (horizontal axis) and R&D usage predicted with the process diagonal-corrected matrix combining transactions and capital flows. The most extreme positive errors are general government,[13] which was also an error outlier in the process-corrected transactions-only analysis, and construction, which was not an outlier in the transactions-based analysis. The reason why construction is an error outlier when predictions include capital flows is straightforward. The capital flows matrix allocates to construction large volumes of kitchen and bathroom appliances, heating and air conditioning apparatus, office partitions and the like, which probably do little to improve productivity on the construction building site, but whose benefits flow largely to those who buy and use the structures constructed. In other words, it is important to implement second-order technology flows, which was done in formulating the original flows matrices but not in the new estimations reported here. The largest negative outlier is air transport, whose estimate in the new matrix undoubtedly understates the dual-use technology contributions of engine and electronic communications and navigation systems producers shared between the defence and air transport sectors.

Among the subset of relatively disaggregated manufacturing industries, the largest positive error outlier is passenger automobiles, whose value in the new estimate includes such innovations as improved disk brakes and electronic ignition controls and more efficient air conditioners whose benefits accrue mainly to vehicle purchasers, but which, without the second-order technology flow adjustments made in the original compilation, are perceived to remain within the vehicle-producing sector. Other large residuals were found for organic fibres and aircraft assembly, both of which were appreciably affected by second-order flows in the original matrix estimation but not in the new estimates described here. These problems suggest that if input–output data were used as the basis of technology flows estimates in the future, selective use of second-order flow adjustments could lead to substantially improved accuracy.

The fifth and sixth entries in Table 1 assess goodness-of-fit when a Leontief inverse total requirements matrix is used as the carrier for technology flows.[14] For reasons stated earlier, no attempt was made to combine capital flows with conventional transactions; the inverse matrix was derived solely from an appropriately aggregated first-order transactions matrix. To avoid excessively 'fat' diagonal values overestimating the importance of what should be process technology, the unit value reflecting deliveries to final demand is subtracted from each diagonal element.[15] The matrix derived in this way is called the 'naive' Leontief matrix. An alternative, in which diagonal elements were replaced by actual industry process usage ratios and all elements were re-normalized to sum to unity, is called the 'process-adjusted Leontief matrix'. By all measures, the fit is much worse than with any of the first-order matrices. The process-adjusted row sums conform slightly more closely than unadjusted values. Since the differences did not appear to be attributable to matrix inversion errors,[16] it seems clear that the total requirements approach characterizes rather different phenomena than those measured in my original effort two decades ago or in my reconstruction emphasizing first-order flows only. The implications of this difference will be addressed subsequently.

The final entries in Table 1 reveal that some of the large mean and quartile estimation errors observed using disaggregated first-order data for some 205 industries are more or less random, cancelling out when the process adjusted-estimates including capital flows are aggregated down to 49 broader sectors.

5.2. Predicting Productivity

An alternative perspective for assessing the success of the new technology flows estimates is to use the column sums as an explanatory variable in regression analyses 'explaining' productivity growth, also taking into account other relevant variables. This was a principal purpose of the original estimation effort two decades ago. At the time, productivity growth data disaggregated by industry were scarcer than they are now. The principal results, reported in Scherer (1982b), emphasized what was at that time a new Bureau of Labor Statistics series using input–output industry definitions and including 87 industries—81 of them from the manufacturing sector along with agriculture, crude oil and gas production, railroads, air transport, communications, and the combined electric-gas-sanitary utilities sector. These data are also used in my new analysis, although because of gaps attributable mainly to the confidentiality of certain Federal Trade Commission Line of Business data, only 80 (or with the Leontief inverse data, 79) industries can be covered here. The dependent variable is the percentage growth of labour productivity, ΔLP, i.e. (estimated) real output per unit of labour input, between 1973 and 1978, both cyclical peak years. An additional variable from the original productivity growth data set was the percentage growth of capital intensity $\Delta(K/L)$ over the same period. Two variables were emphasized in the original paper and are employed again here to measure the contributions of technology: *UsedRD*, which is the appropriately aggregated column sum estimating industry *i*'s R&D usage, either as process R&D or R&D imported from other industries, and *ProdRD*, or

Table 2. Regressions explaining 1973–78 labour productivity growth

	UsedRD	*ProdRD*	$\Delta(K/L)$	R^2	*N*
(1) Original data	0.742	0.289	0.347	0.193	87
	(1.89)	(2.01)	(3.30)		
(2) Original data, matching sample	0.698	0.357	0.332	0.192	80
	(1.74)	(2.27)	(3.03)		
(3) New naive transactions	0.565	0.194	0.314	0.175	80
	(1.19)	(0.81)	(2.85)		
(4) New transactions, process-adjusted	0.352	0.359	0.312	0.164	80
	(0.66)	(2.07)	(2.81)		
(5) New transactions plus capital flows	2.25	−0.073	0.268	0.232	80
	(2.68)	(0.31)	(2.48)		
(6) New transactions plus capital flows, process-adjusted	0.751	0.320	0.299	0.192	80
	(1.73)	(1.96)	(2.72)		
(7) New Leontief inverse	0.302	0.198	0.322	0.183	79
	(1.41)	(0.92)	(2.91)		
(8) New Leontief inverse, process-adjusted	0.232	0.321	0.316	0.169	79
	(0.88)	(1.72)	(2.84)		
	1st Order Inverse				
(9) UsedRD from (6) and (7)	0.815 0.336	0.079	0.302	0.220	79
	(1.88) (1.60)	(0.36)	(2.76)		
(10) UsedRD from (2) and (7)	0.817 0.374	0.088	0.339	0.225	79
	(2.02) (1.77)	(0.40)	(3.12)		

Subscripted parentheses report *t*-ratios.

the amount of non-process R&D performed by industry *i*, virtually all of which was assumed to be exported in embodied form to other using industries. Both are measured as a percentage of relevant industry output values. Ignoring measurement difficulties, *ProdRD* should characterize the benefits appropriated internally by the innovating firm, e.g. area P_MBEP_C in Figure 1, and *UsedRD* the external or exported benefits ABP_M in Figure 1. Our main concern is the contribution of *UsedRD* computed in various alternative ways. Following a proof attributable to Terleckyj (1974), the regression coefficients on the R&D variables can be interpreted, subject to some qualifications, as rates of return on research and development investment.

Table 2 reports the regression results. Regression (1) is drawn from Scherer (1982b) using my original technology flows data. All three variables made statistically significant contributions to the explanation of productivity growth during the (relatively stagnant) middle 1970s (and also, it was shown, with some limitations, in the more dynamic 1964–1969 period). The R^2 value is modest, however, indicating that considerable unexplained noise remains. Regression (2) reports results using the original technology flow estimates for the new sample from which seven industries were removed because of data gaps. It will be taken as the benchmark against which regressions using the new estimates will be compared. In all new regressions (3)–(10), the $\Delta(K/L)$ and *ProdRD* variables are identical to those used in the second, $N = 80$, entry.

There are several surprises. First, regressions (3) and (5), with and without capital goods flows added, but without process diagonal adjustments, outperform their counterparts. Second, the greatest explanatory power (R^2) using first-order carrier matrices is achieved with regression (5), which is based on summed transactions and capital flows, but without process diagonal adjustments. That regression, however, reveals a surprising and indeed implausible constellation of technological impact coefficients. The coefficient on *UsedRD* implies rates of return of 225% on process plus imported R&D—much higher than any estimated with the original data set. At the same time, this strong used R&D effect destroys the impact of internal product R&D, which is also implausible. The reason for this second result is that the regression equation suffers from severe multicollinearity. The simple correlation between the two output-deflated R&D variables is 0.760, and the used R&D variant (without process diagonal adjustment) overwhelms its correlated own-product R&D analogue. Given these anomalies, one is inclined to reject regression (5) and favour regression (6), with the strongest a priori support and the second-highest explanatory power of the new contenders. Indeed, the explanatory power R^2 of regression (6) is identical to that of regression (2) incorporating used R&D data from the original technology flows matrix derived through inspection of 15 112 patents, and the regression coefficients differ only trivially.

Additional surprises materialize when column sums from the Leontief total requirements carrier matrices are substituted to obtain the key used R&D variable. With Leontief inverse estimates both unadjusted for process diagonal values (regression (7)) and process-adjusted (regression (8)), R^2 is less than that of the best two regressions using new first-order technology flow vectors. Neither total requirements-based used R&D variable achieves statistical significance by conventional standards, exceeding the 1.67 t-ratio value delineating 95% confidence in a one-tailed test. As with the first-order estimates, explanatory power is greater without process diagonal adjustments, but at the cost of degrading the product R&D variable's role. A reason for the product R&D impact is that the Leontief estimates without process diagonal adjustments are fairly strongly correlated with the product R&D variable, and this collinearity degrades the product R&D coefficients.[17] Table 3 presents a matrix of correlation coefficients for alternative R&D flow measures defined as a percentage of the value of industry output.

From regressions (7) and (8) in Table 2, the generally similar Pearsonian correlations between 1973–78 productivity growth and alternative R&D flow measures in Table 3, and the Leontief variables' typically low correlations with first-order R&D usage variables, it would appear that the R&D usage variables

Table 3. Correlation matrix for variables used in the regression analysis

	$\Delta(Q/L)$	ProdRD	$\text{UsedRD}_{\text{orig}}$	$\text{UsedRD}_{\text{k,p}}$	UsedRD_{L}	$\text{UsedRD}_{\text{L,p}}$
$\Delta(Q/L)$	1.000	0.264	0.204	0.279	0.259	0.227
ProdRD		1.000	0.186	0.308	0.689	0.532
$\text{UsedRD}_{\text{orig}}$			1.000	0.805	0.009	0.236
$\text{UsedRD}_{\text{k,p}}$				1.000	0.151	0.419
UsedRD_{L}					1.000	0.844
$\text{UsedRD}_{\text{L,p}}$						1.000

Notation:

$\Delta(Q/L)$ Percent annual labour productivity growth, 1973–78
ProdRD Product R&D (percent of industry output value)
$\text{UsedRD}_{\text{orig}}$ Used R&D (original measures) (percent of industry output value)
$\text{UsedRD}_{\text{k,p}}$ Used R&D, new estimates, with capital flows and process diagonal adjustments (%)
UsedRD_{L} Used R&D, Leontief inverse (%)
$\text{UsedRD}_{\text{L,p}}$ Used R&D, Leontief inverse, with process diagonal adjustments (%)

derived from total requirements matrices are characterizing a different dimension of technology flows, but one that has at least some utility in explaining productivity growth. To pursue this insight further, regressions (9) and (10) in Table 2 introduce together two distinct technology usage variables, one derived with an emphasis on first-order transactions and one based on the total requirements matrix without process diagonal adjustment. Higher R^2 values are achieved than in any but regression (5), which is rejected as implausible on a priori grounds. The first-order and total requirements technology usage coefficients exceed 95% statistical significance thresholds in three out of four cases, although, as in equation (7), the power of the product R&D variable is degraded. We observe too that the implied returns on R&D investment are in the range of 70 to 80% with first-order flow measures but only 23 to 37% with total requirements measures. It would appear that the more diffuse usage traced using the Leontief inverse approach yields lower returns than the direct usage in first-order technology embodiments. We conclude more generally that both the first-order and total requirements approaches help explain the links between research and development and productivity growth, with the first-order measures holding a modest edge over those based upon the Leontief inverse.

6. An Alternative Technology Flows Measurement Approach

A promising alternative approach to measuring the inter-industry flow of technology has been pioneered by Robert Evenson and associates at Yale University, using unique data developed in the Canadian Patent Office (CPO). See Evenson & Johnson (1997a, 1997b) and Kortum & Putnam (1997). Beginning in the late 1970s, the CPO began having its staff classify most of the patents it grants (roughly half of which originate from US inventors) according to industry of manufacture, which corresponds to my industry of origin concept, and sector of use, which corresponds to my industry of use concept (see Ellis, 1981). The classifications, discontinued during the 1990s, were typically made at the four-digit industry level in the then-prevailing Canadian Standard Industrial Classification System. Evenson and colleagues have arrayed the Canadian data into technology flow matrices like mine. The comparison made here is from Evenson & Johnson (1997b), whose Table 2 reports flow matrix column sums for counts of issued patents analogous to those analysed in the previous section. My original technology flows matrix was based upon US patents issued to US corporations in 1976 and 1977, to which the patent usage sums reported by Evenson & Johnson for 1978–81 correspond most closely in time.

The Evenson *et al.* data are relatively highly aggregated to the level of 33 sectors. By aggregating my original matrix column sums, combining six of the Evenson sectors into three, and omitting three incompatibly defined sectors, an acceptable match was achieved for 27 sectors, including 19 manufacturing and eight non-manufacturing sectors. Figure 5 arrays their issued patent counts (vertical axis) against my original technology flow matrix column sums (horizontal axis), measured in millions of dollars. Inspection reveals substantial departures from what ought, if the same phenomenon is being measured, to be a linear array of data points. The correlation between the two data sets is 0.560. Especially large discrepancies are observed for the combination of their electrical equipment and electrical machinery groups, which in my analysis had a relatively high incidence of second-order product technology flows; the various transportation services

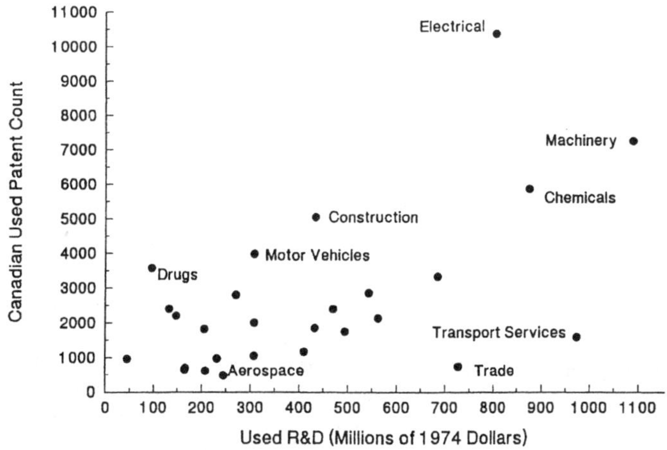

Figure 5. Correlation of Evenson and Scherer Technology Use Data.

industries (which in my analysis were recipients of substantial first- and second-order technology flows); and the wholesale and retail trade sectors, which in my matrix received many general-use inventions apparently not coded by the Canadian Patent Office.[18] It is also possible that the patterns of patenting in Canada, mostly by foreigners, were different than in the United States, in which, during 1977, foreign residents were a distinct minority among patent recipients.

Whatever the reasons, it seems clear that the Canadian patent classifiers and my late 1970s effort measured somewhat different phenomena. Further research on the reasons for discrepancies using more disaggregated column sums and information on second-order technology flows seems in order.

7. Conclusions

From the tolerably good results I have achieved attempting to emulate my original labour-intensive technology flows matrix using mainly input–output data as the carrier matrix, it seems clear that future iterations might be feasible. If the effort is undertaken, combining the transactions and capital flows matrices is essential, since much of the technology flowing between industries is embodied in capital goods. Ideally, capital flows should be capitalized over a larger number of years and depreciated. There is also a persuasive argument for replacing the standard input–output matrix diagonals with information on the share of each industry's R&D effort devoted to internal process improvement. On this, more in a moment. From the results reported here, taking into account total requirements matrix flows is also important. Doing so selectively, as I did two decades ago, yields productivity explanations different from (and somewhat stronger than) those using Leontief total requirements matrices as the technology flow carrier. That the two alternative methods yield superior predictions when used in tandem suggests that both approaches warrant support in future work.

There are several possible ways to obtain the data needed to estimate process diagonals correctly. Ideally, respondents in the joint National Science Foundation—Census Bureau industrial R&D surveys could be asked to provide an estimate for their operations. They may not know the correct fraction precisely, but a knowledgeable approximation is much better than ignorance and manifest error. In diversified corporations, however, a company-wide estimate may conceal wide inter-industry variations. Alternatively, the process breakdowns (averaging 22% for 1996) elicited through a smaller survey conducted periodically by the private-sector Industrial Research Institute could be tapped. See Bean *et al.* (1998). Or at higher cost but greater potential precision, patent applicants might be asked whether their inventions pertain mainly to potential products, internal process improvements, or some mixed or 'other' category.

In my opinion, understanding how technology flows through the economy and enables economic growth is one of the most important matters to which economic analysis can contribute, and therefore it warrants a richer allocation of information-gathering resources. I therefore propose that the US Patent Office emulate its Canadian cousins, but go farther. It would ask each applicant to disclose the NAICS industry category in which the invention originated (with a catch-all for inventions from broad-mandate basic research laboratories and an 'unaffiliated' option for unaffiliated inventors), along with the principal industries in which use of the invention is contemplated, with 'industries to which the originating industry sells products' and 'throughout the economy' as alternatives for general-use inventions. Once patent attorneys became accustomed to asking inventors for such information, the incremental compliance burden would be minute, and valuable information on the structure of the economy would be obtained.

Developing this information as a by-product of invention patenting would go a long way towards solving what was, with my research two decades ago, a minor problem but has now taken on major proportions—the measurement of imported technology flows. My research was focused on patents issued in 1976 and 1977, when the United States was the world's clear leader in most areas of industrial technology and 34.4% of all US invention patents went to foreign residents. At the time, the role of foreign inventors was rising rapidly. From 1966 to 1970, the average share of foreign inventors in total US patenting was only 22%; it rose to 47% in the late 1980s before receding to 44% during the mid-1990s. At the time my original study was conducted, high-technology imports were penetrating the US economy at an accelerating rate (Scherer, 1992). When my research was carried out, one could defend ignoring foreign technology sources, but that is no longer possible. Assuming that imports disseminate technology, or its underlying R&D results, with the same usage patterns as domestic technology sources, as the OECD staff (1996, pp. 26–27, 143) has been compelled by data limitations to assume, provides at best a crude approximation to the contribution of imported technology. A much better estimate could be obtained if foreign inventors, like US inventors, were required to disclose the industry from which their inventions originated (which could then be linked to national R&D statistics) and the industries likely to utilize their inventions.[19]

I have saved for last the most difficult problem to be solved in future technology flows matrix development efforts—obtaining accurate, reliable origin industry research and development data. I began my project during the late 1970s because, for the first time ever, reliable R&D data for highly disaggregated industries became available through the Federal Trade Commission's Line of Business program. Complete reports for four years—1974 through 1977—were obtained before the program was terminated as a result of political pressure orchestrated by US industry.

The closest analogue to the FTC R&D reports covering some 263 industries has been the collection in NSF-Census surveys of applied R&D expenses for some 37 'product fields'. That data collection effort was discontinued following the 1997 survey because of poor response rates.[20] As a result, the only industrial R&D expenditure disaggregations from the NSF-Census surveys are reported for roughly 50 industry groups (expanded through disaggregation of many non-manufacturing groups from the 26 reported in 1997) by the 'principal industry' method. That is, the principal industry in which a company operates is ascertained, and *all* of the company's R&D is thereupon assigned to that industry. For large diversified companies, this method leads to large allocation errors. To illustrate, among the companies included in my study two decades ago, General Electric obtained 706 patents. It is uncertain to which of the many fields in which GE operated at the time, ranging from synthetic resins to aircraft engines, the Census Bureau staff would classify its R&D activities. It was probably 'other electrical equipment,' in which case 57% of GE's patents would be misclassified. If GE were located in the broader (old SIC) two-digit 'electrical equipment' group, the error rate would be 42%. Or to take a less diversified company, 47% of du Pont's 391 patents would be misclassified if its principal industry were deemed to be 'industrial chemicals'. Even at the two-digit SIC level of detail, 24% of du Pont's patents originated outside the broad 'chemicals and allied products' sector. Basing a technology flows matrix on such 'contaminated' R&D data would impart considerable inaccuracy.

The simplest solution to this problem would be to restore line of business reporting in the National Science Foundation—Census Bureau surveys, disaggregating the reporting lines more finely than they have been disaggregated in the past, and exerting strenuous efforts to convince industry participants that the data shed important light on the dynamics of the American economy. Failing that, the principal alternative basis for measuring inter-industry technology flows could be requiring patent applicants to disclose industries of origin and use in their applications. In this case, an average R&D cost per company patent could be estimated using publicly available annual reports on company-financed R&D expenditures. Or, for companies that do not report their R&D figures, the data could be obtained on a confidential basis from NSF-Census filings. For analyses with a longer time perspective, individual patents could be weighted by the citations they subsequently receive—an indicator shown in several studies to reflect patent value better than underlying average research and development expenditures.

I conclude that it is indeed feasible to construct meaningful technology flow matrices using approaches less labour-intensive than those accepted for my effort two decades ago. But substantial progress requires improvements in the data obtained from industry in annual R&D surveys or patent filings.

Notes

1. Compare Jorgenson & Stiroh (2000) and the comment by Gordon (2000, p. 215).
2. With cost-saving process innovations, it is possible but not necessary that all of the benefits are appropriated by the innovator. See Arrow (1962). If the innovations induce price reductions, the benefits are shared.
3. Mansfield *et al.* (1977) estimate that at the median in a sample of 17 innovations, innovators appropriated roughly 44% of the discounted economic benefits from their innovations.
4. Because in deriving gross national product estimates 'real' industry outputs are weighted by industry

price indices, changes in the price index base year can lead to surprisingly large reductions in estimated GNP. This was a special problem in the early 1990s, until GNP weights were chain-linked annually instead of every five years.

5. See Harhoff (1996).
6. Sixty-four percent of these involved a single using industry.
7. The author is grateful in particular to Peter Kubach and Jiemin Guo of BEA.
8. The industries from which second-order flows were computed were weaving mills, fabric knitting mills, organic fibres, tyres and tubes, rubber hose and belting, flat glass, pressed and blown glass, internal combustion engines, pumps, anti-friction bearings, compressors, speed changers and industrial drives, mechanical power transmission equipment, automotive carburettors etc, vehicular lighting equipment, electron tubes, cathode-ray tubes, semiconductors, other electronic components, starter and traction batteries, aircraft engines, and buttons, zippers, etc. Not all transactions, but only those that were preponderantly of a 'component sale to further assemblers' nature, were treated in this way.
9. But see Scherer (1982b, p. 631), in which matrices with second-order flows were emphasized on a priori grounds even though slightly higher Pearsonian correlations with productivity growth were obtained when only first-order flows were measured.
10. On this, see Carter (1970, p. 21), who correctly observes that 'New types of capital goods are at the core of technological change.'
11. An important exception, the construction industry, will be discussed subsequently. Purchasers farther downstream benefit when competition forces cost reductions at the first using stage to be passed on in the form of lower prices. But the change in productivity occurs at the stage using the capital good.
12. The average 1974 R&D cost per consumer good invention patent was \$533 000, as compared to \$594 000 for industrial use inventions. Patents on inventions whose use was considered to be of general use, without any specific industry assignment, had the highest average R&D cost at \$743 000. Patents covering complex system inventions had the highest average R&D cost of \$707 000 among several technological categories; the lowest average was for production processes, at \$450 000.
13. On both dimensions, the most extreme observation is defence and space operations. The R&D values include only company-financed R&D (although some so-called 'independent' R&D reported as company-financed was ultimately reimbursed by the Department of Defense). In addition to the \$1.2 billion of company-financed R&D allocated to defence and space, the original study identified \$4.8 billion of government-financed R&D.
14. The author is indebted to Aubhik Khan and Robert Hunt of the Federal Reserve Bank of Philadelphia for inverting the 207 × 207 transactions matrix, which was too large for the author's ancient computer. Two using industries—guided missile production and the government's defence and space operations—had to be omitted.
15. A somewhat different procedure to avoid overestimation of diagonal effects was adopted by the OECD group (1996, pp. 142–143).
16. Some sectors of the 207 square matrix were at the same level of disaggregation as the 487 × 487 matrix published in US Department of Commerce (1979, volume II), so a comparison to verify the accuracy of our inversion effort was possible. All diagonal elements had unadjusted values of unity or greater, as is required. Most of the compared cells differed by no more than 10%. A few larger deviations were expected (and found) in inverting matrices of such disparate aggregation.
17. That the Leontief inverse measures without process diagonal adjustment are associated with advanced product technology is also suggested by an examination of the three most extreme values, all taken as a percentage of industry output value: computers, 10.04%; optical and ophthalmic instruments, 9.36%; and other office machinery, 6.15%. The median value for all industries was 1.36%. With my original used R&D indices, the largest three values occurred in electronic components (3.39%), air transportation (3.02%), and synthetic fibres (2.56%). Two of the three are high process technology users; the third (air transport) a major importer of embodied technology.
18. See Kortum & Putnam (1997, p 174, note 13). For those who might wish to replicate the comparison, it should be noted that in Scherer (1984, p. 451), which presents disaggregated used R&D matrix column sums, the value for coal mining should be 72.4 rather than 35.1.
19. For a pioneering effort using Canadian patent data, see Hanel (2000). Several studies have found that patents sought outside one's home market tend to be of greater economic value on average than patents received at home.
20. Communication from Raymond Wolfe of the National Science Foundation, 23 May 2002. From a mini-conference held 23 April 1998, on 'R&D and Innovation Statistics' under the auspices of

the Census Bureau's Advisory Committee of Professional Associations, one of the strongest recommendations to emerge was that more effort be devoted to obtaining industrial R&D expenditure data broken down by disaggregated originating lines of business.

References

Arrow, K. J. (1962) Economic welfare and the allocation of resources for invention, in: National Bureau of Economic Research conference report, *The Rate and Direction of Inventive Activity* (Princeton, Princeton University Press).

Baumol, W. J. (2002) *The Free-Market Innovation Machine: Analyzing the Growth Miracle of Capitalism* (Princeton, Princeton University Press).

Bean, A. S., Russo, M. J. & Whiteley, R. L. (1998) Benchmarking your R&D: results from IRI/CIMS Annual R&D Survey for FY '96, *Research Technology Management*, 41, pp. 21–30.

Carter, A. P. (1970). *Structural Change in the American Economy* (Cambridge, Harvard University Press).

Cloutier, L. M., De Bresson, C. & Dietzenbacher, E. (eds) (2003) *Changement Climatique, Innovation Technologique et Finances: Nouvelles Directions d'Anlayses Entrée-Sortie* (Montreal, Presses de l'Université du Quebec, forthcoming).

Denison, E. F. (1979) *Accounting for Slower Economic Growth* (Washington, Brookings Institution).

Ellis, E. D. (1981) The philosophy, construction, and uses of the Canadian Patent Data Base PATDAT, *World Patent Information*, 3, pp. 13–18.

Evenson, R. E. & Johnson, D. (1997a) Introduction: Invention Input–output Analysis, *Economic Systems Research*, 9, pp. 149–160.

Evenson, R. E. & Johnson, D. (1997b) Innovation and Invention in Canada, *Economic Systems Research*, 9, pp. 177–192.

Flamm, K. (1987) *Targeting the Computer: Government Support and International Competition* (Washington, Brookings Institution).

Gordon, R. J. (2000) Comment, *Brookings Papers on Economic Activity* 2000:1, pp. 212–222.

Griliches, Z. (1979) Issues in assessing the contribution of research and development to productivity growth, *Bell Journal of Economics*, 10, pp. 91–116.

Griliches, Z. (1995) R&D and productivity, in: P. Stoneman (ed) *Handbook of the Economics of Innovation and Technological Change* (Oxford, Basil Blackwell).

Gustafson, W. E. (1962) Research and development, new products, and productivity change, *American Economic Review*, 52, pp. 177–185.

Hanel, P. (2000) R&D, interindustry and international technology spillovers and the total factor productivity growth of manufacturing industries in Canada, *Economic Systems Research*, 12, pp. 345–361.

Harhoff, D. (1996) Strategic spillovers and incentives for research and development, *Management Science*, 42, pp. 907–925.

Jaffe, A. B. & Trajtenberg, M. (2002) *Patents, Citations & Innovations: A Window on the Knowledge Economy* (Cambridge, MIT Press).

Jorgenson, D. W. (1990) Productivity and economic growth, in: E. R. Berndt & J. Triplett (eds), *Fifty Years of Economic Measurement* (Chicago, University of Chicago Press), pp. 19–118.

Jorgenson, D. W. & Stiroh, K. (2000) Raising the speed limit: U.S. economic growth in the information age, *Brookings Papers on Economic Activity*, 1, pp. 125–211.

Kortum, S. & Putnam, J. (1997) Assigning patents to industries: tests of the Yale technology concordance, *Economic Systems Research*, 9, pp. 161–175.

Mansfield, E., Rapoport, J., Romeo, A., Wagner, S. & Beardsley, G. (1977) Social and private rates of return from industrial innovations, *Quarterly Journal of Economics*, 91, pp. 221–240.

Mokyr, J. (1990) *The Lever of Riches* (Oxford, Oxford University Press).

OECD (Organization for Economic Cooperation and Development) (1996) *Technology and Industrial Performance* (Paris, OECD).

Scherer, F. M. (1982a) Industrial Technology Flows in the United States, *Research Policy*, 11, pp. 227–245.

Scherer, F. M. (1982b) Inter-industry technology flows and productivity growth, *Review of Economics and Statistics*, 64, pp. 627–634.

Scherer, F. M. (1984) Using linked patent and R&D data to measure inter-industry technology flows, in: Z. Griliches (ed) *R&D, Patents, and Productivity* (Chicago, University of Chicago Press), pp. 417–461.

Scherer, F. M. (1992) *International High-Technology Competition* (Cambridge, Harvard University Press).

Scherer, F. M. (1993) Lagging productivity growth: measurement, technology, and shock effects, *Empirica*, 20, pp. 5–24.

Schmookler, J. (1966) *Invention and Economic Growth* (Cambridge, Harvard University Press).

Schumpeter, J. A. (1942) *Capitalism, Socialism, and Democracy* (New York, Harper).

Solow, R. A. (1957) Technical change and the aggregate production function, *Review of Economics and Statistics*, 39, pp. 312–320.

Terleckyj, N. E. (1974) *Effects of R&D on the Productivity Growth of Industries: An Exploratory Study* (Washington, National Planning Association).

US Department of Commerce, Bureau of Economic Analysis (1979) *The Detailed Input–Output Structure of the U.S. Economy: 1972*, two vols. (Washington, USGPO).

US Federal Trade Commission (1981) *Statistical Report: Annual Line of Business Report: 1974* (Washington, US Federal Trade Commission).

Young, A. H. (1989) BEA's measurement of computer output, *Survey of Current Business*, July, pp. 109–115.

Appendix

Industry Names for Table A1

1 Agriculture
2 Mining
3 Food and Tobacco
4 Textiles
5 Apparel and Leather
6 Wood Products
7 Furniture
8 Paper Products
9 Publishing
10 Basic Chemicals
11 Organic Chemicals
12 Pharmaceuticals and Toiletries
13 Agricultural Chemicals
14 Miscellaneous Chemicals
15 Petroleum Extraction & Refining
16 Rubber and Plastic Products
17 Stone, Clay, Glass Products
18 Ferrous Metals
19 Non-Ferrous Metals
20 Fabricated Metal Products
21 Engines & Turbines
22 Farm Machinery
23 Construction Equipment
24 Metal-Working Machinery
25 Other Machinery
26 Office Equipment
27 Industrial Electrical Equipment
28 Electrical Appliances
29 Misc. Electrical Products
30 Communications Equipment
31 Electronic Components
32 Motor Vehicles
33 Aircraft
34 Guided Missiles and Ordnance
35 Other Transportation Equipment
36 Instruments
37 Misc. Manufacturing
38 Trade
39 Communications
40 Construction
41 Services
41.1 Electric, Gas, and Sanitary Services
41.2 Medical Services
41.3 Other Services
41.4 Finance, Insurance, and Real Estate
42.1 Surface Transportation
42.2 Air Transportation
42.3 Water Transportation
43.1 Government Functions
43.2 Defense and Space Operations
44 Personal Consumption

PART III

WHAT DO THE MEASURES MEASURE?

World Patent Information, Vol. 4, No. 1, pp. 12–17, 1982.
Printed in Great Britain.

0172-2190/82/01 0012-06 $02.00/0
Pergamon International Information Corp.

The Office of Technology Assessment and Forecast Industry Concordance as a Means of Identifying Industry Technology Origins

*F. M. Scherer, Northwestern University**

Summary

This paper examines the difficulties encountered when one attempts to link patents to data on such variables as R & D expenditures in the industries where the patents originated. A comparative experiment using U.S. OTAF industry assignments based upon patent class codes only and a 'manual' procedure using much richer information reveals a high rate of disagreement as to origin industries. Alternative means of obtaining better information on origin industries are considered.

Introduction

Invention patents provide a potentially rich source of information on the contours of industrial technology. To tap this potential, one may among other things wish to trace patents back to the industries in which they *originated*. This might be done, for example, because one is studying the relationships between such industry characteristics as output growth or seller concentration and invention, or because one seeks to relate patents as a measure of inventive output to R & D expenditures as an index of input. For these and other purposes, it would be convenient to have a simple, inexpensive means of linking patents to appropriate industries.

The Office of Technology Assessment and Forecast (OTAF) in the U.S. Patent and Trademark Office has developed a computerized procedure for making such links.[1] Its basis is a concordance between detailed patent class and subclass codes and industry codes. When a U.S. invention patent is issued, one piece of information accompanying the technical specification is a set of one or more patent class and subclass codes to which the invention's subject matter most closely relates or to which it is cross-referenced. To utilize this information, OTAF has compiled a concordance assigning each patent subclass to a corresponding segment of the U.S. Standard Industrial Classification. The industry categories are for the most part defined at the two- or three-digit level.[2] In its most detailed version, the OTAF concordance contains a total of 41 non-overlapping SIC industry categories. Once the patent class and subclass are ascertained, the patent is linked through a simple computer look-up procedure in the concordance. When multiple subclass codes are referenced, multiple industry assignments may result.

Unfortunately, serious inaccuracies may occur under this procedure. What is 'inaccurate' or 'accurate' in this context is relative; it depends upon the uses to which one wishes to put the industry classifications. If all one wants is to assign patents to more or less arbitrarily defined but consistent industry categories and not link the resulting patent tallies to other data sets, the OTAF procedure is probably quite acceptable. An example would be certain tables included in the biennial U.S. National Science Board *Science Indicators* report. Where problems arise it is in attempting a link to other industry data sets. The specific context in which these problems came to light was a link made by the author between patent data and statistics on research and development expenditures subdivided by industry codes.

Linking Problems

The basic difficulty is that in making such a link, one must attempt to emulate the organization of the data to which the patents are being related. In the case of research and development expenditures, this depends upon individual company structures and reporting conventions, which in turn depend upon the specific reporting context. Consider for example the General Motors Corporation. Among other things, it does R & D on, and obtains patents covering, turbojet

* The research reported here was conducted under U.S. National Science Foundation grant No. PRA-7826526. This report was written while the author was a visitor at the Max-Planck-Institut für ausländisches und internationales Patent-, Urheber-, und Wettbewerbsrecht in München, with a fellowship from the Max-Planck Gesellschaft. The assistance of patent classification team members Mary Gianos, Chun Yue Lai, Brett Spencer, and Pin Tai is also gratefully acknowledged.

aircraft engines. To what industry should such patents be classified? If one is interested in establishing a link to R & D statistics compiled under the main definition used in National Science Foundation surveys, the patents should be classified to the motor vehicles industry, because that is GM's *primary* industry and the one to which *all* of its R & D program outlays would be assigned under the applicable survey reporting conventions. No matter how well one does classifying patents on the basis of their subject matter, serious errors will result if no heed is taken of company organization, since 'conglomerateness' is a prominent feature of U.S. corporate structure. In 1972, 33% of U.S. manufacturing corporations' employment was in industries other than the reporting enterprises' primary industry, with the industry locus being defined at approximately the three-digit level of disaggregation.[3]

The research and development data to which the link emphasized here occurred were largely free of this conglomerateness problem, since they resulted from the Federal Trade Commission's 1974 Line of Business survey. That survey asked 443 reporting companies to break down their applied R & D outlays into some 276 industry categories according to the specific divisional or establishment locus at which the R & D was performed. The average company reported its activities in 9.65 different 'line of business' categories. With such 'pure' R & D data, one might suppose, the problems of arbitrary company organization and reporting conventions should fade and a link based upon patent classes should be more trouble-free. Nevertheless, matters are not quite so simple. The OTAF concordance (or in all likelihood, any patent class-oriented concordance emphasizing intrinsic subject matter) can generate industry codes quite different from those adopted by companies in either their internal divisional structure or their responses to government information requests. There are three main reasons why discrepancies materialize. They can be illustrated through an extension of the previous General Motors example.

For one, a subject matter-oriented classification will often yield wrong conclusions about internal production process inventions, which account for roughly 25% of all U.S. corporation patents. Suppose, for example, that GM receives a patent covering an apparatus for welding turbojet engine fan blades to the rotor assembly. An analysis of the patent's subject matter will probably lead to the conclusion that U.S. SIC code 362 (including electric welding apparatus) is the proper industry code, when in fact the relevant GM R & D would be reported not by process orientation, but in *product line* SIC 372 (aircraft and aircraft equipment). Using the concordance, one would choose the wrong two-digit industry sector — a major error by any standard.

Second, a classification system looking only at the patent's subject matter may miss the forest for the trees on component inventions. Thus, GM engineers might come up with an improved high temperature alloy for use in jet engine fan blades. This invention would probably be classified to SIC code 331 (steel products) or perhaps SIC 333 (nonferrous metals). Again, the two-digit code in which the R & D is actually reported would be missed.

Third, a mismatch might occur because of peculiarities in the way a company's reports are organized. GM's turbojet engines are produced by the Allison Division, which also develops gas turbines for land vehicle use and generator set applications. Such engines happen to be classified under U.S. SIC code 351 (engines and turbines). If General Motors chose to report both turbojet engine and motive gas turbine R & D in code 351 because, say, the two activities were integrated and sales of non-aeronautical turbines were greater, even a technically correct assignment of turbojet engine patents to code 372 would fail to link the inventions to their underlying R & D. Or because gas turbines have been used experimentally in GM trucks, R & D data for *all* turbine operations might conceivably be integrated into the company's motor vehicle activity reports. In such instances, there is no way of determining how to effect a match with the relevant R & D dollars without knowing exactly how the company actually organizes its reports.

A Comparative Test

To ascertain how serious these and other problems might be, a test of two alternative patent classification methods was conducted. One was the computerized OTAF concordance system, as modified in 1980. The other was a 'manual' procedure used by the author and others to classify some 15 112 U.S. invention patents as part of a larger project studying the relationships between industrial R & D and productivity growth.[4] The proximate objective in the latter effort was to link patents to the 'lines of business' in which companies would report their 1974 R & D outlays under the FTC's Line of Business survey. With this in view, the classification team (consisting of four engineering and chemistry students) had extensive information on the product lines and publicly-announced Standard Industrial Classification codes into which sample company operations were subdivided. Also valuable to the matching effort was information on what specific activities companies had at particular geographic locations, since inventions could often be linked to locations from the inventor's address, especially when multiple inventors had the same city of residence.[5] Obscure cases were resolved through telephone calls to the companies. Despite this intensive effort, the team members' classifications failed to match any line of business under which companies actually reported for 7.3% of the sample patents, and on an additional 10.4% one or more of multiple coded origin industries did not match the companies' actual Line of Business program reports.

From the 15 112 patents classified in this way, a sample of 100 (later reduced to 99 because of information gaps) was drawn. The sampling procedure was not exactly random but is probably representative. In effect, the patents were arrayed in numerical order and every 150th patent from the larger sample was included. OTAF kindly provided a set of industry classifications based upon its updated 1980 concordance. The OTAF list contained all industry categories to which a class reference or cross-reference was made, so multiple assignments were possible. Indeed, for 17 patents, assignments were made by OTAF to different, non-overlapping two-digit industry sectors, and in 13 more cases, there were non-overlapping multiple assignments at the three-digit level within a two-digit category. These OTAF industry assignments were compared with those obtained under our manual approach.[6] A match was considered made whenever one of our industry codes agreed with or (since the OTAF categories were broader) nested within any OTAF category, whether or not there was disagreement on one or more other multiple codings. In other words, the comparison was biased on the side of minimizing the amount of disagreement found between our codings and those of OTAF.

Despite this bias, disagreements materialized at the two-digit level on 37 patents, or 37% of all cases.[7] In 49% of all cases, there was disagreement between our classifications and the finer (e.g. sometimes three-digit) industry codes generated by the OTAF procedure.

An appendix lists the sample patent numbers on which there was agreement at the two-digit level and, for the patents on which there was disagreement, provides details from which the nature of the disagreement can be discerned and/or rechecked. An attempt was made to characterize the two-digit code disagreements in terms of four qualitative problem types. Dual explanations were plausible for two patents, in which case each explanation was counted as one-half. For 13 patents, disagreement occurred because the OTAF classification emphasized the industry that might provide specialized process equipment when in fact the invention was made internally by a firm operating in (and presumed to be reporting in) a different *product* line. Twelve classifications differed because the invention involved a component, detailed feature, or material that OTAF associated with an industry other than the broader system or subsystem product line sold by the reporting company. Eight disagreements resulted because the OTAF classification, although probably correct in a technical sense, did not correspond with the company's divisional structure, as reported in annual reports, SIC codes published in the *Poor's Register of Corporations*, Dun & Bradstreet plant activity reports, etc. Of these, two involved inventions that were rather clearly by-products branching off from an established line of endeavor. Four disagreements appeared to come from anomalies or errors in the OTAF concordance.

Conclusion

It is evident that substantial difficulties are encountered in attempting to link patents to originating industries on the basis of information covering only the invention's subject matter and not the peculiarities of patent-receiving companies' internal organization. I know of only three ways to escape the problems identified here. One would be to repeat the procedure used in our research, manually inspecting and classifying each individual patent. This is very labor-intensive, costly, and (after a while) tedious. Members of our classification team averaged 6.7 patents per hour to extract various information, the most difficult of which was a proper origin industry code. Second, one might extend the OTAF computerized approach and take into account not only intrinsic subject matter (as reflected in patent classes and subclasses), but also the SIC codes in which companies report their operations and perhaps a matching of inventor addresses with the SIC codes of establishments in nearby postal zones. This would require extensive information updated regularly as well as quite complex computer programming, and it is far from clear that it would be feasible. Or third, one could turn to the individuals in the best position to have the desired information: the persons (either inventors or their attorneys) applying for a patent. This seems by far the simplest and most reliable approach, and it would add insignificantly to the cost of obtaining patent protection. If full advantage is to be taken of the rich information on technology flows that comes as a by-product of the patent system, such a change in patent application procedures should be seriously considered.

References

(1) A. C. Marmor, W. S. Lawson and J. F. Terapane, The technology assessment and forecast program of the United States Patent Office. *World Patent Information* **1**, 15–23; 1979.

(2) At the two-digit level, the U.S. SIC contains 20 manufacturing groups. At the three-digit level, manufacturing includes a total of 140 industry categories.

(3) F. M. Scherer, *Industrial Market Structure and Economic Performance* (rev. ed., R. McNally, 1980), pp. 76–77.

(4) For a methodological overview, see F. M. Scherer, 'Using Linked Patent and R & D Data To Measure Inter-Industry Technology Flows,' paper presented at a National Bureau of Economic Research conference on Research and Development, Patents, and Productivity, October 1981.

(5) For details, see the author's interim progress report under National Science Foundation grant PRA-7826526, 'Research and Development, Patenting, and the Micro-Structure of Productivity Growth,' February 28, 1980.

(6) No account is taken of reclassifications based upon the actual link to FTC Line of Business data, which are confidential.

(7) The alternative two-digit categories into which OTAF classifications were made comprised food and kindred products, textile mill products, chemicals, petroleum extraction and refining, rubber and plastic products; stone, clay and glass products; primary metals, fabricated metal products, non-electrical machinery, electrical equipment, transportation equipment, and scientific and professional instruments.

Appendix

A. Patents with Agreement at Both the Two- and Three-Digit Levels

3962430	3964094	3964844	3965209	3966429
3968092	3969651	3970452	3972046	3972706
3973873	3975073	3975528	3976278	3976718
3978446	3979557	3980321	3980852	3981497
3981928	3982958	3983346	3985586	3985968
3987107	3987600	3988649	3989366	3989771
3991313	3992432	3992953	3997007	3997519
3999690	4001693	4002375	4004416	4005027
4005524	4006140	4007205	4008376	4009031
4009448	4010097	4011711	4012328	4015077

B. Patents with Agreement at the Two-Digit Level but Disagreement at the Three-Digit Level

3960872	3963001	3966965	3984421	3984967
3986614	3994917	4001237	4003909	4006666
4007814	4011213			

C. Patents with Disagreement at the Two-Digit Level

Patent Number	Company	OTAF Classification (2-Digit SIC)	Our Classification (3-Digit SIC)	Explanation
3961367	RCA	35	365	Method for recording digital computer data signals. OTAF assignment to computer industry is technically correct. But RCA exited from the computer industry and is likely to report relevant R & D under communications equipment codes.
3962092	Dow Chemical	35	282, 307	Screen for plastic resin extruder. Dow sells the resins and products therefrom and is not primarily in the process machinery business.
3963644	Union Oil	28	291	Cracking catalyst for petroleum refining. This is a process invention for Union Oil as a petroleum refiner. Union is not primarily in the business of selling catalysts.
3965847	General Motors	35	371	Mileage odometer. Under Line of Business reporting, GM would integrate such component work into its motor vehicles category.
3967532	Caterpillar	34	353	Hydraulic valve for scraper equipment. Although some valves (but not this type) belong in SIC 34, the invention pertains to Caterpillar's construction equipment business.
3968624	GTE	35	344	Fastener for metal building construction of GTE's Unistrut subsidiary. Both metal buildings and fasteners are in SIC 34, so OTAF assignment appears to reflect a concordance error.
3969223	UOP	28	739	Refining process using zeolite catalyst. UOP is in the catalyst business, but this invention appears more closely related to their plant engineering business. This is a particularly close call.
3970876	Burroughs	36	357	Logic circuit using semiconductors. For Burroughs' computer systems business; they are not primarily in the semiconductor or other components business.
3971575	Combustion Engineering	35	344	Nuclear reactor fuel assembly locking device. For Combustion Engineering's reactor and boiler business. Patent subclass is pipe joints and couplings, so OTAF assignment to SIC 35 appears to be a concordance error.
3973133	General Tire	35	286, 376	Ozone generator from Aerojet General. OTAF assignment is right technically, but no reason to believe Aerojet General reports in the machinery industry.
3974305	Westinghouse	28	364	Treatment of chemical used in fluorescent lamps. Westinghouse sells the lamps, not the chemical.
3977217	Ingersoll Rand	35	396	Knitting needle element from Torrington, which characterizes its business as needles, not textile machinery.
3977963	Gulf Oil	28	291	Method to avoid poisoning hydrocracking catalysts. Reduces costs in Gulf's oil refining business. Patent class is mineral oil processes; not clear why assigned to chemicals, even though catalyst-related.

Patent Number	Company	OTAF Classification (2-Digit SIC)	Our Classification (3-Digit SIC)	Explanation
3979179	Anheuser-Busch	28	204, 208	Gelling organic liquids in starch processing. It is a chemical process, but Anheuser-Busch reports its activities not in chemicals but in foods.
3982371	Scovill Mfg.	34	243	Panel (e.g. glass window) holding structure. Although components of this invention are metallic, they pertain to the wooden windows and doors in which Scovill's Caradco Division specializes.
3983870	MMM	22	384	Surgical leg support stocking. Part of 3M's surgical supplies business, rather than conventional textile products.
3988216	National Steel	35	331	Method for making galvanized strip. National is in the steel business, not primarily in making machinery for steel processors.
3990191	Burlington Industries	35	221, 225	Yarn dyeing process. Burlington is in the fabrics business (with yarn production integrated with weaving and knitting), not primarily in the textile machinery field.
3990906	Goodyear	35	301	Process for cleaning tire molds. Goodyear sells tires, not tire-making machinery.
3991968	Chrysler	34, 35	371	Retaining device for automobile seat. Fasteners do go into SIC 34 if detached, but under FTC reporting guidelines, Chrysler would vertically integrate such parts work into auto business.
3993661	Polaroid	28	386	Photo developing chemical. Polaroid is primarily in the photo equipment business, whose SIC code in any event also includes photo processing chemicals.
3994244	Shell Oil	28	291	Method for treating petroleum refinery sludge wastes. Shell responding here to EPA mandates with respect to refining business. Patent class involves furnaces, so OTAF assignment to chemicals is surprising.
3995426	Caterpillar	37	353	Control device for fluid drive vehicle. The vehicle characterization misled the computer; Caterpillar's vehicles originate from the machinery industry, not motor vehicles.
3996032	PPG	35, 38	322	Method of making glass fibers. The invention involves glass-making machinery, but PPG is improving its fiber business processes, not selling machinery for its own sake.
3996466	Xerox	36	386	Corona discharge device for copying machine. The patent class is radiant energy and the invention does involve electricity, but Xerox's primary business here is copiers.
3998031	W. R. Grace	35	264, 282	Method for making foamed tampons. Grace was developing the tampons here, not machinery for sale. The line was new and it was unclear whether it would be reported under sanitary paper products or as an extension of Grace's resin business at the same laboratory.
3998667	Owens-Illinois	36	322, 329	Method for making glass ceramics for semiconductor doping. Owens-Illinois sells the materials to semiconductor makers and is not in the semiconductor business *per se*.
3999062	IBM	38	357	Spectrophotometer for color analysis used with computers. OTAF classification is technically correct, but IBM is not primarily in the instrument business. This invention enhances computer sales.
4000206	Phillips Petroleum	28	291	Process to make gasoline, benzene, hydrogen, and other by-products. Does provide petro-chemical feedstocks, so a close call, but used in petroleum refinery.
4000703	Pullman	34	374	Train hopper car hatch locking mechanism. Locking mechanisms are in SIC 34, but Pullman's business is railroad cars.
4002829	W. R. Grace	36	307, 355	Optical scanning device for image-making. For Grace's new Letterflex plastic blanket printing process, whose coding was uncertain, but was clearly not in electrical equipment.

cont. overleaf

Patent Number	Company	OTAF Classification (2-Digit SIC)	Our Classification (3-Digit SIC)	Explanation
4003427	Grumman	34	376	Heat pipe fabrication. OTAF classification technically correct, but Grumman is not listed in pipe business. This is a spillover from space work under Department of Transportation contract.
4010640	American Cyanamid	36	289	Apparatus to measure structural stiffness of adhesives. Cyanamid is not recorded as operating in electrical equipment, but does sell high-performance adhesives, to which this invention pertains.
4012699	AT & T	38	366	Laser-based optical amplifier. Bell System developed this as part of its communications equipment work, not as an instrumentation subsystem in its own right.
4013442	Exxon	34	287	Encapsulation of fertilizer particles. Part of Exxon's fertilizer making business. Unclear how OTAF concordance put metal carbonate subject matter from fertilizer chemistry class into fabricated metal products.
4013872	Mobil	36	291	Temperature control device for heating of oil. Improves Mobil's petroleum refining processes, not primarily for sale as electrical machinery.
4014638	GTE	34, 38	364	Photoflash lamp. GTE-Sylvania makes the lamps, which the SIC puts under electric lamps rather than photo equipment (SIC 38). Probable concordance error.

[16]

EXPLORING THE TAIL OF PATENTED INVENTION VALUE DISTRIBUTIONS°

Dietmar Harhoff*, Frederic M. Scherer**, Katrin Vopel***
** University of Munich, Munich School of Management, Munich, Germany*
*** Harvard University, John F. Kennedy School of Government, Cambridge MA, USA*
**** Centre for European Economic Research (ZEW), Mannheim, Germany*

Chapter contents:

Abstract: We explore the tail of patented invention value distributions by using value estimates obtained directly from patent holders. The paper focuses on those full-term German patents of the application year 1977 which were held by West German and U.S. residents. The most valuable patents in our data account for a large fraction of the cumulative value over all observations. Several tests are conducted to pin down more precisely the nature of the high-value tail distribution. Among the Pareto, Singh-Maddala and log normal distributions, the log normal appears to provide the best fit to our patented invention value data.

Keywords: Patents, skew distributions
JEL Classification: O34, O31

° The research underlying this paper was supported by grants from the Sloan Foundation and the ZEW. The authors are also grateful for the invaluable research assistance of Jörg Kukies, Marc Dehoust, Teri Elniski, Renate Häßler, Sascha Karrassek, Stephan Seeberger, Stefan Schmidt, Suzanne Sheldon, and Jesus Viejo Gonzalez, and for helpful comments from Bronwyn Hall and Jenny Lanjouw.

O. Granstrand (ed.), Economics, Law and Intellectual Property, 279-309.

Economics, Law and Intellectual Property

12.1 Introduction

Many industrialized nations charge invention patent holders periodic maintenance or renewal fees. Failure to pay causes the patent permanently to lapse. This phenomenon has been discovered (or rediscovered[1]) by economists and exploited to estimate the economic value of patent protection.[2] The lower the incremental fee at which payment is discontinued, the smaller is the patent right's estimated value. Recent patent renewal studies reveal a highly rightward-skewed distribution of values, with the weight of evidence favoring a log normal distribution of individual patent values, although an even more skew Pareto-Levy distribution has not been ruled out conclusively. However, for the sizeable fraction of patents that remain in force to their full term, the distribution parameters must be derived by extrapolation from patterns for patents allowed to expire earlier. This paper draws upon two new surveys to estimate directly the value of right-hand tail inventions whose patent renewal fees were paid to full-term expiration.

Exploring the distribution properties of the tail is important, because with skew distributions, outlying tail values account for a large fraction of the cumulative value over all observations. Lanjouw *et al.* (1998), for example, estimate that within four technology groups, the most valuable five percent of German patents – all presumably renewed to full term – contributed 22 to 31 percent of the total value of all cohort members. The longer and thicker the distribution's right-hand tail, the more difficult it is for business managers to achieve stable average returns through portfolio strategies. Indeed, if the distribution of values is Pareto-Levy and the distribution parameter $\alpha < 1$, the distribution has neither finite mean nor variance asymptotically, which implies that sample means fail to converge (under the weak law of large numbers) toward a stable value.[3] Extreme skewness of invention values (as distinguished from the value of covering patents) can also have macroeconomic implications. Assuming a Pareto distribution of high-value inventions with $\alpha = 0.8$, Nordhaus (1989) found through a simulation study that, aggregating across the entire U.S. economy, invention-based productivity effects fluctuated widely over time as a result of unusually favorable or unfavorable Monte Carlo simulation draws. His simulated patterns corresponded remarkably closely to actually observed productivity growth fluctuations.

[1] For an earlier but less rigorous analysis, see Dernburg and Gharrity (1961).
[2] See e.g. Pakes and Schankerman (1984), Schankerman and Pakes (1986), Pakes (1986), Lanjouw (1998), and Lanjouw, Pakes and Putnam (1998).
[3] See e.g. Johnson *et al.* (1994), Chapter 20.

12.2 Methodology

12.2.1 Sampling and Survey Questionnaire

The German patent renewal fee system provided for us a crucial wedge into the full-term patented invention value question. Germany was chosen because its patent examination system is unusually rigorous, screening out applications of low inventive content, and because its renewal fee schedule is one of the most progressive in the world. Figure 12-1 shows how renewal fees escalated for patents on which applications were filed in 1977. Beginning with DM 100 for the third year, the fees keeping the patent in force over incremental years rose to DM 2,700 for the 18th year. Over the maximum possible term of 18 years, cumulative renewal fees of DM 16,075 had to be paid.[4]

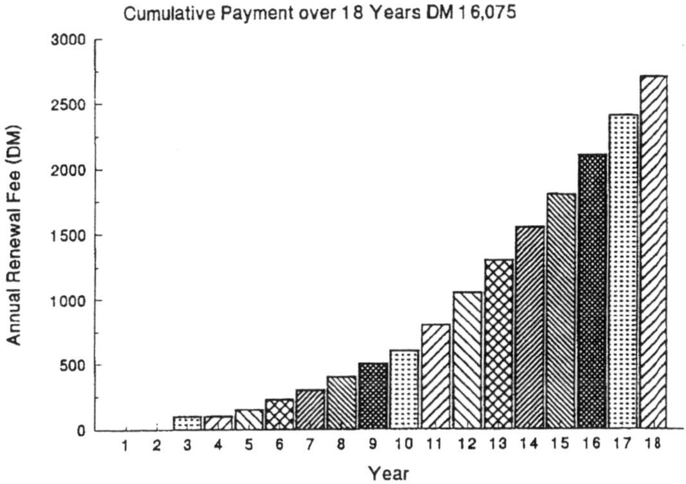

Figure 12-1 Progression of German patent renewal fees over time

[4] The maximum term was extended to 20 years after 1977. For patents declared available for non-exclusive licensing, renewal fees were reduced by 50 percent. Interviews revealed that firms sometimes chose not to pay their fee for the last year because patents are cancelled for non-payment only after half the year has passed.

[5] Most of our U.S. sample patents had 1976 U.S. priority dates, but their German counterpart patent expiration dates were geared to 1977 German applications. Some German patents in our sample had application dates (Nachmeldungen) later than 1977, but were entitled to a 1977 German priority.

Our survey was targeted toward the Class of 1977 – i.e. applications with a 1977 priority date, leading to patents expiring at full term during 1995.[5] Since the survey was conducted in 1996, only a year after the final year's fees were paid, memories concerning survey patents' value remained reasonably fresh. Beginning in 1978, inventors seeking patent protection in Germany enjoyed the option of filing patent applications with the European Patent Office. The emergence of two parallel systems complicates the tracking of inventions, and so the Class of 1977 was the last one for which a relatively clean research design could be implemented.

The total number of German patent applications filed in 1977 was 57,782, of which 27,954 were from applicants residing in West Germany and the remaining 29,828 from foreign applicants.[6] The lag from application to issue was bimodal, with mean of 6.5 years, median of 6 years, and modes at three years (16.9 percent of all issues) and nine years (11.2 percent of all issues). Of the 22,903 patents that were issued and not revoked in subsequent opposition proceedings,[7] 4,349, or 19.0 percent, were renewed to full term. For the six leading invention-originating nations, the number of patent applications, full-term patents, and full-term renewal rates as a percentage of applications are summarized in Table 12-1.

Table 12-1: Patent applications and full-term renewals by nation, 1977

Country of Origin	Applications	Full-Term Patents	Full-Term Patents as Share of Applications
Germany	27,954	1,431	5.12%
United States	10,207	900	8.82%
Japan	4,469	936	20.94%
France	2,675	221	8.26%
Switzerland	2,509	214	8.53%
Great Britain	2,451	166	6.77%
Other	7,517	481	6.40%
All Nations	57,782	4,349	7.53%

[6] Approximately 6,900 of the 57,782 applications were withdrawn by applicants before the 18-month initial publication deadline.

[7] See Harhoff, Scherer and Vopel (2003) for a discussion of opposition proceedings and their possible use as an indicator of patent value.

Full-term renewal rates are considerably higher for foreign patent recipients than for German residents, consistent with the finding of Putnam (1996) that inventors seek patent protection outside their home jurisdictions mainly for their most promising inventions. The extraordinarily high renewal rate for Japan suggests a more aggressive approach to maintaining patent portfolios; see IVA (1993).

The research for this paper focused on the full-term patents held by West German and U.S. residents. For the German patents, a two-stage methodology was pursued. A preliminary telephone and telefax screening elicited patent value estimates and identified the most valuable patents, about which on-site interviews were then sought to develop more detailed historical information. For the U.S. patents, only an extended first stage was executed.

Of the 1,431 full-term German patents, it was possible to trace 1,325, or 92.6 percent, to surviving firms or individual inventors. Most of the untraced patent owners were individual inventors or firms too small to be located through standard business directories. The 684 firms or individuals owning the 1,325 traced patents were contacted by telephone to identify the most appropriate respondent, and a questionnaire was dispatched, usually by facsimile, on each relevant patent. Its essence was a single substantive counter-factual question,[8] translated as follows:

> "If in 1980 you had known how its contribution to the future profitability of your enterprise would unfold, what is the minimum price for which you would have sold the patent, assuming that you had a good-faith offer to purchase?"

This asset value approach (consistent with the view of patents as intellectual property) was chosen over a question asking for the accumulated flow of profits following pilot interviews with German patent-holders testing both approaches. The team contacting patent owners had a more detailed written protocol answering the most common queries posed by recipients.[9] Respondents were asked to place their patent within one of five broad minimum

---

[8] On the use of counter-factual questions, see Tetlock and Belkin (1996).

[9] On the basis of early experience, the question posed in the U.S. survey instrument was altered slightly, conforming to the guidelines used in the German contact team's protocol: "If in 1980 you knew what you now know about the profit history of the invention abstracted here, what is the smallest amount for which you would have been willing to sell this patent to an independent third party, assuming that you had a bona fide offer to purchase and that the buyer would subsequently exercise its full patent rights?"

selling price ranges: less than DM 100,000, DM 100-399,999, DM 400-999,999, DM 1-5 million, and more than DM 5 million.

Our approach adopts a distinctly different counter-factual assumption than the one implied by previous patent renewal studies. Failure to renew a patent means that the covered subject matter will move into the public domain, allowing it to be used by both the original patent holder and others. Erosion of the original inventor's profits due to non-renewal is often constrained by first-mover advantages shown by Levin *et al.* (1987) to be more potent on average than patent protection. In our approach, on the other hand, sale of all rights in a patent to a third party implies that the buyer can prevent the original patent holder from practicing the subject invention or demand license fees approaching the value of profits foregone. Conversely, in studies using the renewal value of patents, the patent that is not renewed lapses into the public domain so that the former patent holder can still employ the formerly patented technology. For broad patents that covered key features of products or processes – a case found in interviews to be common for the most valuable patents – the sale of rights to a competitor puts at risk the whole stream of quasi-rents realizable from a product or process. Or at minimum, the sale of rights can impose upon the seller the profit sacrifice from foregoing key patented features or the cost of inventing around them. Therefore, higher valuations are expected under our approach than under the renewal approach.[10]

12.2.2 Response Rates

After extensive follow-up, positive responses were obtained on 772 patents (58.3 percent of the total 1325 on which contacts were initiated) held by 394 organizational entities and individual inventors (including ten unaffiliated inventors). Table 12-2 reports probit regressions of a binary survey participation variable on the characteristics of respondents and non-respondents for 684 full-term patent-holding firms and individuals. Three explanatory variable groups are introduced: firm size (or a dummy variable for independent inventors), industry or (for independent inventors) organizational affiliation, and the professional specialty of the principal contact person. The base (intercept) case is a firm with 50 or fewer employees in the machinery industry, with a patent counsel as the principal contact person. The firm size ef-

---

[10] We elaborate on this issue in a companion paper where we use a simple formal model to contrast the renewal and the asset approaches to defining the value of a patent. See Harhoff, Scherer and Vopel (2003).

fects are significant and generally rising with employment when they are considered in isolation, but fail to be significant with respondent and industry effects included. The respondent effects are highly significant, with responses being particularly likely when an inventor or patent attorney was the principal contact. Pseudo-r^2 values reveal that the three sets of characteristics explain at most only modest amounts of response variance.

Table 12–2: Probit analysis of German patent-holder response behavior*

Characteristic	Number of Cases	Firm Size Model	Full Model
Individual inventor	55	-0.282 (0.078)	-0.441 (0.080)
Firm Size (employees):			
<50	115	(reference group)	(reference group)
50-249	129	0.056 (0.061)	0.041 (0.066)
250-499	77	0.010 (0.072)	-0.010 (0.076)
500-999	87	0.083 (0.066)	0.055 (0.074)
1000-4999	112	0.192 (0.058)	0.104 (0.072)
5000 or more	52	0.122 (0.075)	-0.002 (0.097)
No. of employees unknown	57	-0.038 (0.080)	-0.131 (0.087)
Firm Size Dummies χ^2		14.4	8.2
p-value (df)		p=0.025 (6)	p=0.223 (6)
Respondent Function:			
Patent counsel	179		(reference group)
R & D Executive	142		-0.056 (0.064)
Inventor	63		0.074 (0.082)
General management	181		-0.250 (0.061)
Legal counsel	13		-0.057 (0.155)
Marketing / PR	21		-0.371 (0.097)
Function unknown	30		-0.490 (0.063)
Respondent Type Dummies χ^2			51.5
p-value (df)			p=0.001 (6)
Industry Group Dummies χ^2			14.6
p-value (df)			p=0.101 (9)
Number of observations		684	684
χ^2 (df)		38.82 (7)	107.2 (22)
Pseudo-R^2		0.042	0.116

* Standard errors are reported in parentheses. The dependent variable is participation in the survey (0/1).

A separate analysis reported in Harhoff, Narin, Scherer and Vopel (1999) revealed that the economic valuations elicited in our survey were positively and significantly associated with the number of citations in German patents subsequent to the survey patents. Companies holding four of the most highly cited Class of 1977 full-term patents chose not to participate in the survey. Inspection suggests that those non-responding patents would have reported relatively high economic values. However, a formal test was unable to reject the hypothesis that citation frequency did not differ between responding and non-responding patents. Thus, there appears to be no evidence of selection biases in our first-stage survey responses.

The U.S. survey, added as a by-product of the German survey, used a stratified sampling approach. All U.S.-resident holders of two or more full-term German patents (totalling 630) were included in the starting sample, as were 89 holders of a single full-term patent, including 70 patents held by U.S. corporations known to remain in existence and 19 by New England-based individuals and small companies. A manual search of these 719 German patents was conducted to identify U.S. patent application priority serial numbers for the U.S. patent applications on the basis of which, under international rules, a priority filing date is accepted in foreign jurisdictions. From those 719 German patents, 632 could be linked to valid U.S. application serial numbers – sometimes several per patent. Eighty-seven others dropped out because the priority claims were to nations other than the United States (e.g., when the invention came from a foreign subsidiary of a U.S. multinational enterprise) or because no non-German priority claims were reported. The U.S. application serial numbers for the remaining 632 German patents were linked to 485 U.S. patents, which comprised the basis of our U.S. sample.[11] For the 23 percent of cases with no link to a U.S. patent, the most likely explanation was rejection of applications by the U.S. Patent Office, although there are gaps in the underlying records, and a few parallel U.S. patents may simply have been missed.[12] For 20 of the 485 U.S. sample patents, no questionnaires were dispatched, in 11 cases because no address could be found for the patent holder and in nine other cases because linking

[11] The U.S. serial numbers for entities with two or more German patents were linked by the U.S. Patent Office, for which the help of Jane Myers and William Brown is gratefully acknowledged. The link for the single German patent subsample was accomplished using the Boston Public Library's computerized patent data base.

[12] Roughly two-thirds of patent applications filed in the United States during the 1970s resulted in issued patents. That only 23 percent of the linked applications failed suggests a positive correlation between the probability of issuance in the two jurisdictions.

errors were discovered (e.g., the U.S. application was filed too early or too late to support a 1977 German application).

Of the 465 U.S. questionnaires sent, usually after a preliminary telephone contact, 225, or 48.4 percent, were completed by a total of 78 companies or (in three cases) individual inventors. [13] There was a clear bias in the returns. For 12 companies with eight or more linked patents, the response rate was 37 percent of relevant patents despite intensive follow-up; for firms with fewer than eight patents, the rate was 54 percent, even though follow-ups for the companies with fewer than four patents were less thorough. Two main reasons were given for non-response by the companies with extensive patent holdings: unacceptable compliance burden, and fear that response would reveal too much about the profitability of individual company divisions.

12.3 Results

12.3.1 The First -Stage German Survey

Table 12–3 reports the distribution of estimated values for 772 German patents on which first-stage responses were obtained. Ninety-nine patents were valued in the highest range, at more than DM 5 million. Figure 12–2 plots the cumulative distribution function on doubly logarithmic coordinates. If the distribution of values were Pareto-Levy of the simplest form, it would plot as a straight line on the coordinates chosen, i.e., with $\log N = \log k - \alpha \log V$, where V is the reported patent value (in thousands of German Marks) and N is the number of patents with value V* or greater. An ordinary least squares regression fitted to the data is (with standard deviations in brackets): [14]

$$(1) \qquad \underset{[0.116]}{\log N = 3.481} - \underset{[0.043]}{0.385} \log V; \qquad R^2 = 0.963,\ n = 5.$$

[13] For some extensively restructured corporations, divisions were treated as separate companies.

[14] Since we are particularly interested in the right-hand tail of the patent value distribution, we use in this preliminary characterization an unweighted regression of group geometric mean values and ranks, rather than weighting by group patent counts. The slope from observation-weighted maximum likelihood estimate (e.g., in Table 4, Part A) is smaller than the unweighted estimate of regression (1). The lower-bound value for the "less than DM 100,000" group was set at DM 23,000, which approximates the cost of initial patent application plus full-term renewal payments. Logarithms are taken to base ten here.

[15] Similar results emerged with ordered probit regressions. That is not surprising given that the log value thresholds were chosen such as to induce interval scaling in the variable VALUE.

Adding a quadratic value term increases R^2 by 0.034, which is significant at the 5 percent level in an F-ratio test. If a log linear regression is fitted to the three highest-value classes (with one residual degree of freedom), the estimated slope is 0.518. Both slope estimates fall into the range of α values at which Pareto distributions lack asymptotically finite means and variances. Thus, a high degree of skewness is suggested.

Table 12-3: Distribution of full-term German patent values

Estimated Value Range	Number	Percent
More than DM 5 million	99	12.8 %
DM 1-5 million	121	15.7 %
DM 400,000 - 999,999	149	19.3 %
DM 100,000 - 399,999	200	25.9 %
Less than DM 100,000	203	26.3 %
Total	772	100.0 %

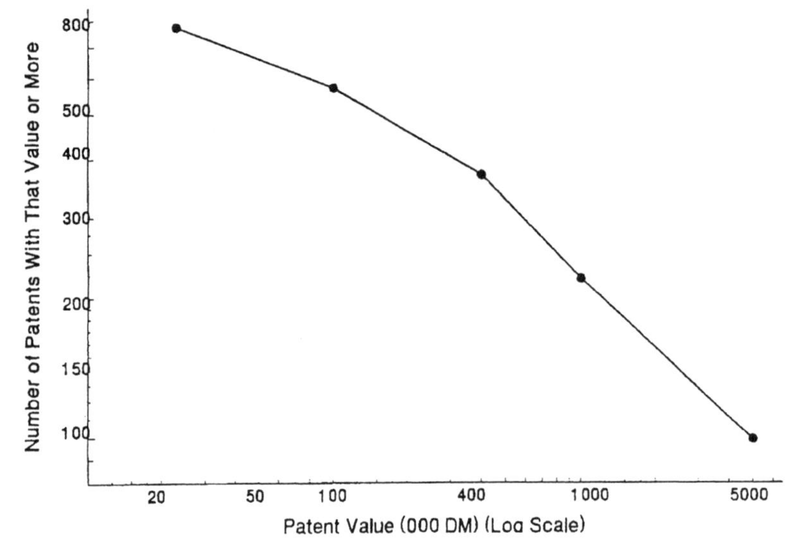

Figure 12-2: Pareto plot of surveyed German patent values (N = 772)

In a disaggregated regression over 772 patent responses, the five value reporting classes were characterized by a dependent variable VALUE, ranging from 1 (for the least valuable inventions) to 5 (for the most valuable). Zero-one dummy variables were defined for patents originating from electrical industry firms (ELEC, 226 cases), chemical industry firms (CHEM, 131 cases), and machinery makers (MACHY, 203 cases); and for responses from patent counsel (PAT, 426 cases), research and development executives (RDEXEC, 128 cases), and inventors (INVENT, 77 cases), with all other industries and responses from other occupational specialties serving as the base or intercept case. Other variables were the logarithm of company employment (EMPLOY) and a one-zero dummy GAP for 64 cases with no employment data. The resulting regression, with standard errors in brackets, was as follows: [15]

$$\text{(2)}\quad \underset{[0.167]}{\text{VALUE} = 2.491} + \underset{[0.133]}{0.004\ \text{ELEC}} + \underset{[0.159]}{0.626\ \text{CHEM}} + \underset{[1.475]}{0.200\ \text{MACHY}}$$

$$- \underset{[0.136]}{0.372\ \text{PAT}} - \underset{[0.147]}{0.157\ \text{RDEXEC}} + \underset{[0.165]}{0.358\ \text{INVENT}}$$

$$+ \underset{[0.050]}{0.063\ \log \text{EMPLOY}} - \underset{[0.250]}{0.010\ \text{GAP}};$$

$$R^2 = 0.0441,\quad n = 772.$$

The value responses of patent counsel are found to be on the conservative side, ceteris paribus, while inventors lean toward relatively generous estimates of their contributions. Chemical industry inventions are more valuable on average by nearly two-thirds of a choice interval, possibly because (as Levin *et al.* (1987) show) patents provide a particularly strong appropriability regime. The analysis reveals no significant firm size effects. Adding a quadratic employment variable (not separately reported) left both size variables insignificant. Industry, respondent, and company size variables together explain only a small fraction of the variation in patent value responses, suggesting that most of the variation was invention-specific.

12.3.2 The German Interview Survey

Interviews were sought with all 69 firms owning the 99 patents whose value was estimated at DM 5 million or more in the first-stage questionnaires. Fifty-five on-site interviews were conducted, supplemented by one telephone interview and five cases in which patent holders completed a detailed questionnaire without a face-to-face interview.[16] The interviews covered a wide range of topics, including the history of the inventions patented, the number of nations in which patents were sought, the extent of licensing (domestic and foreign), the incidence of infringement disputes if any, and most importantly for the present paper, a detailed exploration of the subject patents' value. Interviewees were asked to supplement the minimum patent-rights sale-price range answers to the questionnaire's counter-factual question with a point estimate, surrounded when appropriate by a range of uncertainty. In addition, they were asked to estimate the annual sales (including foreign and export sales) and profits, cost savings, and/or royalty receipts attributable to the patented invention, the fraction of profits that would not have been achieved without the patent (usually, the source of the greatest estimation uncertainty), and the role patent protection per se played in sustaining the profits.[17] Interviewees often had detailed records on which to base their answers because of a German law requiring employers to compensate employee inventors for their inventions using a formula including revenues, profits, the fraction of sales attributable to subject inventions, and the degree of initiative exercised by the employee.[18] The estimates analyzed here focus in most cases on the profit sacrifice that would have resulted if the firm were precluded from using the patented invention, not on the role patent protection played per se, although in some cases the two questions were disentangled only with difficulty. Some interviews were able to cover in detail only one high-value invention among several held by a respondent. Careful analysis of one (usually the most valuable) was considered preferable to superficial examination of several.

Useable value responses were obtained on 69 patents originally valued at DM 5 million or more. Estimates for ten other patents were considered too unreliable to include in the formal analysis.[19] Among the 69, profit flow estimates were available in 54 cases; counter-factual asset sale price estimates were available in 38 cases; and in 23 cases, both flow and asset sale estimates were obtained. In those 23 cases, the geometric mean value was DM 11.1 million per invention with flow estimation and DM 19.1 million in the

---

[16] The interviews were conducted primarily by Jörg Kukies and Katrin Vopel.

[17] The interviews explicitly tallied the value of both domestic and foreign rights. On average, patents covered by interviews were patented in 10 nations other than Germany, with a standard deviation of 9.9.

[18] See *Patent- und Musterrecht* (1995, pp. 92-109).

[19] The estimated average value per patent in those cases was DM 22.6 million.

asset sale counter-factual. Although the differences appear sizeable, they are not significant at the 0.05 level (t = 1.21) in a test of logarithmic means.

The flow value estimates were computed from annual sales and profit ratio data, with all values converted to 1995 price levels and then discounted back to the year 1980 at a 0.05 real discount rate. In what follows, flow value estimates are used in all but the 15 cases in which they were unavailable and counter-factual estimates were substituted. In 57 cases, best-estimate values were accompanied by an error range, whose average single-side magnitude was 25.1 percent of the best estimate, with a median value of 22.7 percent.

Thirteen of the best estimates, all based upon flow data, were below DM 5 million. In five of those cases, counter-factual asset value estimates were elicited also at the interview stage. All five were greater than or equal to DM 5 million. Discrepancies resulted in these and other cases for complex reasons – e.g., because a high counter-factual value was placed on the blocking role of patents or on ultimately unrealized, but in 1980 plausible option values, because it was difficult to disentangle the role of the invention in a complex systems product or because multiple patents covered the invention, and because respondents estimated surprisingly low profit margins on product sales. We have opted to err on the side of accepting the more conservative flow values when conflicts existed.

Altogether, the best-estimate values ranged from less than DM 1 million (in five cases) to well over DM 1 billion. The mean value was DM 38.8 million if all 69 observations are averaged and DM 15.1 million if the most lucrative invention, with a point estimate roughly 15 times that of the second most valuable invention, is excluded. Figure 12–3 introduces the values for individual high-value patents into the cumulative distribution function of Figure 12–2, multiplying each interview patent's rank by 1.45 to reflect the fact that only 69 of the 100 first-stage high-value patents yielded useable estimates, and aggregating interview estimates below DM 5 million into appropriate lower-value groups. Relative to Figure 12–2, the overall distribution becomes considerably more concave downward. The straight-line fit associated with a Pareto distribution is plainly absent. However, the large jump between the largest and second-largest patent values leaves considerable uncertainty concerning the distribution's form. If a function linear in the logarithms is forced by ordinary least squares on all 56 observations of DM 5 million or more, the slope is -0.874, with standard error of 0.033. If the billion-DM outlier is excluded, the estimated slope steepens to -1.040, with standard error of 0.027.

Several tests were conducted to pin down more precisely the nature of the high-value tail distribution. A graphical test attributable to Bryson (1974) solidly rejected the hypothesis that the distribution for patents valued at DM 5 million or more was negative exponential, Weibull, or any less skew variant. See also D'Agostino and Stephens (1986, Chapter 2).

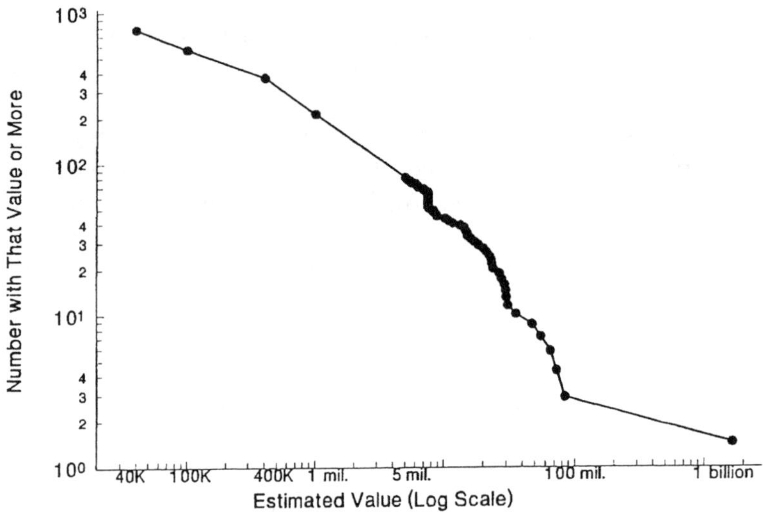

Figure 12–3: Plot of all renewed German patent values on Pareto coordinates

A difficulty with any test is that our data relate only to the right-hand tail of a wider patented invention value distribution. Figures 12-4a and 12-4b provide a simple graphic test of log normality attributable to D'Agostino (1986, pp. 23-29). Where V is mean log patent value and *s* its standard deviation, we plot the standardized log normal variate $Z_i = (V_i - V)/s$ actually observed against a theoretical log normal distribution (from published tables) with the same mean and standard deviation. A necessarily arbitrary assumption is that the total number of observations in the high-value half of the theoretical distribution consists of the 69 patents with originally reported values exceeding DM 5 million plus 69/100^{ths} of the lower-valued full-term observations.[20] Thus, the high-value cases with useable point estimates are

[20] Because the complete distribution is not known, the means and standard deviations are ambiguous. Where Z* is the theoretical standardized normal variate derived on the basis of a

assumed to be a sample from the hemi-distribution. Not surprisingly, the theoretical log normal variates over-predict the high-value distribution extremity when the single billion-DM outlier is excluded (Figure 12-4a) and under-predict it when that outlier is included (Figure 12-4b). For a quantitative comparison with the Pareto-Levy distribution, theoretically predicted values of V_i were derived from the relationship $V_i^* = V + s\, Z_i$. The sum of squared deviations SSD between those predictions and the actual V_i was computed and compared to the equivalent SSD from fitting a one-parameter Pareto-Levy regression.[21] The resulting SSDs (in logarithms to base 10) are as follows:

	Log Normal	**Pareto-Levy**
All patents with V > DM 5 mil.	0.6634	0.7215
Largest outlier excluded (n = 55)	0.2345	0.3128

In both cases, and especially when the largest outlier is excluded, the fit is better with a log normal distribution than with the Pareto-Levy.

Maximum likelihood techniques provide a more general approach to the goodness-of-fit problem, permitting *inter alia* extrapolation of the fitted distribution function to all patents, including those not renewed to full term, and hence excluded from our survey. Three highly skew alternatives were evaluated by maximum likelihood techniques: the two-parameter log normal, the one-parameter Pareto-Levy, and the three-parameter Singh-Maddala distribution. The third of these alternatives, whose high-value tail approximates the Pareto-Levy but does not impose log-linearity on low-value observations, has been used with success in characterizing income distributions. See Singh and Maddala (1976), McDonald (1984), McDonald and Ransom (1979), and Esteban (1986). The likelihood functions used to estimate distribution parameters are summarized in an appendix. Here we focus on the results and their interpretation.

patent's value rank, best-fitting values were estimated using the transformed regression log $V_i = V + s\, Z_i + \varepsilon_i$, with V and s as parameters.

[21] For comparability, and unlike regression (1), the sum of squared deviations was minimized in the patent value direction.

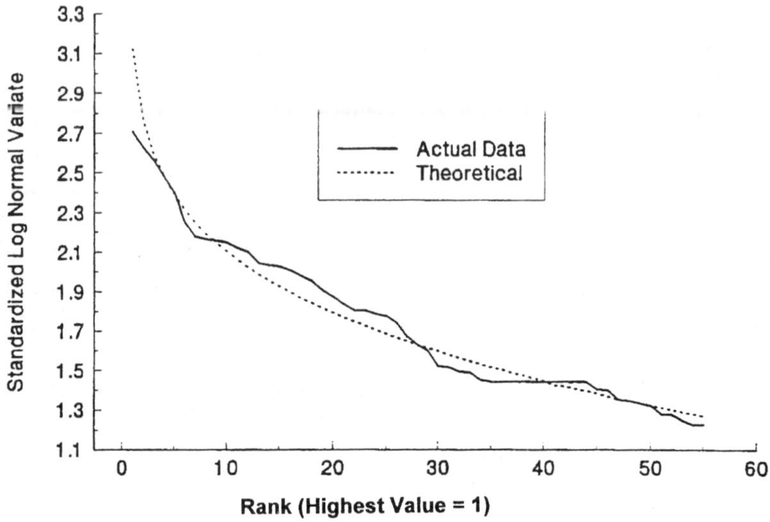

Figure 12–4a: Log normal plot of high-value German patent values (highest value excluded, N=55)

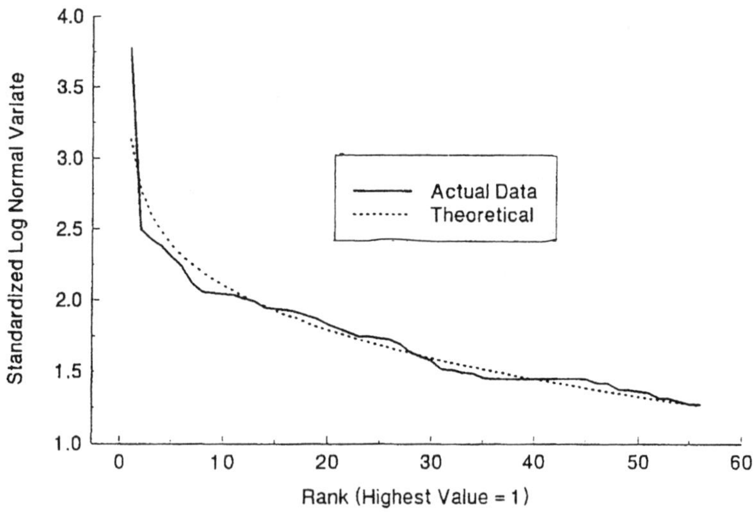

Figure 12–4b: Log normal plot of high-value German patent values (all values included, N=56)

Table 12–4 summarizes the estimation results and goodness-of-fit tests. Part A focuses only on the patents in our response group, i.e., excluding German-origin patents not renewed to full term and truncating the lower-bound value of the full-term patents at DM 23,000.[22] Patents valued above DM 5 million were divided into three discrete groups. Both with and without the most valuable outlier, the log normal form provides a distinctly superior fit. Part B reports maximum likelihood estimates when the distribution includes 9,282 non-renewed German-origin patents, assumed to be generally less valuable than the renewed patents but constrained to have positive values.

Table 12–4: Maximum likelihood estimates for three alternative distributions

Table 12–4 Part A: German patent survey responses only*

Distribution	Pareto	Pareto	Log Normal	Log Normal
Sample	N = 772	N = 771 (largest value excluded)	N = 772	N = 771 (largest value excluded)
Parameters	$\alpha = 0.341$	$\alpha = 0.342$	$\mu = 12.09$ $\sigma = 2.428$	$\mu = 12.14$ $\sigma = 2.366$
χ^2 (df)	175.3 (7)	179.6 (6)	22.4 (6)	26.0 (5)

* Lower-bound patent value truncation at DM 23.000

Table 12–4 PartB: Including non-renewed German-origin patents

Distribution	Pareto	Pareto	Log Normal	Log Normal	Singh-Maddala	Singh-Maddala
Sample	N = 772 + 9282	N = 771 + 9282 (largest value excluded)	N = 772 + 9282	N = 771 + 9282 (largest value excluded)	N = 772 + 9282	N = 771 + 9282 (largest value excluded)
Parameters	$\alpha = 0.216$	$\alpha = 0.216$	$\mu = 4.028$ $\sigma = 5.463$	$\mu = 4.061$ $\sigma = 5.433$	a = 10.02 b = 62.39 q = 0.034	A = 10.19 b = 63.83 q = 0.037
χ^2 (df)	565.6 (9)	573.0 (8)	192.7 (8)	199.5 (7)	325.2 (7)	333.5 (6)

[22] Estimates for the Singh-Maddala distribution could not be obtained with truncation, since the numerical optimization routine did not converge. For the log normal and Pareto distributions, we experimented with other truncation points, e.g., by analyzing only patents with reported values greater than DM 1 million or DM 5 million. In all cases, the log normal distribution provided a better fit, but the differences between the two became quite small and sensitive to omission of the most valuable patent.

Both with the most valuable outlier included and excluded, the log normal distribution provides by far the best fit, with the Singh-Maddala distribution placing second and the Pareto form third. Chi-squared values rise dramatically relative to those for estimates confined to our right-hand tail sample observations, reflecting the larger number of groups and the difficulties of extrapolating outside the range of high-value observations.[23] Estimated means and standard deviations for the log normal form also deviate considerably between the truncated and all-patent analyses.[24] However, graphic inspection reveals that the shape of the right-hand log normal tail changes little between the truncated and all-patent cases. It seems clear that the log normal distribution provides a quite good fit, especially for the high-value tail.[25]

12.3.3 The U.S. Survey

While the German data can be considered reasonably representative for the 1995 group of patents, our U.S. survey does not allow us to test for the overall distributional properties with the same confidence as with the Ger-

---

23 Formally, the Chi-squared values are sufficiently high to reject even the log normal distribution hypothesis. The rejection of distributional assumptions is a common result in large samples, e.g., in the income distribution literature. See McDonald (1984) and Kloek and Van Dijk (1978).

24 The parameter estimates here, unlike other parts of the paper, are in logarithms to base e.

25 We conducted additional tests to determine how Paretian slope estimates and coefficients of skewness (estimated in the logarithms) were affected by plausible individual patent valuation errors. The estimates are fairly insensitive to changes in patent values within the reported range of uncertainty.

26 Inflating for the less than full sampling of entities holding only one full-term German patent, the total number of U.S. counterpart patents must have been approximately 604, or 1.5 percent of the 41,252 U.S. patents issued to U.S. residents in 1978 – the median year of issue for linked U.S. patents.

man data. Nonetheless, the U.S. data can be used to generate some interesting insights.

The 485 linked U.S. patents with which our U.S. survey began comprised an even smaller fraction of their applicable universe – less than two percent – than the patents in our German survey.[26] They were in effect thrice-screened – once by being applied for both at home in the United States and in Germany, once by having applications approved in both relatively stringent jurisdictions, and again by being renewed to full term in Germany.[27]

Because field interviews were not contemplated, the range of values among which respondents were asked to choose was elaborated and expanded to include at the upper bound a category, "more than $100 million". Table 12–5 reports the distribution of outcomes. One respondent placed all three of his company's patents in the "more than $100 million" group and added explanatory comments (invited on all questionnaires) showing a clear misinterpretation of the requested counter-factual question. Those three patents are excluded from subsequent analyses, leaving 222 valid responses.

Table 12–5: Distribution of full-term U.S. patent values

Estimated Value Range	Number	Percent
More than $100 million	22	9.8
$50-100 million	6	2.7
$20-50 million	10	4.4
$5-20 million	34	15.1
$1-5 million	41	18.2
$500,000 to $1 million	30	13.3
$100,000 to $499,999	45	20.0
Less than $100,000	37	16.4
Total	225	100.0

Figure 12–5 plots the cumulative value distribution on doubly logarithmic coordinates, assuming the minimum value of the least profitable inventions to have been $40,000. The ordinary least squares log linear regression equation is given by:

$$(3)\quad \text{Log N} = \underset{[0.074]}{2.945} - \underset{[0.021]}{0.322} \log V; \qquad R^2 = 0.976,\ n = 8.$$

[27] Among other things, they were cited much more frequently in subsequent U.S. patents. See Harhoff, Narin, Scherer, and Vopel (1999).

where estimated patent value V is scaled in thousands of dollars, standard errors are given in brackets, and logarithms are to the base 10. Modest concavity is evident in the plotted points. When a quadratic value term is added, the 0.022 increase in R^2 is significant in an F-ratio test at the 0.01 level. When a linear regression is fitted only to the top four value categories (i.e., $5 million and higher), the estimated slope is -0.429, consistent with the tail of a highly skewed Pareto-Levy distribution. However, if that right-hand tail segment is extrapolated, it implies that the single most valuable patented invention would be valued at approximately $90 billion.[28] This is an order of magnitude greater than plausible values associated with the most profitable "blockbuster" drugs, although close to the 1997 stock market value of Microsoft (whose fortunes were based upon multiple innovations, none covered by patents in our sample). It seems probable that within the "more than $100 million" open interval, the distribution function must become more concave, as it does within the "more than DM 5 million" interval for the German patent data.

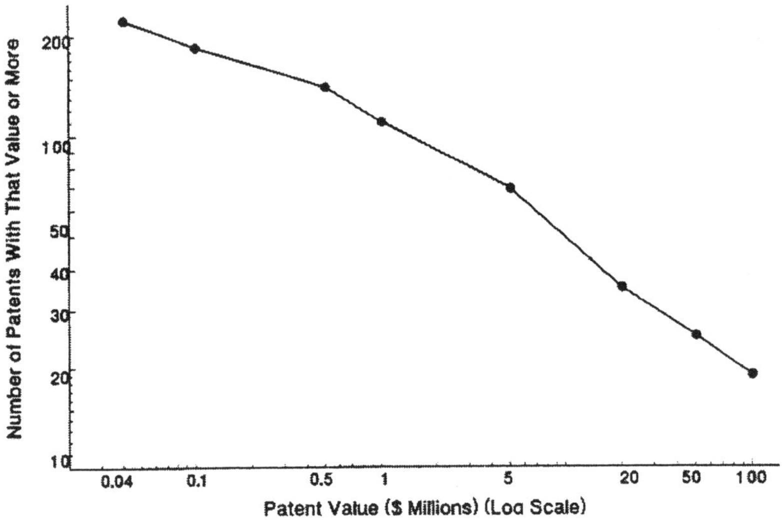

Figure 12–5: Pareto plot of U.S. patent values (N = 222)

While we do not apply the distributional tests to the U.S. data that we have used before, we can combine the patent value data from our U.S. sam-

[28] For the second to fifth most valuable patents, the predicted values are $17.3 billion, $7.3 billion, $3.57 billion, and $2.12 billion, respectively.

ple with data on R&D expenditures and other variables. For 48 sample companies with 165 linked full-term patents, it was possible to obtain the 1976 company-financed research and development expenditures which presumably gave rise to the surveyed patented inventions and their estimated value, aggregated to the company level.[29] For the closed-end intervals, values were estimated as the geometric mean of the interval bounds. For the open "more than $100 million" interval, the average patent value was assumed to be $250 million.[30] This seems a conservative estimate, considering the evidence from our German study that the average value of the most valuable patents was three to seven times the top-value interval's threshold.[31] Letting VALSUM be the estimated value of a company's patented inventions and R&D its 1976 R&D outlays, the resulting OLS regression for companies with data of acceptable quality was:

$$4)\ \text{Log VALSUM} = \underset{[0.553]}{-1.174} + \underset{[0.314]}{1.193}\ \text{log R\&D}; \qquad R^2 = 0.239,\ n = 48;$$

with logarithms taken to base 10 and standard errors given in brackets. The relationship between value and R&D is positive as expected, and highly significant statistically. A ten percent increase in R&D outlays was accompanied on average by a 12 percent increase in the value of linked patented inventions. There is considerable "noise" in the relationship, however. The standard error of estimate is 1.07, which means that individual company profits deviated upward or downward by as much as two orders of magnitude from central tendencies implied by the regression equation. Lowering the assumed average value of the above-$100 million inventions to $200 million reduced the regression coefficient to 1.170; the t-ratio remained at 3.80. In a multiple regression including the logarithm of 1976 company sales along with R&D outlays, the sales variable had a negative sign, suggesting that larger companies do not realize higher profits from their inventions, ce-

---

29 The source of R&D data was "What 600 Companies Spend for Research", *Business Week*, June 27, 1977, pp. 62-83.

30 However, in one case, a lower $100 million figure was used as a result of additional information provided by the company.

31 Inflating for sample response rates, one might predict from our results approximately 157 inventions valued at more than $5 million in the United States and 167 inventions valued at more than DM 5 million in Germany. Adjusting for market size differences and purchasing power parity exchange rates, the two most valuable cohort counts do not appear wildly inconsistent.

teris paribus[32], but both the sales variable and the increment in R^2 fell far short of statistical significance.

For the 48 companies with complete data, company-financed research and development outlays in 1976 totalled $4.82 billion, 27.6 percent of the reported $17.44 billion aggregate for all of U.S. industry in that year. Assuming an average value of $250 million for inventions valued at more than $100 million, the total value of those companies' inventions was $5.16 billion. That the estimated value of linked full-term renewed inventions, which comprised only a small fraction of all the inventions patented by responding companies, exceeded (or at least approximated) total company-financed R&D outlays suggests that on average the profit rewards to industrial R&D were appreciable.[33]

Going beyond the response limitations of our sample, 1976 R&D expenditure data were available for a total of 97 U.S. companies with U.S. patents linked to full-term renewed German patents. Those 97 corporations spent $10.69 billion on company-financed R&D in 1976 and obtained approximately 11,087 U.S. patents in 1978, including 381 patents (3.4 percent of their total 1978 patent receipts) linked to full-term German patents.[34] The OLS regression of all patents received in 1978 by the companies on their R&D spending, with standard errors in brackets, was as follows:

$$\text{(5)}\quad \text{Log ALLPATS} = \underset{[0.107]}{0.570} + \underset{[0.060]}{0.726} \log \text{R\&D}; \qquad R^2 = 0.604,\ n = 97.$$

The more R&D the companies performed, the more patents they received. However, the relationship between U.S. patents linked to German full-term renewed patents and R&D exhibits much more noise (standard errors in brackets):

$$\text{(6)}\quad \text{Log LINKPATS} = \underset{[0.112]}{-0.090} + \underset{[0.063]}{0.298} \log \text{R\&D}; \qquad R^2 = 0.191,\ n = 97.$$

32 See also Cohen and Klepper (1996).

33 That substantial benefits are realized from unpatented technology and run-of-the-mill patented inventions is shown by Hollander (1965).

34 The total patent counts were derived using the U.S. Patent Office's annual Index of Patents for 1978. Since the counts for subsidiaries receiving patents in other than the parent's name were undoubtedly incomplete, the 11,087 estimate is biased downward.

Indeed, the relationship is of about the same strength as regression (4), relating (for a sub-sample of companies) the *value* of linked patented inventions and R&D.

We return now to survey data disaggregated to the level of individual linked patents. For most of the patents, information was available on the professional specialty of the questionnaire respondent, the industry from which the invention originated, and the nature of the invention. These data allowed us to explore three questions that have been of considerable interest in the economic analysis of patents and innovations.

We first considered the question whether patented products are more valuable than patented processes. Among other things, 202 inventions could be coded as pertaining primarily to a product sold by the company or to a production process used by the company. Of those 202, 27.7 percent were process inventions – a figure closely comparable to estimates from earlier studies.[35] We could not detect any strong evidence that process inventions are less valuable – the differences were not statistically significant. Being subject to less market risk than product inventions, process inventions may also have less variable values, although the group variances fall just short of significance at the 0.05 level in an F-ratio test.

Secondly, as with the German data, we also tested for respondent biases and industry effects. We summarize the main results here without presenting the full regression tables. Controlling for industry effects, process inventions are found to be two-thirds of a value group less profitable on average. Drug and surgical supply inventions are significantly more profitable; electronic inventions nearly a category group less profitable. As in the German sample, patent attorneys appear to make relatively conservative estimates, all else equal. Unlike the German results, however, inventors reported relatively less valuable inventions, probably because few of the inventor respondents were employed by companies with a strong commercial orientation.

Finally, we considered the impact of patent protection on the profitability of intrinsically valuable inventions. To determine whether the strength of patent protection was contributing importantly to our survey results, indices from the so-called "Yale Survey" of the effectiveness of patents in protecting the competitive advantages from product and process inventions in narrowly defined industries were linked to our patent data, depending upon the line of

---

[35] See e.g. Scherer (1982).

business in which an invention arose.[36] Since separate Yale survey indices existed for products and processes, the coding reflected that distinction. To avoid collinearity with the variables used in the Yale coding, the PROCESS dummy was excluded, as were industry-specific dummy variables. Where YALE is the matched Yale survey appropriability index, the resulting regression was (standard errors in brackets):

$$(7)\quad \underset{}{\text{VALUE}} = \underset{[0.731]}{3.245} + \underset{[0.139]}{0.276}\ \text{YALE} + \underset{[0.639]}{0.301}\ \text{RDEXEC}$$

$$+ \underset{[2.120]}{0.636}\ \text{INVENT} - \underset{[0.398]}{0.991}\ \text{PAT};$$

$$R^2 = 0.080,\ \ n = 194.$$

Greater effectiveness of patent protection in appropriating the benefits from innovation is found to increase the reported value of our survey inventions, although only by about one-fourth of a value group. Evidently, most of the value differences are specific to the inventions and not the result of generalized patent strength variations. The conservative bias in patent attorneys' responses is also apparent in regression (7).

12.4 Implications

Previous studies have found the distribution of patented invention values to be highly skew. We confirm that conclusion, but also contribute a new insight: when patent values are measured relative to the counter-factual of being deprived of an invention's use, rather than having the invention lapse into the public domain (as in patent renewal studies), much higher extreme values emerge. Analyzing the renewal of German patents applied for between 1952 and 1972, Pakes (1986, p. 777) estimated the *maximum* patent value to be roughly $420,000 (in 1980 dollars). Our surveyed maxima are several hundred times greater. Subdividing German patent data into industry groups, Lanjouw (1998, Table 4 estimated an average value per patent for the top 0.1 percent of patents to lie between DM 191,000 and DM 433,000 (at 1975 price levels). Since resident German patent recipients obtained

[36] See Levin *et al.* (1987). The indices varied from 6.88 (in agricultural chemicals) and 6.53 (in pharmaceuticals) down to 2.00 (in electric batteries). Higher values denote greater effectiveness of patents in appropriating the benefits from innovation.

9,826 patents in 1980, the comparable top 0.1 percent cohort for our sample period includes ten patents. We find the average value of those patented inventions to have been DM 48 million if the most extreme observation is excluded or DM 193 million if it is included – at least two orders of magnitude larger.

Among several plausible alternative skew distributions, a two-parameter log normal distribution appears to provide the best fit to our patented invention value data. Although the log normal distribution has better-behaved asymptotic properties than the Pareto-Levy alternative, the observed distribution remains highly skew. The most valuable five percent of all renewed German patents accounted for 46 to 61 percent of total sample patent value, depending upon whether the largest outlier is excluded or included. In the U.S. sample, the 19 patents valued at more than $100 million (8.5 percent of the sample by number) accounted for 76 to 80 percent of total sample value, depending upon whether the average value of those patents is assumed to be 2.0 or 2.5 times the threshold value. The facile convergence to stable means sought in R&D risk-hedging through portfolio strategies operates at best imperfectly. Investing in technological innovation must be viewed as intrinsically risky, even if (as our U.S. value sums suggest) profitable in the aggregate.

Given the highly skewed nature of patent values, it has also become clear that simply counting patents in an attempt to measure the "intellectual capital" of corporations is a futile undertaking. Mere counts are likely to reveal very little of the underlying value of a patent portfolio.[37] In related research (Harhoff, Scherer and Vopel 2003), we have investigated an alternate route – we discuss correlates of patent value that can be used to weight patents in order to approximate the value of patent portfolios. It turns out that the researcher's problem of assessing patent value *ex post* can be resolved to some extent; it should not be surprising that those correlates which help to identify patent value typically have a highly skewed distribution themselves. In the absence of such *ex post* measures, assessing patent value *ex ante* remains a daunting task.

In combination with other evidence on new pharmaceutical compounds and the outcomes of high-technology venture investments (Scherer 1998), we now possess a clear picture of the range of profit or quasi-rent distribution shapes resulting from investments in innovation. Those distributions are

[37] Griliches (1990) also comments on the measurement question and is sceptical about counts of patents as measures of knowledge capital. See also Lanjouw, Pakes and Putnam (1998).

the result of stochastic processes played out as the menu of technological possibilities is tapped, market prospects are tested, and competitive forces impinge. In related research[38], the data employed in this paper have been used to identify likely stochastic processes consistent with the observed profit distributions.

A final conclusion is noteworthy. In a world with skew-distributed outcomes of the R&D and innovation process, there is very little room for "picking winners" – either at the business or the public policy level. First, managers in large corporations seeking to reach some degree of risk-hedging within the firm will be forced to build relatively large portfolios of R&D projects. Second, society at large will be better off conducting contests among many projects, e.g. in the form of newly founded technology-oriented firms, than with trying to assess the value of an invention or a startup firm early on. We can also learn from these results that a high rate of failing projects is not necessarily a signal of failure of policy – one cannot escape the fundamental distributional properties of R&D and inventions.[39] Hence, a large share of failures or low-value projects is the natural concomitant of having a few outstanding successes.

[38] See Scherer, Harhoff and Kukies (2000).

[39] Other policy implications of the skewed distribution of innovation returns are discussed in Scherer and Harhoff (2000).

Technical Appendix

Estimation of Distribution Parameters

Using the data described in the body of the paper, we estimate the parameters of Pareto-Levy, log-normal and Singh-Maddala distributions. The distribution function for the Pareto-Levy is given by:

$$\psi(x|\alpha)=\left(\frac{X}{X_0}\right)^{-\alpha}=x^{-\alpha}$$

where $x = X / X_0$ is defined over the interval $[1,\infty]$. X is the actual valuation for the patent, and X_0 corresponds to the minimum valuation for a patent in the sample. The parameter α determines the slope of the distribution function in a log(value)-log(rank) plot.

The Singh-Maddala function is a three-parameter distribution with cumulative distribution function

$$\Sigma(a,b,q)=1-\frac{1}{\left(1+(y/b)^a\right)^q} \quad .$$

where b is a scale parameter and a and q determine the shape of the distribution function.

The third function under consideration is the log-normal whose probability function is given by the standard normal cumulative function evaluated at $(\ln X-\mu)/\sigma$ where μ is the mean of the log-normal distribution and σ is the corresponding standard deviation.

To obtain the estimates in Part A of Table 12–4, we used the 772 patents for which we obtained valuations in our survey to estimate parameters from truncated distributions. We assume that the value of the patent right is bounded from below at DM 23,000, i.e. the group for which respondents gave a minimum counter-factual sales price of less than DM 100,000 is assumed to have values in the interval [23000,100000]. The choice of the lower bound is somewhat arbitrary, but approximates the discounted sum of renewal fees in 1995 currency units (in order to be consistent with our patent value estimates).

Each observation enters the likelihood function as the probability of observing a patent right's value to be in a known value interval. The interval observation comes either from the first-stage survey, or it has been obtained in the second-stage interviews. In cases where the interviews elicited only a point estimate, we assume that the interval is defined by the point estimate minus or plus 10 percent of the point estimate.

Thus, instead of observing the actual values X_i, we actually observe an upper bound B_{Ui} and a lower bound B_{Li} such that $B_{Ui} \leq X_i < B_{Li}$. Hence, the likelihood function (neglecting constant terms) is given by

$$log\, L(\theta) = \sum_{i=1}^{N_1} log\left(\frac{F(B_{Ui},\theta) - F(B_{Li},\theta)}{1 - F(T,\theta)}\right) + \sum_{i=N_1+1}^{N_1+N_2} log\left(1 - \frac{F(B_{Li},\theta)}{1 - F(T,\theta)}\right)$$

Here $F(\cdot)$ is the cumulative distribution function, θ denotes a vector of parameters, and T is the truncation point, assumed to be equal to DM 23,000. The values of observations $(1,\ldots,N_1)$ fall into intervals that are bounded from above and below, while the values of observations $(N_1+1,\ldots,N_2)$ are bounded from below, but no upper bound has been established. The latter case occurs whenever we do not obtain detailed interview information on patents with first-stage valuations of greater than DM 5 million.

In the case of the Pareto-Levy distribution, the above expression simplifies considerably, since the truncated Pareto-Levy is again a Pareto-Levy distribution. Therefore, the likelihood is given by

$$log\, L(\alpha) = \sum_{i=1}^{N_1} log\left(b_{Li}{}^{-\alpha} - b_{Ui}{}^{-\alpha}\right) + \sum_{i=N_1+1}^{N_1+N_2} log\left(b_{Li}{}^{-\alpha}\right)$$

where $b_{Ui} = B_{Ui}/X_0$ and $b_{Li} = B_{Li}/X_0$ and constant terms are again neglected.

We estimate the parameters by implementing the likelihood function in a non-linear optimization routine, using Newton-Raphson numerical techniques to obtain numerical solutions. The standard errors are computed from a variance-covariance matrix obtained by multiplication of the score vectors. For each distribution, we perform the estimation with and without the most valuable patent in our sample in order to explore the sensitivity of our results to that particular observation. This is an informal test of robustness, since the estimated parameters of the distribution should not change much if a particular observation is taken from the sample, assuming that the distributional assumption is correct.

In Part B of Table 12–4, we take into account that the actual number of German-owned patents not renewed to full term is actually known. For the

Singh-Maddala and the log-normal distribution, we assume that patent values are non-zero and positive. The likelihood function is now given by

$$\log L(\theta) = \sum_{i=1}^{N_1} \log(F(B_{Ui},\theta) - F(B_{Li},\theta)) + \sum_{i=N_1+1}^{N_1+N_2} \log(1 - F(B_{Li},\theta)) +$$

$$+ \sum_{i=N_1+N_2+1}^{N} \log(F(B_{Ui},\theta))$$

where the third group describes the patents that were not renewed to full term – in their case, we only know the upper bound of DM 23,000. For estimating the parameters of the Pareto-Levy distribution, we assume that patent values in the latter group range from DM 1 to DM 23,000. Note that the lower bound can in principle be estimated, but we have relied thus far on the assumptions described here.

For computing a goodness-of-fit measure, we use up to 11 value intervals which are defined by the threshold values DM 23000, DM 100000, DM 400000, DM 1 million, DM 5 million, DM 10 million, DM 20 million, DM 40 million, DM 80 million and DM 120 million. The Pearson Chi-Square is computed as

$$N\sum_{j=1}^{J}\left(\frac{n_j}{N} - P_j(\hat{\theta})\right)^2 \Big/ P_j(\hat{\theta})$$

where *j* denotes the interval, n_j/N is the actual share of patents observed to be in interval *j*, and $P_j(\hat{\theta})$ is the predicted share of observations in the interval. This statistic is distributed as a Chi-Squared random variable with *J-k-1* degrees of freedom, where *k* is the number of distribution parameters estimated.

12.5 Literature References

Bryson, M.C. (1974). 'Heavy-tailed Distributions: Properties and Test', *Technometrics*, Vol 16, pp. 61-68.

Cohen, W.M. and Klepper, S. (1996). 'A Reprise of Size and R&D', *Economic Journal*, Vol. 106, pp. 925-951.

D'Agostino, R.B. and Stephens, M.A. (1986). *Goodness-of-Fit Techniques*. New York: Marcel Dekker.

Dernburg, T. and Gharrity, N. (1961). 'A Statistical Analysis of Patent Renewal Data for Three Countries', *Patent, Trademark, and Copyright Journal*, Vol. 5, pp. 340-361.

Esteban, J.M. (1986). 'Income Share Elasticity and the Size Distribution of Income', *International Economic Review*, Vol. 27, pp. 439-444.

Griliches, Z. (1990). 'Patent Statistics as Economic Indicators: A Survey', *Journal of Economic Literature*, Vol. 28, pp. 1661-1707.

Harhoff, D., Scherer, F.M. and Vopel, K. (2003). 'Citations, Family Size, Opposition and the Value of Patent Rights', *Research Policy*, Vol. 32, No. 8, pp. 1343-1363

Harhoff, D., Narin, F., Scherer, F.M. and Vopel, K. (1999). 'Citation Frequency and the Value of Patented Inventions, *Review of Economics and Statistics*, Vol. 81, No. 3, pp. 511-515.

Hollander, S. (1965). *The Sources of Increased Efficiency: A Study of DuPont Rayon Plants*. Cambridge: MIT Press.

IVA (1993). *Profit from Innovation: A Comparison of Swedish and Japanese Intellectual Property Management*. Stockholm: Royal Swedish Academy of Engineering Science.

Johnson, N.L., Kotz, S. and Balakrishan, N. (1994). *Continuous Univariate Distributions*. Vol. I, 2nd ed.. New York: Wiley.

Kloek, T. and Van Dijk, H.K. (1978). 'Efficient Estimation of Income Distribution Parameters', *Journal of Econometrics*, Vol. 8, pp. 61-74.

Lanjouw, J.O. (1998). 'Patent Protection in the Shadow of Infringement: Simulation Estimations of Patent Value', *Review of Economic Studies*, Vol. 65. pp. 671-710.

Lanjouw, J.O., Pakes, A. and Putnam, J. (1998). 'How to Count Patents and Value Intellectual Property: Uses of Patent Renewal and Application Data', *Journal of Industrial Economics*, Vol. XLVI, No. 4, pp. 405-433.

Levin, R.C., Klevorick, A., Nelson, R.R. and Winter, S.G. (1987). 'Appropriating the Returns from Industrial Research and Development', *Brookings Papers on Economic Activity*, No. 3, pp. 783-820.

McDonald, J.B. (1984). 'Some Generalized Functions for the Size Distribution of Income', *Econometrica*, vol. 52 (May 1984), pp. 647-663.

McDonald, J.B. and Ransom, M.R. (1979). 'Functional Forms, Estimation Techniques and the Distribution of Income', *Econometrica*, Vol. 47, pp. 1513-1525.

Nordhaus, W.D. (1989). 'Comment', *Brookings Papers on Economic Activity: Microeconomics*, pp. 320-325.

Pakes, A. (1986). 'Patents as Options: Some Estimates of the Value of Holding European Patent Stocks', *Econometrica*, Vol. 54, pp. 755-784.

Pakes, A. and Schankerman, M. (1984). 'The Rate of Obsolescence of Patents, Research Gestation Lags, and the Private Rate of Return to Research Resources', in Zvi Griliches, ed., *R&D, Patents, and Productivity*. Chicago: University of Chicago Press, pp. 73-88.

Patent- und Musterrecht (1995). 3rd edition; Munich: Beck-Texte.

Putnam, J. (1996). 'The Value of International Patent Protection', Unpublished Ph.D. Dissertation, Yale University.

Schankerman, M. and Pakes, A. (1986). 'Estimates of the Value of Patent Rights in European Countries during the Post-1950 Period', *Economic Journal*, Vol. 97, pp. 1-25.

Scherer, F.M. (1982). 'Inter-Industry Technology Flows in the United States', *Research Policy*, Vol. 11, pp. 227-245.

Scherer, F.M. (1998). 'The Size Distribution of Profits from Innovation', *Annales d'Economie et Statistique*, Vol. 49/50, pp. 496-516.

Scherer, F.M. and Harhoff, D. (2000). 'Technology Policy Implications for a World with Skew-Distributed Outcomes', *Research Policy,* Vol. 29, pp. 559-566, 2000.

Scherer, F.M., Harhoff, D. and Kukies, J. (2000). 'Uncertainty and the Size Distribution of Rewards from Technological Innovation', *Journal of Evolutionary Economics*, pp. 175-200.

Singh, S.K. and Maddala, G.S. (1976). 'A Function for Size Distribution of Incomes', *Econometrica*, Vol. 44, pp. 963-970.

Tetlock, P.E. and Belkin, A. (1996). 'Counterfactual Thought Experiments in World Politics', *Social Science Research Council Items*, Vol. 50 (December).

[17]

CITATION FREQUENCY AND THE VALUE OF PATENTED INVENTIONS

Dietmar Harhoff, Francis Narin, F. M. Scherer, and Katrin Vopel*

Abstract—Through a survey, private economic value estimates were obtained on 964 inventions made in the United States and Germany and on which German patent renewal fees were paid to full-term expiration in 1995. A search of subsequent U.S. and German patents yielded counts of citations to those patents. Patents renewed to full-term were significantly more highly cited than patents allowed to expire before their full term. The higher an invention's economic value estimate was, the more the patent was subsequently cited.

I. Introduction

THE availability of rich machine-readable data sets has led during the past decade to considerable use of patent citations as an instrument for measuring the impact and fine structure of technological innovations. Through citation counts, it is argued, one can achieve a better approximation to the economic value of inventive activity than using only unweighted patent counts or data on research and development expenditures. Tapping unique survey data on individual patents, this paper tests whether more valuable patents are in fact more extensively cited.

II. Citations as a Measuring Instrument

When an invention patent is issued by government patent offices, its text normally includes citations to prior patents (and also to relevant prior scientific and technological literature). To be patentable, an invention must be novel, i.e., not anticipated in the already practiced state of the art. Citations placed in the patent text, usually by either the inventor's attorneys or by patent office examiners (depending upon national practice), serve to show how the claimed invention differs from the "prior art" and hence to help stake out the metes and bounds of the new claims. In a sense, citations identify the shoulders upon which the new claims stand in their attempt to advance technology. Given this, patent citations can serve as measures of economic value in two quite different ways (Griliches, 1990, pp. 1688–1689).

For one, it is reasonable to suppose that the prior inventions cited in new patents tend to be the relatively important precursors that best define the state of the art. The broader the shoulders, the more likely they are to be cited. From this follows the hypothesis to be tested here—that patents of relatively high economic value are cited more frequently than are low-value patents. This measurement approach is analogous to using journal article citation counts to identify scholars whose research has been particularly seminal and to assess the quality of the institutions where they are employed.[1]

Second, because prior inventions set the stage for new inventions, citations are used to measure a potentially important economic externality, i.e., the impact the knowledge embodied in prior inventions has in stimulating new contributions. See, for example, Caballero and Jaffe (1993), Jaffe et al. (1998), and the survey by Jaffe (1998).

Our research here emphasizes the first of these measurement approaches, hypothesizing that more valuable patents are more frequently cited. In this, it builds upon Trajtenberg (1990)—who found that medical diagnostic imaging patents of greater social value were more-frequently cited—and Hall et al. (1998), whose preliminary results suggest that companies with more frequently cited patents have higher stock market values, other relevant variables held equal. Unlike Trajtenberg, we focus on the private value of our survey patents and their underlying inventions to patent holders, not on their social value (including positive externalities).

III. The Surveys

The sampling approach underlying our analysis was based upon two phenomena identified in earlier work with patents. See, for example, Lanjouw et al. (1996). First, many nations require patent holders to pay periodic renewal or maintenance fees to keep their patents in force. The longer fees are paid, the higher a patent's implied value is. Second, inventions originating in one national jurisdiction are in a sizeable fraction of cases also patented in other nations. The greater the invention's anticipated value, the more likely patenting abroad is.

To exploit these two properties, we focused on German patents resulting from applications filed in 1977, and hence with a full-term expiration date (given the eighteen-year life span prevailing at that time) falling within 1995.[2] Germany was chosen because it is one of the leading industrialized nations, because its patent examination system is particularly rigorous, and especially because its patent renewal-fee system is one of the most progressive in the world. Fees keeping a patent in force rose from DM 50 for the third year to DM 2,700 for the eighteenth year, cumulating to DM 16,075 ($8,856 at average 1980 exchange rates) over the statutory eighteen-year life. From 57,782 applications in the 1977 cohort, 22,898 patents were issued and not subsequently revoked. Of these, 4,349 (19%) were renewed to full-term expiration in 1995.

The samples analyzed here began with 1,431 full-term German patents issued to residents of Germany (mostly to companies operating in Germany, but also to a few independent inventors) and 900 full-term German patents issued to

Received for publication September 24, 1997. Revision accepted for publication October 27, 1998.

* University of Munich, CHI Research Inc., Harvard University, and Center for European Economic Research, Mannheim, respectively.

[1] For a survey of the extensive literature pursuing this approach, see Medoff (1996).

[2] The maximum patent life was increased from eighteen to twenty years beginning with 1978 priority applications.

The Review of Economics and Statistics, August 1999, 81(3): 511–515

U.S. residents. Telephonic contacts were achieved in Germany with the holders of 1,352 patents from German-resident inventors. To form the U.S. sample, an attempt was made to link the German patents back to counterpart U.S. patents. Attrition occurred because not all German patents disclosed prior U.S. applications,[3] because the U.S. entities holding a single full-term German patent were included in the linking effort only when there was reason to believe they could be located,[4] and because only 61% of the remaining U.S. applications were successful in yielding U.S. patents. Thus, the starting point of the U.S. survey comprised 485 parallel and linked U.S. patents.

Each surveyed patent owner was contacted by telephone and facsimile during 1996 (the year following patent expiration) and asked to answer a single counterfactual question, which, in the U.S. survey version, was phrased as follows:

> If in 1980 you knew what you now know about the profit history of the invention abstracted here, what is the *smallest* amount for which you would have been willing to sell this patent to an independent third party, assuming that you had a bona fide offer to purchase and that the buyer would subsequently exercise its full patent rights?

This asset-value approach (consistent with the view of patents as intellectual property) was chosen following pilot interviews over a question asking for the accumulated flow of profits. Because selling full rights meant that the patent holder could be enjoined from using that invention or forced to pay royalties reflecting the invention's value, the survey responses elicited a private value that corresponds most closely to the discounted present value of profits that would be foregone by not having the invention *and* its accompanying patent protection.

Respondents in the German survey were asked to locate their patent's value within five categories ranging from "less than DM 100,000" ($55,000 at 1980 exchange rates) to "more than DM 5 million" ($2.75 million). Electronic contacts were supplemented by face-to-face interviews for 69 German patents valued at more than DM 5 million. U.S. respondents, contacted first by telephone and then by facsimile or mail, were given a broader range of eight categories, from "less than $100,000" to "more than $100 million." Altogether, there were 772 useable responses covering the inventions of German origin and 192 usable responses in the survey of U.S. patents.[5] The value distribution of responses was highly skewed. Among the German patents, 12.9% occupied the highest value category. Among the linked U.S. patents (thrice winnowed—once for yielding a U.S. patent, again for yielding a German patent, and a third time for being renewed to full term in Germany), 18 (9.4%) were placed in the "above $100 million" category. Statistical distribution properties of the survey responses are analyzed by Harhoff et al. (1997).

IV. U.S. Patent Citation Link Results

A search of citations to the 192 U.S. sample patents contained in other U.S. patents subsequently issued (i.e., between 1978 and 1996) was made using the extensive database of CHI Research, Inc. (See, for example, Narin (1995).) The number of recorded citations ranged from zero (nine cases) to 169, with a mean of 15.91, median of 9.0, and standard deviation of 20.83. The average number of citations to all U.S. patents issued to U.S. corporations, weighted by issue year counts of patents in our sample, was 6.83. (Citation rates to patents held by independent inventors and foreign corporations were lower. Universe citation counts were approximately 15% higher on average in the U.S. Patent Office classification groups into which our sample fell than in the citation universe generally.) With or without an adjustment for sample composition, the differences between mean citation ratios for our sample and weighted universe means are significant at the 0.001 level. Thus, U.S. patents selected on the basis of full-term German counterpart renewal are substantially more frequently cited than run-of-the-mill patents.

Figure 1 displays average citation frequencies for the eight economic value groups into which respondents classified their patents, with the number of observations per group recorded in parentheses. In every value group, the average citation frequency for full-term German counterpart patents exceeds the weighted universe average of 6.83. Within our full-term sample, two broad groupings appear. The most profitable patents—those valued at $20 million or more—are visibly more highly cited than less-valuable patents, i.e., 29.6 citations on average for the former, compared to 13.1 for the latter. An F-ratio test reveals the two groups to be significantly different; $F(1, 190) = 18.66$. A less-structured test for differences among the eight classes is also highly significant; $F(7, 184) = 3.12$.

Although the group citation averages in figure 1 do not rise monotonically with estimated patent value, regression analysis of the number of citations (in natural logarithms) on

[3] When one seeks patents both at home and abroad, it is customary to file first at home, disclosing the home application serial number to the foreign patent office as part of establishing the legal date on which priority of invention occurs. Foreign patents then cite the home application serial number when it is available.

[4] Lost by this last criterion were independent inventors and firms without publicly traded securities located outside the New England area and patents issued to firms that disappeared by merger or other corporate control transactions.

[5] Of the 222 U.S. patents analyzed elsewhere by the authors, 20 U.S. patents could not be included in the analysis here because a large corporate respondent provided for the sake of confidentiality only a size distribution of patent values detached from individual patent numbers, and ten replies arrived too late. Attrition between the original linked sample of 485 and the 222 with usable replies occurred mainly because of company decisions not to participate, but also because some patents went to sold-off company divisions or individuals and could not be traced to knowledgeable respondents.

FIGURE 1.—U.S. PATENT CITATION FREQUENCIES BY VALUE GROUP

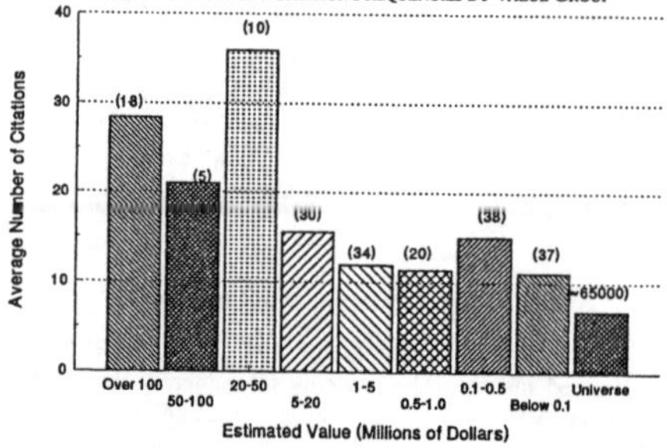

TABLE 1.—REGRESSIONS OF (LN) U.S. PATENT CITATIONS ON VALUE (N = 192)*

	OLS		Negative Binomial	
Variable	1a	2a	1b	2b
Constant	2.076	1.845	2.682	2.267
	(0.089)	(0.273)	(0.074)	(0.228)
ln VALUE	0.119	0.137	0.105	0.136
	(0.033)	(0.034)	(0.027)	(0.028)
7 IPC Dummies	No	Yes	No	Yes
Chi-squared for dummies (df)		12.4 (7)		18.49 (7)
R^2 or pseudo R^2	0.063	0.123	0.064	0.100
Chi-squared (df)	12.3 (1)	24.4 (1)	15.3 (1)	34.0 (8)

Note: * Standard errors are given in subscripted parentheses.

survey value estimates (also in natural logarithms) supports the existence of a generally positive relationship.[6] Six of the eight value groups were assigned values equal to the geometric mean of their bounds; the mean value of the most valuable patents was estimated (on the basis of telephonic interviews) to be $250 million; and the mean of the least valuable patents was assumed to be $50,000. The regressions, with and without dummy variables controlling for the eight main classes of the International Patent Classification (IPC), are presented in table 1.[7] (Standard errors are enclosed in parentheses.) In OLS regressions, zero values of the citation count dependent variable are assumed to be 0.3. Negative binomial regressions avoid this arbitrary zero-value convention and also permit closer comparison with later German patent results. In all cases, the economic value coefficients (interpretable as elasticities) are statistically significant, with t-ratios of 3.6 or higher.[8] The relationships exhibit strongly diminishing returns. In regression 1a, for example, patents valued in 1980 at $1 million elicited 7.97 predicted citations on average (with a two standard error range of from 0.7 to 90 citations); patents valued at $10 million, 10.49 citations; and patents valued at $100 million, 13.70 citations (with a two-sigma range of 1.2 to 156). Among the seven dummy variables, only the coefficient for the international patent class corresponding approximately to the scientific and professional instruments sector was statistically significant, with a positive sign.[9]

The relatively low R^2 values for all regressions and the wide confidence bounds reveal that the citation-value relationship is quite noisy. To the extent that errors in both the citation and value observations are random, pooling of observations to the firm level can enhance explanatory power and reduce errors-in-variables biases. For 43 companies holding in total 139 of the 192 sample patents, the survey responses were complete, covering all U.S. patents linked to full-term German patents, and, in addition, it was possible to obtain data on total company-financed research and development outlays in 1976 (the year before German applications were filed).[10] The analogue of regression 1a in table 1 for those 43 companies is:

$$\text{Ln CITES} = \underset{(0.22)}{2.62} + \underset{(.058)}{0.314} \text{ ln VALUE}; \; r^s = 0.413, \quad (1)$$

with standard errors reported in subscripted parentheses. The increases in the estimated elasticity and r^2 relative to table 1 suggest that some averaging-out of random errors has occurred.

It is conceivable that the amount of R&D funds expended by a company contributes to the technological quality (and hence citation-attracting power) of its patents independently of the patents' economic value. A second variable, RDPAT, measuring 1976 R&D expenditures per linked patent, was therefore added to regression (1), yielding:

[6] Our maintained hypothesis is that more valuable patents elicit more subsequent citations; thus, value is the appropriate righthand-side variable. Referees interested in using citations as a predictor of value suggested the reverse formulation. With either formulation, the righthand-side variable is measured with error—for monetary values, because survey responses are imperfect and available only in eight groups; and for citations, because choices by patent agents and/or examiners of which prior patents to cite have substantial random components. In either case, regression coefficients are likely to be biased downward.

[7] The International Patent Classification groupings are roughly (A) food, shelter, and clothing; (B) production processes and transportation; (C) chemicals (including drugs); (D) textiles and paper; (E) construction and mining; (F) machinery, machine elements, heating, cooling, and weapons; (G) instruments and control devices and atomic energy; and (H) electrical and electronic equipment.

[8] A referee suggested that by confining the regression analysis to full-term patents, we limit the variance of patent values and hence the strength of value-citation relationships. This is true, but what is selected out is mainly the great mass of low-value patents. Among the full-term survivors, there remain value variations spanning at least four orders of magnitude.

[9] In the U.S. citation universe, patents with that IPC code also had unusually high citation frequencies.

[10] The patents linked to full-term German patents of those companies comprised only approximately 2.5% of all the patents issued in 1978 to the same companies. R&D cost was of course incurred to generate the nonlinked patents, which undoubtedly contributed to company profits. Thus, the value estimates are biased downward.

$$\begin{aligned} \text{Ln CITES} &= 3.29 + 0.334 \ln \text{VALUE} \\ &\quad\ (0.53)\quad (.060) \\ &\quad - 0.235 \ln \text{RDPAT}; \qquad (2) \\ &\quad\ (.167) \\ R^2 &= 0.441. \end{aligned}$$

The coefficient for RDPAT has an unexpected negative sign, but its t-ratio of 1.41 falls below conventional statistical significance thresholds. Thus, R&D outlays appear to have no independent explanatory power once private economic value is taken into account.

V. German Patent Citation Link Results

Using the databanks of the Fraunhofer Institute for Systems Technology and Innovation Research in Karlsruhe, Germany, a similar link was effected between full-term patents issued to German residents on which economic value estimates were obtained through facsimile and interview methods and the German patent citations to those patents. As with the linked U.S. data, the German patent values had a highly skewed distribution that most closely approached the log normal (Harhoff et al., 1997).

Successful citation links were achieved for all 772 of the patents on which German survey responses were secured. A tally of citation counts was also made for all 10,780 German patents issued to German residents on the basis of 1977 applications, including the 1,431 patents renewed to full term. This research yielded the initially surprising finding that German patents cite prior German patents much less frequently than U.S. patents cite prior U.S. patents. For the universe of all German patents with 1977 application dates issued to German residents, the average number of subsequent citations was 0.504 per patent (compared to more than six for corresponding U.S. patents). The maximum number of recorded citations was 15; 69.5% of the 10,780 German universe patents received no citations at all.

The difference in citation rates between Germany and the United States is attributable in part to the much larger size of the U.S. patent base. During the 1980s, there were roughly four times as many U.S. patents as German patents to cite any given prior patent. In addition, there are material differences in patent laws and practice.[11] In the United States, patent applicants are required to list relevant prior patents. This is not mandated under German or European Patent Office practice; citations are introduced mainly by patent office examiners. In the United States, examiners may add to or delete the applicant's citations, but the tendency of both parties is toward comprehensive referencing to highlight an invention's multiple facets. In the German and European patent offices, examiners must confine their citations to prior patents essential in evaluating an invention's

FIGURE 2.—GERMAN PATENT CITATION FREQUENCIES BY VALUE GROUP

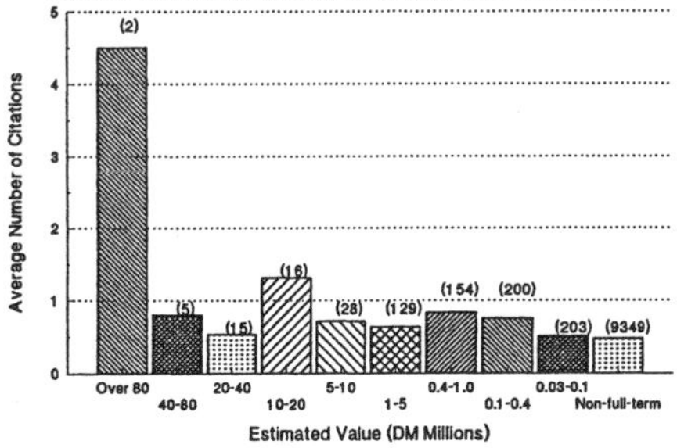

claim of novelty. The consequence is a lower citation rate in Europe, with all else equal.[12]

Patents renewed to full term were more likely to be cited and had more citations on average than patents allowed to lapse before their maximum eighteen-year term. The average number of citations per full-term patent was 0.732, 47% more than the nonfull-term German resident cohort average of 0.470. The difference in means is highly significant, with $t = 9.14$. The full-term patents on which economic value estimates were obtained through our German survey had 0.703 citations on average. The difference between survey and non-full-term patents is again highly significant ($t = 6.52$).[13] Among the full-term (survey) patents, 61.7% (60.4%) had zero citations, compared to 70.7% for the non-full-term cohort. In a binomial proportions test, the difference is significant at better than the 0.01 level.[14]

Figure 2 displays the average citation frequencies (including zero-citation cases) for survey response patents grouped by their estimated economic value. For patents originally valued at more than DM 5 million, finer disaggregations from subsequent interviews are used. Also shown is the average for nonfull-term cohort patents. The two most

[11] This interpretation is based upon discussion with Bernd Bettels of the European Patent Office, The Hague, and a communication from him to Francis Narin.

[12] Because the applicant contributes less to generating citations, citation counts may be less useful as indicators of technological spillovers for European patents than for American patents.

[13] When average citation rates for non-full-term patents in specific International Patent Classification categories were weighted to reflect the composition of our survey sample, the adjusted measure (0.478 citations) was virtually identical to the unweighted universe total.

[14] Four of the eight universe patents with ten or more citations had been renewed to full term, and hence were included in our survey frame, but their owners chose not to participate in the survey. From inspection of those patents, there is reason to believe that appreciable economic values would have been reported. Among the total of 659 full-term patents on which no response was obtained, 63.3% had zero citations, compared to 60.4% for responding patents. The difference is not statistically significant in a binomial test; $t = 1.13$. The mean number of citations conditional upon nonzero citation was 2.083 for nonresponding patents and 1.775 for responding patents, with a t-ratio of 4.32 in a test of mean differences. In a test of unconditional means, $t = 1.67$. Thus, the responding and nonresponding samples were different, but mainly with respect to citation rates for the highly cited patents. In a probit analysis of survey participation by full-term patent holders, the number of citations did not have significant explanatory power at the 0.10 level, but response rates were significantly related to differences in International Patent Classification categories.

TABLE 2.—NEGATIVE BINOMIAL REGRESSIONS OF (LN) GERMAN PATENT CITATIONS ON VALUE

	All Full-Term Patents		Full-Term Patents Valued Above DM 5 mil.	
Variable	1a	2a	1b	2b
Constant	−1.534	−0.703	−4.606	−6.618
	(0.435)	(0.475)	(2.329)	(2.170)
ln VALUE	0.090	0.105	0.272	0.345
	(0.033)	(0.034)	(0.139)	(0.129)
7 IPC Dummies	No	Yes	No	Yes
Chi-squared for dummies (df)		18.13 (7)		10.87 (7)
N	752	752	66	66
Chi-squared (df) (1)	7.43 (1)	24.45 (8)	3.82 (1)	11.80
Log L	−859.2	−850.6	−86.7	−82.8
Pseudo R^2	0.012	0.030	0.176	0.309

valuable inventions are grouped together for reasons of confidentiality; disaggregation would not change the implications. As with the U.S. patents, early-expiration universe patents are less cited than all but the least valuable full-term patents. By far the highest citation frequencies are for the two patents with the greatest reported economic value.

This visual implication is confirmed by regression analysis of individual patent citation counts on reported economic values. Due to the high incidence of zero values and the low citation counts, ordinary least squares regression is inappropriate. Poisson regression is also ruled out, since the variance of the citation counts is roughly twice the mean count. Therefore, a negative binomial formulation is used, with economic VALUE measured (in millions of DM) as the geometric mean of reporting class interval bounds for first-stage survey responses (with a minimum value of DM 30,000 for the "less than DM 100,000 group"[15]) and as the mean point estimate for second-stage interview responses. The resulting regressions, with the number of citations as dependent variable and the economic value estimates (along with seven IPC dummy variables) as explanatory variables, are presented in table 2.

Several implications stand out. First, in every case the regression coefficients on ln VALUE (interpretable as elasticities) are significantly positive, with *t*-ratios in the full-sample regressions of 2.72 and 3.09. Second, the full-sample elasticities lie within a standard error of those obtained for the completely different U.S. sample of table 1, despite the much smaller average number of German citations. Third, when the sample is limited to 66 patents with value estimates exceeding DM 5 million, higher elasticities with substantially larger standard errors emerge.[16] Fourth, the seven International Patent Classification dummy variables together are statistically significant in a *chi*-square test. As in the U.S. data, the instruments group had particularly high citations, as did (unlike the U.S. results) the class including textiles and paper goods. But finally, the pseudo-R^2 values are even more modest than in the U.S. regressions, revealing again that the value-citation relationship is quite noisy.[17]

V. Conclusion

In sum, our results from two surveys, one in Germany and one in the United States, reveal that patents reported to be relatively valuable by the companies holding them are more heavily cited in subsequent patents. Our evidence suggests at least a two-stage relationship between economic values and citation counts. First, patents that are renewed to full-term expiration in environments such as Germany with highly progressive annual maintenance fees are more highly cited than run-of-the-mill patents allowed to expire before running to full term. Other research reveals such full-term patents to be more valuable on average than patents allowed to lapse at midterm. Second, within the relatively select cohort of full-term patents, citation frequency rises noisily with reported economic value.

REFERENCES

Caballero, R. J., and Adam B. Jaffe, "How High Are the Giants' Shoulders: An Empirical Assessment of Knowledge Spillovers and Creative Destruction in a Model of Economic Growth," in Olivier Blanchard and Stanley Fischer (eds.), *NBER Macroeconomics Annual,* vol. 8 (Cambridge: MIT Press, 1993).

Griliches, Zvi, "Patent Statistics as Economic Indicators: A Survey," *Journal of Economic Literature* 28 (December, 1990), 1661–1707.

Hall, Bronwyn H., Adam B. Jaffe, and Manuel L. Trajtenberg, "Patent Citations and Market Value: A First Look," paper presented at a National Bureau of Economic Research meeting (March, 1998).

Harhoff, Dietmar, F. M. Scherer, and Katrin Vopel, "Exploring the Tail of Patent Value Distributions," working paper (July, 1997).

Jaffe, Adam B., "Patents, Patent Citations, and the Dynamics of Technological Change," *NBER Reporter* (Summer, 1998), 8–11.

Jaffe, Adam B., M. Fogarty, and B. Banks, "Evidence from Patents and Patent Citations on the Impact of NASA and Other Federal Labs on Commercial Innovation," *Journal of Industrial Economics* 46 (June, 1998), 183–206.

Lanjouw, Jean O., Ariel Pakes, and Jonathan Putnam, "How To Count Patents and Value Intellectual Property; working paper (August, 1996).

Medoff, Marshall H., "A Citation-Based Analysis of Economists and Economics Programs," *The American Economist* 40 (Spring, 1996), 46–59.

Narin, Francis, "Patents as Indicators for the Evaluation of Industrial Research Output," *Scientometrics* 34 (November–December, 1995), 489–496.

Trajtenberg, Manuel, "A Penny for Your Quotes: Patent Citations and the Value of Innovations," *RAND Journal of Economics* 21 (Spring, 1990), 172–187.

[15] Presumably, the least-valuable full-term patents were expected to recover their renewal fees of DM 16,075 plus application and legal fees of roughly the same magnitude.

[16] This was also true for the U.S. sample. When binomial regressions were confined to 97 patents valued at $1 million or higher, the VALUE coefficient was 0.207 (0.059) without IPC dummy variables and 0.182 (0.063) with dummy variables. For regression 2b with 66 high-value patents, one could not confidently reject the hypothesis that the citations were Poisson-distributed. The ln VALUE coefficient in a Poisson regression with IPC dummies was 0.387 (0.100).

[17] Pseudo-R^2 values were computed by squaring correlations between the actual and regression-predicted numbers of citations (in absolute levels, not logarithms).

PART IV

THE ROAD AHEAD

[18]
THE INNOVATION LOTTERY

F. M. SCHERER*

During the past several years Dietmar Harhoff (University of Munich) and I have been collecting and analyzing quantitative data on the size distribution of rewards realized by individuals and organizations carrying out technological innovations. From our new evidence and insights by other scholars on the economics of individuals' gambling behavior, it is possible to propose a theory of incentives for innovation in technology and other creative endeavors. Our findings provide support for a half-century-old obiter dictum by the renowned Austrian (and later Harvard University) economist Joseph A. Schumpeter. Characterizing the admixture of chance, ability, and energy that sustains innovation and technological progress in a capitalistic economy, Schumpeter wrote:

> Spectacular prizes much greater than would have been necessary to call forth the particular effort are thrown to a small minority of winners, thus propelling much more efficaciously than a more equal and more 'just' distribution would, the activity of that large majority of businessmen who receive in return very modest compensation or nothing or less than nothing, and yet do their utmost because they have the big prizes before their eyes and overrate their chances of doing equally well.[1]

Schumpeter's conjecture was based upon little more than casual empiricism. In this chapter I summarize the relevant new evidence and tread, perhaps too incautiously, into the realm of theory-building.

* Aetna Professor of Public Policy Emeritus, John F. Kennedy School of Government, Harvard University; visiting professor, Princeton University.

[1] Joseph A Schumpeter, Capitalism, Socialism, and Democracy 73–74 (1942).

I. Profits and their Possible Size Distributions

During the early 1960s, while seeking to understand the links between market structure and technological innovation, I happened onto some data from a small survey asking holders of US invention patents about the profitability of their inventions. Analyzing the profit data, I found their statistical distribution to be 'skew' and indeed highly skew, with most of the observations lying in the range of low profit values but 'with a very long tail into the high value side'.[2] A simple test suggested that the data might conform to what is called a Pareto distribution, to whose peculiar properties the mathematician Benoit Mandelbrot was at the time directing attention.[3]

A statistical size distribution is essentially a systematic array of numerical observations drawn through sampling, conventionally shown by plotting on the horizontal axis of a graph the range of observation values (eg, profits) and on the vertical axis the relative frequency with which those values occur in the sample. Figures 1.1(a) and 1.1(b) illustrate two such distributions created using my computer's random number generating program. The array in Figure 1.1(a) is typical of a 'normal' distribution. It resembles a symmetric bell-shaped curve, with the largest and smallest observation values occurring with much lower frequency than those in the middle of the distribution (ie, near the sample mean, or average value). Figure 1.1(b), on the other hand, reflects a particular kind of skew distribution known as a log normal distribution.[4] By far the greatest mass of observations occurs at relatively low values, but the distribution has a long thin right-hand 'tail' of large values occurring with very low frequency.

The length and thickness of the tail determine the degree of skewness characterizing a distribution. The longer the tail, the more skew the distribution is. Some general classes of skew distributions are more skew than others. The Pareto distribution I believed I had found in the 1960s for patented invention profits is more skew, ie, with a longer and fatter tail, than the typical log normal distribution illustrated in Figure 1.1(b). The differences between skew distributions can be characterized through a simple graph (Figure 1.2) popularized by Benoit Mandelbrot. The coordinates are logarithmic. On the

[2] F M. Scherer, *Firm Size, Market Structure, Opportunity, and the Output of Patented Inventions*, 55 AM. ECON. REV. 1098 (1965).

[3] Benoit Mandelbrot, *New Methods in Statistical Economics*, 71 J POL. ECON. 421 (1963).

[4] It is called log normal because if one takes logarithms of the observation values, the logarithmic values exhibit a symmetric normal distribution pattern. Indeed, Figure 1.1(b) was created by taking the 1,000 values used in Figure 1.1(a) (from a random number program generating normally distributed values with mean zero and a standard deviation of 1) and using them as exponents in the expression 10^n, where n is the value of a Figure 1.1(a) observation. Since the logarithm to base 10 of 10^n is n, the distribution of n is normal while 10^n is skew-distributed. The mean (average) value of the distribution in Figure 1.1(b) is 10.50; the median value is 1.11.

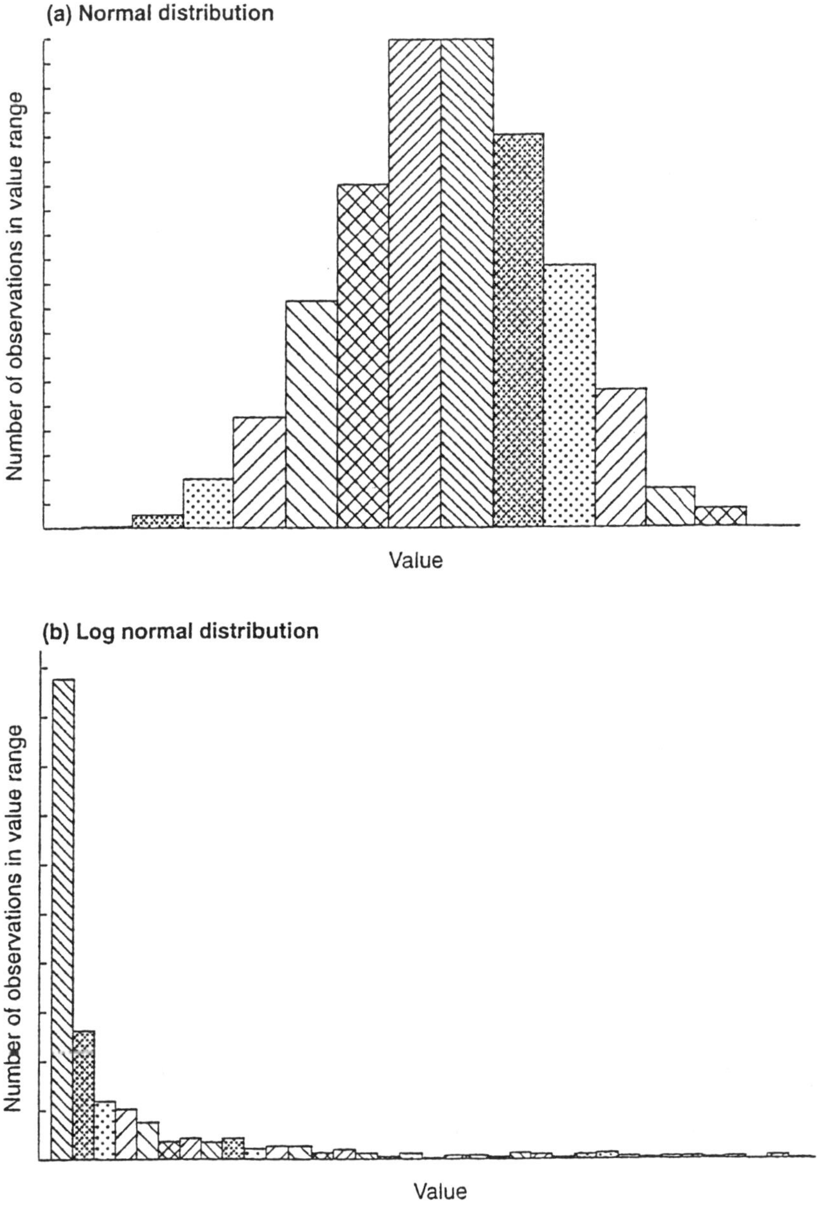

FIG. 1.1: Comparison of normal and (skew) log normal distributions

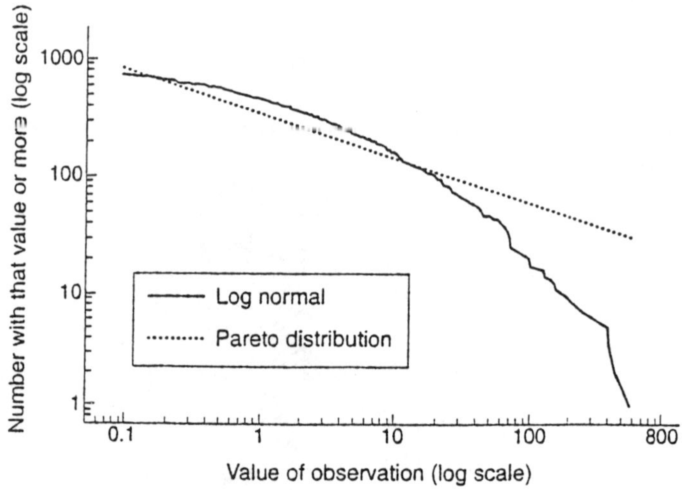

FIG. 1.2: Pareto plot of simulated log normal and Pareto distributions

horizontal axis, one identifies the values of the observations (eg, profits). The observations are ranked from most valuable to least valuable, and on the vertical axis, one plots the number of observations with values equal to or greater than any given horizontal axis value.

The solid curve in Figure 1.2 traces the log normal distribution plotted in Figure 1.1(b). The most valuable observation (number 1 on the vertical scale) has a value of 570; 18 observations have values of 100 or more; and 172 observations have values of 10 or more. The dotted straight line shows a Pareto distribution fitted to the data underlying Figure 1.1(b).[5] The simplest Pareto distributions plot as straight lines on doubly logarithmic graphs, which is why such graphs are called Pareto graphs. The dotted Pareto distribution line conforms fairly closely to the solid log normal distribution curve over the less valuable observations ranked from 100 to 1,000. However, it diverges ever more sharply from the downward-bending log normal distribution in the right-hand tail of high-value observations. Indeed, the most valuable log normal observation of 570 is only the 30th most valuable observation on the Pareto line. If the Pareto line were extrapolated linearly, one would find the value of the most valuable observation (number 1 on the vertical axis) to be 2.94 million, and the second most valuable observation (number 2 on the vertical) 520,000. True Pareto distributions have extremely large 'outliers' relative to the mass of observations.

[5] Specifically, a straight line was fitted by least squares to the observations, yielding the equation log rank = 2.585 – 0.3996 (log value of observation). The slope value 0.3996 is called the Pareto alpha or the Pareto slope coefficient.

Skew distributions exhibit ill-behaved sampling properties. Statisticians' 'law of large numbers' states that the mean (average value) of a sample of observations should converge ever more closely to some 'true' average value as one draws ever larger samples. But with skew distributions, this convergence operates at best slowly, and it may not happen at all. Pareto distributions are extreme in their non-conformity to the law of large numbers. Indeed, Pareto distributions with high but plausible degrees of skewness do not conform at all.[6] Rather, as one draws ever larger samples, there is a chance that one will come up with an extremely large observation value (such as the 2.94 million value extrapolated from the Pareto distribution of Figure 1.2) that is so large relative to all previous observations that it 'blows' the sample mean to a much higher value. For the Pareto distribution illustrated in Figure 1.2, the sample average is 1,004 if the largest observation is excluded but rises to 2,947 if it augments the other 999 observations. What this means in practical terms is that it is very hard or maybe even impossible to secure stable average profit returns by pursuing portfolio strategies eg, pooling many research and development projects into a portfolio.

II. The Profits from Technological Innovation

My discovery that the distribution of profits associated with patented inventions might exhibit sampling instabilities like those described in the previous paragraph has been gnawing at me for three decades.[7] My interest was sustained in part by others' findings, from studies of the rates at which fees are paid to keep patents in force in some national jurisdictions, that the distribution of patent values is indeed highly skew. Returning to the scene of the crime, I joined Dietmar Harhoff to conduct surveys ascertaining through direct queries the estimated value of 776 German and 222 US inventions on which German patent protection was maintained for a full 18-year term following applications filed in 1977.[8] We confirmed that even for patents considered valuable enough to warrant paying German renewal fees escalating over time to a cumulative total of DM 16,075 (roughly $6,825 at 1977 exchange rates), most of the inventions are of only modest value, while a few yield blockbuster rewards. Among the 776 German patents—already self-selected from the seven times larger cohort of contemporary patents not

[6] This is true when the so-called Pareto slope coefficient has an absolute value less than 10. The slope coefficient for the Pareto distribution in Figure 1.2 is shown in n. 5 above to be 0.3996. Thus, asymptotically (as ever larger samples are drawn, and the straight dotted line in Figure 1.2 shifts in parallel upward), the sample mean diverges toward infinity.

[7] Acquaintances of long standing might observe that quite a lot of me has been gnawed away in the interim

[8] Dietmar Harhoff et al., *Exploring the Tail of Patent Value Distributions* (working paper), Center for European Economic Research, Mannheim, Germany (1998).

renewed to full term—the most valuable 78 (ie, the top ten percent) accounted for an estimated 88 percent of all 776 patents' economic value to their holders. For the smaller but even more select group of US inventions, which were patented both in Germany and the United States, the top 22 accounted for 81 to 85 percent of the 222 patents' value. The best of the best conferred economic prizes measured in the billions of dollars or Deutschmarks.

Figure 1.3 plots the German patent value distribution on doubly logarithmic (ie, Pareto) coordinates. If the size distribution were Paretian, it would plot as a straight line on the chosen coordinates. In fact, downward concavity reveals that there is less skewness than for the Pareto case. A statistical analysis provides the strongest support for a log normal distribution like the solid-line distribution in Figure 1.2.[9]

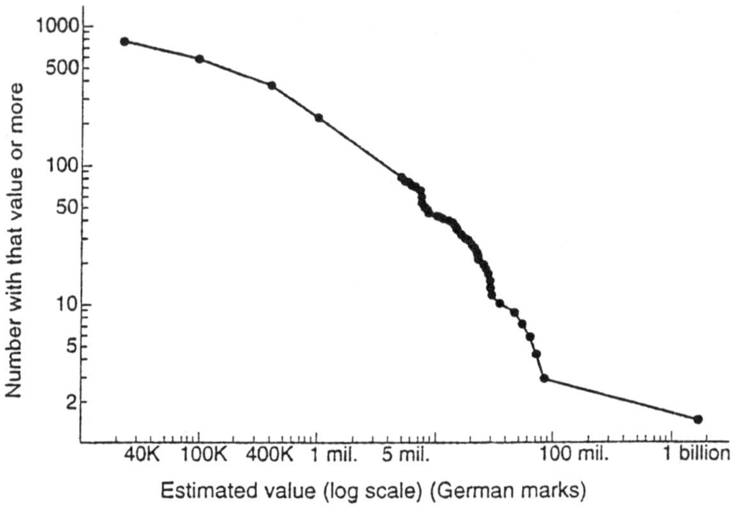

FIG. 1.3: Plot of German renewed patent values on Pareto coordinates

In a parallel survey of the license royalties obtained by six research-oriented US universities on their patent portfolios during four years of the early 1990s, I found that a single bundle, comprising the three Cohen-Boyer gene splicing patents, yielded 24 to 33 percent of the total royalties obtained from 350 to 486 individual bundles of licensed technology. The top six bundles, ie,

[9] For the log normal data underlying Figure 1.2, the 100 largest observations (ie, the top 10 percent) accounted for 76.9 percent of total sample values. By this concentration measure, the Figure 1.2 data are somewhat less skew than the German patent data.

making up one to two percent of the sample members by number, generated from 66 to 76 percent of total sample patent royalties.[10]

Confirmation of the rewards skewness phenomenon comes from information on the discounted present value of quasi-rents (ie, the discounted surplus of sales revenues over estimated production and marketing outlays) realized by companies from new pharmaceutical chemical entities marketed during the 1970s and early 1980s in the United States following Food and Drug Administration approval.[11] The most profitable 10 percent of those product introductions contributed 48 to 55 percent of total sample quasi-rents. Among eight sets of data on which reward size distribution data were obtained, the two drug quasi-rent samples exhibited the least skewness. Their distributions were much less skew than Pareto distributions and somewhat less than the log normal distribution, but more than plausible alternative skew distributions such as the negative binomial or the Weibull.[12]

Further insights emerge from evidence on the outcomes of investments in that most dynamic of US industrial sectors, new high-technology ventures. Surveys of 1,053 investments by venture capital funds in individual venture capital targets reveal a highly skew distribution of outcomes. Some 59 to 62 percent of the ultimately realized returns came from the most successful 10 percent of investments by number.[13] A Federal Reserve Board staff study of 225 venture capital partnerships formed between 1980 and 1986 revealed a weighted-average internal rate of return on investment as of 1993 averaging 7.95 percent, with a weighted-average median of 5.58 percent.[14] Finding the mean in excess of the median or mid-sample value, which occurred for every individual year of fund formation as well as in the aggregate, is another indication of distribution skewness. So also is the high variability over time of median rates of return, ranging from 1.6 percent (for partnerships organized in 1981) to 13.2 percent (for 1980-vintage funds). When the distribution of outcomes is skew, as we have seen, atrophied operation of the law of large numbers makes it hard to diversify away random sampling variability by forming portfolios of numerous investments.

To sharpen insights on the performance of new high-technology ventures, an exhaustive sample was drawn of venture fund-backed companies that floated initial public common stock offerings (IPOs) in the 1983–86 period and that operated in specified high-technology fields.[15] Typically, new

[10] F. M. Scherer, *The Size Distribution of Profits from Innovation*, 49/50 ANNALES D'ECONOMIE ET DE STATISTIQUE 495 (1998).

[11] See Henry Grabowski & John Vernon, *A New Look at the Returns and Risks to Pharmaceutical R&D*, 36 MGMT SCI. 804 (1990); *Returns on New Drug Introductions in the 1980s*, 13 J. HEALTH ECON. 383 (1994).

[12] See Scherer, n. 10 above.

[13] Ibid.

[14] GEORGE W. FENN ET AL., THE ECONOMICS OF THE PRIVATE EQUITY MARKET (1995). The weighted-average standard deviation was 9.60 percent.

[15] A more complete analysis is presented in F M. Scherer et al., *Uncertainty and the Size Distribution of Rewards from Technological Innovation*, 10 J. EVOLUTIONARY ECON. 175 (2000).

companies supported by venture capital partnerships attempt IPOs only after they have made appreciable progress toward achieving technical and market success, and so they tend to be more mature investments than those discussed in the previous paragraph. A hypothetical first-day investment of $1,000 was made in the common stock of the 110 sample companies whose stock continued to be traded on a public market after the IPO date.[16] Dividends were reinvested and stock splits were tallied. By the end of 1995, 52 of the 110 companies remained as independent entities. The value of their stocks as of 31 December 1995, along with proceeds (reinvested temporarily in the NASDAQ index) from the liquidation of drop-out company stocks, was as follows:

52 surviving companies	$417,002
23 acquired companies	96,400
35 delisted companies	21,178
Total terminal value	$534,580

Had the same $110,000 been invested in the NASDAQ index in January 1983, investors would have had $501,908. Thus, investors in our sample of high-technology companies increased the value of their initial investment by a substantial multiple, earning on average a 12.2 percent return on their investment, but fared only slightly (and not statistically significantly) better than investors in a complete portfolio of NASDAQ company stocks.

For our present purposes, the key point is this: as the years progressed from the time of the initial investments analyzed here, the distribution of outcomes became more and more skew. By the end of 1995, the best-performing 11 companies (10 percent by number of the initial sample) contributed 62 percent of the investors' total portfolio value. Figure 1.4 shows how the value of individual company investments changed over time. It is limited to 10 companies, including the five most successful full-term survivors and five others selected randomly. The share values for the random choices cluster so tightly below $2,000 that they are largely indistinguishable. Had one invested $1,000 in Adobe Systems, Concord Computing, and Amgen, on the other hand, one would have had shares valued at $77,565, $74,130, and $55,980 respectively by the end of 1995. The cross-sectional distribution of full-term survivors as of December 1995, plotted in Figure 1.5 on doubly logarithmic coordinates like those used in Figures 1.2 and 1.3, exhibits a concave log normal scatter of values not unlike the one observed for German patents, except that there is no single extremely large outlying value. Skew-distributed samples commonly exhibit such erratic behavior in their right-hand tail.

[16] Twenty-one IPO companies left no discernible trace of stock trading in stock exchange records To standardize the investments for timing differences, $1,000 was 'parked' in the NASDAQ index as of 1 January 1983, with the proceeds being invested on the date of any sample company's IPO.

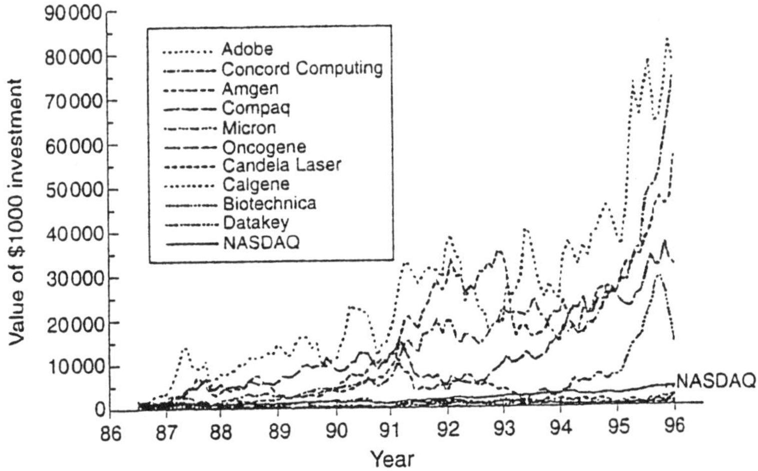

FIG. 1.4: Evolution of the value of $1,000 investments in 10 IPOs

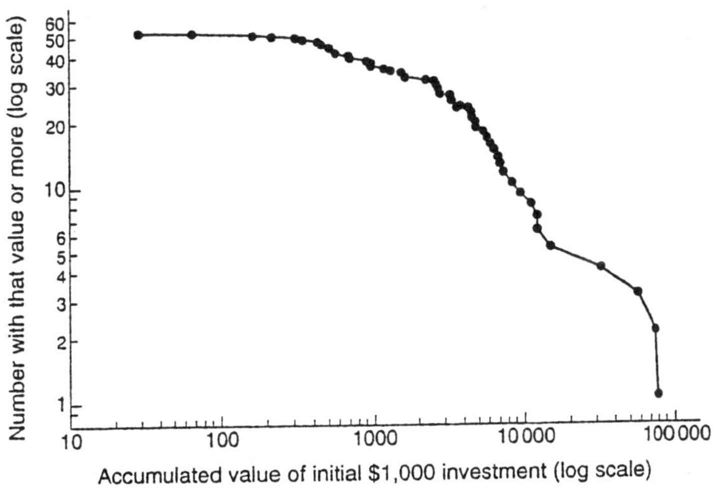

FIG. 1.5: Pareto plot of 52 1983–86 IPO investments' value in December 1995

Thus, we conclude that there are striking regularities in the distribution of rewards to technological innovation. A minority of 'spectacular winners' appropriate the lion's share of total rewards, as Schumpeter predicted. The size distribution of rewards is highly skew, with a long right-hand tail. The distributions do not appear to be Paretian, contrary to my 1965 hypothesis; the log normal distribution characteristically provides a better fit. The rewards from individual patents exhibit more skewness, ie, the top 10 percent capture a higher total share of payoffs, than the rewards from innovations such as new drugs, which may be covered by numerous product and process patents, or investments in high-technology startup companies, which sometimes market multiple innovations and hence average out some of the variability associated with individual innovations.

III. Innovation in Popular and Not-So-Popular Culture

Not all innovation is technological. Innovation also occurs in cultural domains, eg, in the composition or performing of music, the production of motion pictures, and the writing of books. We supplement our insights here by analyzing fragmentary data from the first two of these branches.

From *Billboard* magazine, data were obtained on the US sales of the 70 best-selling popular music albums and single recordings in 1997.[17] The best-selling album, with sales of 5.3 million estimated by sampling, was by the Spice Girls; the best-selling single, with sales of 8.1 million units, was Elton John's 'Candle in the Wind', commemorating the death of Britain's Princess Diana. It is known (also from *Billboard*) that approximately 613 million music albums and 134 million singles records were sold in 1997. The best-selling album accounted for 0.86 percent of total album sales; the best-selling 70 albums (all with sales of 1 million or more) for 21.0 percent. The best-selling single recording gained 6.1 percent of singles sales; the 70 best-sellers (with sales of 500,000 or more) 55.0 percent of singles sales. It is not known how many albums and singles were offered in total during 1997. A crude extrapolation from the (nearly log normal) distribution of data on the top 70 albums suggests that 1,000 albums accounted for 60 percent of total album sales, with average sales for individual albums in the bottom decile of this best-selling group averaging approximately 123,000. A similar extrapolation reveals (with somewhat larger uncertainty) that 800 records sufficed to account for 98 percent of total singles sales, with the least successful hundred of those 800 averaging sales of roughly 10,000 each. Even though individual winners did not capture the lion's share of sales, the top 10 winners surpassed the sales of records ranked 800th in the distribution by approximately 25 times on average for albums and 250 times for singles. Clearly, it is much more lucrative to be a top winner.

Figure 1.6 plots on doubly logarithmic coordinates the size distribution of sales for the top 70 records of 1997. For both albums and singles, some concavity is evident, although the extreme 'Candle in the Wind' outlier reverses the curvature for singles, just as an extreme value does in the distribution of German patents (Figure 1.3).

[17] *Best-Selling Records of 1997*, BILLBOARD MAGAZINE, January 31, 1998, at 76.

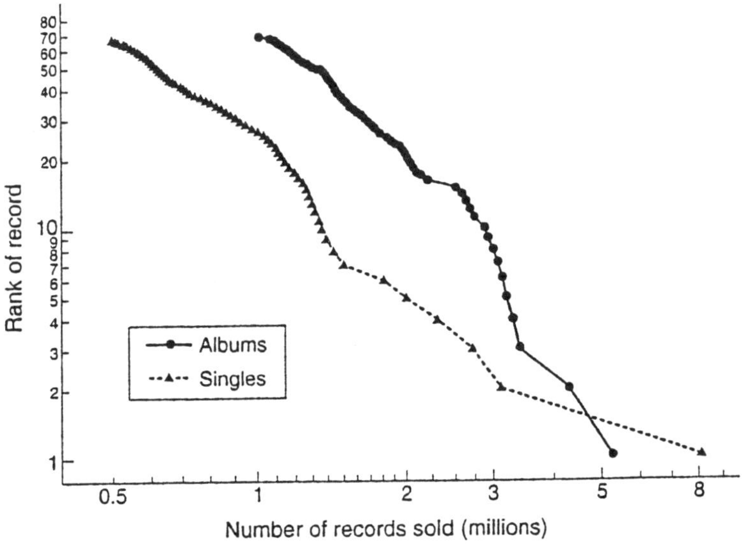

FIG. 1.6: Pareto plot of leading popular record sales in 1997

These statistics are for individual records. Another article in *Billboard* arrays the cumulative total record sales outside the United States over 50 years through December 1997 for the 48 most successful recording groups affiliated with one of the leading popular record companies, Atlantic Records.[18] The highest-selling group, Led Zeppelin, achieved career sales of 29.6 million records, more than twice those of the second-ranked group and 34 times those of the 48th-ranked artist. Led Zeppelin's sales comprised 17.6 percent of the sales of all 48 leading groups. This and other evidence reveals more skewness in career record sales than for the distributions across individual records, but less than what was observed for any group of technological innovations.

The success of classical music composers, tallied not in lifetime financial rewards but in the attention posterity has paid them, also appears to follow a skew distribution. From the Schwann *Reference Guide to Classical Music* for Fall 1996, the number of records currently available was estimated for 686 composers born between 1650 and 1840.[19] The measure used was linear centimeters of record listings, with five lines of type per centimeter on average and the typical record entry running from one to three lines. A list of the eight leaders yields no major surprises:

[18] *Atlantic's International Best-Sellers*, BILLBOARD MAGAZINE, January 17, 1998, at A–28.

[19] A posthumous record count is analogous to the publication citation counts used to assess the influence of scholars. See DEREK J. DE SOLLA PRICE, LITTLE SCIENCE, BIG SCIENCE (1963).

W. A. Mozart	1,656 cm
L. Beethoven	1,262 cm
J. S. Bach	1,190 cm
J. Brahms	644 cm
P.I. Tchaikovsky	568 cm
F. Schubert	553 cm
F. Chopin	497 cm
F. J. Haydn	461 cm

Mozart alone commands 11.3 percent of the recordings; the top eight composers 46.8 percent, and the 69 composers comprising the top 10 percent by number 87.8 percent.[20] These values, reflecting a high degree of skewness, are broadly consistent with those observed for the profits or royalties from technological innovations.[21] Figure 1.7 plots the distribution function on doubly logarithmic coordinates. For the vast majority of composers, it is close to a

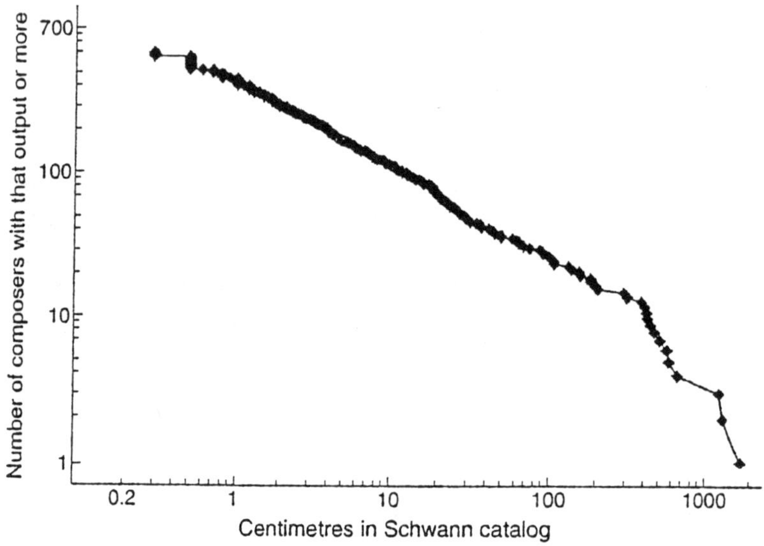

FIG. 1.7: Pareto plot of 686 composers' recordings available in 1996 (composers born between 1650 and 1840)

[20] Mozart's leadership is even stronger when composers are arrayed in order of records per year of working life, with the working life defined (somewhat inaccurately for the case of Mozart) to begin at age 16. The top eight composers by this criterion accounted for 57.1 percent of the total distribution.

[21] In ongoing research, I am attempting to assemble data on the the size distribution of composers' earnings during their lifetimes. On Mozart, for which the documentation is particularly rich, see W. J. Baumol & Hilda Baumol, *On the Economics of Musical Composition in Mozart's Vienna*, 18 J. CULTURAL ECON. 171 (1994).

straight line, consistent with a Pareto distribution, departing from the Pareto form and bending downward only for 13 leaders.

Arthur DeVany and W. D. Walls compiled statistics on the distribution of revenues for motion pictures (most of them readily forgotten) appearing among the Top 50 list in *Variety* magazine between May 1985 and January 1986.[22] Over the entire nine-month period surveyed, the top 10 percent of market-leading films by number generated approximately 60 percent of total sample revenues. Again, a skew distribution is revealed. The distribution was strongly concave relative to doubly logarithmic coordinates and hence inconsistent with a Pareto law, but more skew than a log normal distribution.

IV. An Ideal Reward System?

From the compilation of evidence reviewed here, it would appear that Schumpeter was correct at least statistically: the big prizes from innovation are thrown to a small minority of winners, while the majority of innovative efforts confer only modest rewards. For new drug chemical entities and investments in high-technology startups, on which the evidence is most complete, the median project yields less than the capitalized cost of capital funds invested, but losses on the majority of projects are more than offset by gains from the most successful projects. We ask now whether Schumpeter might also have been correct in his conjecture that a skew distribution of rewards motivates innovative activity 'more efficaciously' than a more 'just' system of rewards, which presumably would bestow returns more or less closely proportional to the investments seeking them.

Skew reward distributions are particularly risky because it is difficult to make random deviations cancel each other out and converge on some stable average value by forming diversified portfolios containing numerous investments.[23] Most of received investment theory assumes that investors are risk-averse. If innovation requires investment and investors are risk-averse, how can a highly skew distribution of rewards be conducive to innovation?

The assumption that investors are risk-averse is difficult to square with the widespread and long-standing popularity of sweepstakes lotteries in the

[22] Authur DeVany & W D. Walls, *Bose-Einstein Dynamics and Adaptive Contracting in the Motion Picture Industry*, 106 ECON. J. 1493 (1996).

[23] How difficult it is was shown by computer simulations in which hypothetical random samples were drawn repeatedly from the skew distribution of new drug quasi-rents estimated by Grabowski and Vernon. Even when new product introductions were pooled to the full US industry-wide level (ie, with a single entity introducing 18 new products annually, each yielding profits over a 21-year span), annual profit swings as large as plus-or-minus 25 percent of long-run industry averages remain. See F. M. Scherer & Dietmar Harhoff, *Technology Policy for a World of Skew-Distributed Outcomes*, 29 RES. POLICY 559 (2000).

United States and other nations.[24] The larger in absolute value the prize, captured at infinitesimal odds, the more enthusiastic individuals' participation appears to be. With an actuarial value of payoffs well below the sum of players' bets—eg, for state sweepstakes lotteries, roughly half the stakes on average—such lotteries, like most other forms of gambling, are not 'fair' gambles.[25] Yet people flock to play.

Investing in high-technology startup companies is similar to (but less extreme than) lotteries in the skewness of rewards and the relatively low probability (well below 0.1) of a really big payoff. But they are strikingly different in the sense that venture investment tends, given reasonable diligence, to yield discounted returns exceeding actuarially the value of the stakes invested. Thus, at least historically, they have been 'fair' gambles. Perhaps that is difference enough to distinguish between buying sweepstakes tickets and investing in the innovation lottery. (Whether high-technology stock returns will continue to exceed investments in the future as more and more naive investors plunge into the high-technology game remains to be seen.) But let us probe further to see whether more can be discerned.

That people embrace high-stakes gambles but engage more or less simultaneously in the risk-averse behavior associated with paying an administrative premium for insurance has long fascinated economists. In a seminal paper, Milton Friedman and L. J. Savage showed that a consumer may rationally choose both to buy insurance against risks that would reduce her wealth and accept unfair gambles on the off chance of a sizeable wealth increase. Their argument is illustrated in Figure 1.8, which assumes that 'utility' (whose measurement problems need not detain us) is derived from consumption and that consumption is facilitated by wealth. Friedman and Savage postulated that the utility function, ie, the mathematical relationship between an individual's utility and her wealth, has the peculiar ogive shape shown by the heavy solid line U(W) in Figure 1.8. Suppose the consumer's initial wealth position is W_0, yielding (read horizontally over to the vertical axis) utility realization U_0. Now suppose the consumer can invest in a 'fair' lottery ticket whose cost, if with relatively high probability she fails to draw the winning combination, leaves her with lower wealth W_L. But if she wins the lottery (with low probability), she is propelled to much higher wealth W_W (yielding correspondingly higher utility U_W). The straight line segment ACB shows a locus of expected utility values given by the probability of winning P_W times the utility U_W from winning plus $(1 - P_W)$ times the utility U_L from being at the loser's

[24] See, eg, Charles T. Clotfelter & Philip J. Cook, Selling Hope: State Lotteries in America (1989); Matthew Breuer et al., *State Lotteries: The Determinants of Ticket Sales*, 8 Wagner Rev. 15 (1997); Gregory Bresinger, *The Lottery Racket*, 16 Free Market 1 (1998). The irrationality of lotteries appears to be one of the few things on which economists of all ideological persuasions agree.

[25] A fair gamble is one in which the actuarial value of the prize, ie, the value of the prize times the probability of winning it, equals the value of the bets.

wealth level W_L. If the odds are such that the probability-weighted expected wealth outcome is $\bar{W}$, the expected utility (read horizontally from point C on the expected value line ACB to the vertical axis) of that outcome is exactly the same as the utility U_0 from not entering the lottery and remaining with certainty at wealth level W_0, then our consumer will be indifferent between playing and not playing. But if the 'cost' of a lottery ticket (eg, in the form of less than 'fair' odds) is less than $W_0 - \bar{W}$, so that the probability-weighted expected value of the consumer's post-lottery wealth is greater than $\bar{W}$, the consumer will realize higher expected utility participating in the lottery than not participating. This happens because substantial increases in wealth yield disproportionately large increases in utility (ie, marginal utility is increasing with increases in wealth), outweighing the actuarial value of the modest decrease in utility associated with buying a ticket and losing. By similar reasoning, it can be shown that from an initial wealth position W_0, the consumer faced with some small probability of losing much of her wealth (eg, because her house burns down) but able to insure against that contingency by paying a premium for insurance will, within some range of insurance costs, choose to buy the insurance. The reason is that, in the concave-downward (low-wealth) segment of the consumer's utility function, wealth losses entail disproportionately large sacrifices of utility, the avoidance of which is worth paying more than the actuarial value of insurance.

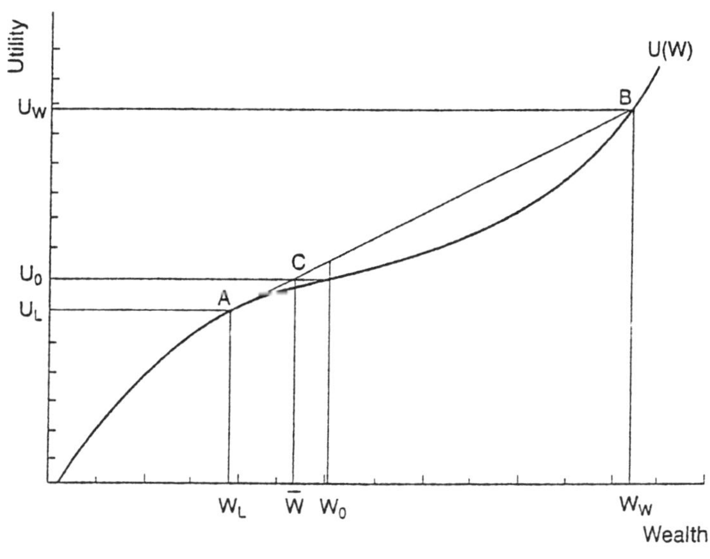

FIG. 1.8: Utility function consistent with buying insurance and betting in lotteries

Friedman and Savage speculate that real-world individuals' utility functions often have upward-bending curvature at wealth levels much higher than those currently attained, implying increasing marginal utility of wealth, because a big wealth increase moves the consumer to a new, much higher material and social status. Such a large change in effect fulfills their optimistic, or perhaps wildly optimistic, dreams. Certainly, something like that explains the propensity of lower-income citizens to spend disproportionate amounts of their income playing the lottery, in effect gambling on an escape from poverty.[26] It is also consistent with anecdotal evidence about the motives of high-technology firm entrepreneurs, who see a successful startup as the principal means of becoming truly wealthy. In this respect, sweepstakes players and technological entrepreneurs may be more similar than their conventional demographic characteristics imply.

Lotteries, of course, are not the only form of gambling to offer the possibility of large payoffs at low odds. Horse racing provides a rich environment for analyzing risky choices. There is evidence that the actuarial returns from bets on 'long shots' tend to be lower than on better-rated horses, suggesting (given the way parimutuel odds are set) that some bettors have a positive preference for long shots. Noting that the same individuals place bets on multiple races and often bet on more than one horse per race—portfolio strategies that pool and hence reduce risks, conventionally defined—Golec and Tamarkin question previous studies' inference that such bettors are risk lovers. They argue that one must distinguish between two commonly confounded aspects of risk: the variance (technically, the second moment of a statistical distribution), which measures the average variability of sample observations around their mean, and the skewness (the third moment), which, as we have seen, measures asymmetries associated with very long tails on one side of the distribution.[27] Golec and Tamarkin test the hypothesis that it is skewness (the third moment), not variance (the second moment, and the more conventional measure of risk) that the long-shot bettors embrace.[28] In a statistical analysis of 2,309 races, they find persuasive evidence that bettors are simultaneously variance-averse and skewness-loving.

This discovery, which is consistent with the utility function shape exhibited in Figure 1.8, might equally well explain the behavior of high-technology

[26] See CLOTFELTER & COOK, n. 24 above, at 95–104. It is for this reason that their book is titled 'Selling Hope'.

[27] Statisticians distinguish four main moments, or mathematical characterizations, of sample distributions. The first moment, or mean, characterizes the sample's central tendency; it is the sum of observation values divided by the number of observations. The second moment, or variance, characterizes the degree to which observations deviate from the mean; the third moment, the skewness of the distribution of observations, was clarified earlier; and the fourth moment, the 'peakedness' or height of the distribution compared to the length of its tails.

[28] Joseph Golec & Maurry Tamarkin, *Bettors Love Skewness, Not Risk, at the Horse Track*, 106 J. POL. ECON. 205 (1998).

entrepreneurs, inventors, and (to the extent that financial motives play a role) creators of popular culture. In high technology, those who commit their fortunes and most of their waking hours to pioneering a new venture forego the option of placing multiple bets enjoyed by horse race bettors. But high-technology 'angels', like horse players, simultaneously bet on long shots and pursue variance-reducing multi-investment portfolio strategies. At this stage, the proposition that those who invest in high technology derive positive utility from the skewness of rewards is offered as hypothesis, not as a demonstrated phenomenon. The most one can say is that it is intuitively plausible. Testing it rigorously is likely to be more difficult than with horse racing, since all of the skewness evidence presented in this chapter was derived ex post, and it is unclear whether the ex ante skewness potential of specific high-technology investments or classes of bets can be measured empirically.

V. Intellectual Property Implications

In the evidence that the rewards from innovation are skew-distributed and that at least some risk-takers are skewness lovers there are implications for intellectual property policies. However, several caveats need to be recognized.

For one, especially in the arts but to some extent also in the realm of technology, creative activity is often driven by non-pecuniary motives. To be sure, an artist, author, or inventor must keep body and soul together, but as long as that condition is met and the incremental costs of creation are not large, the uncertain prospect of spectacular payoffs may be more of a 'nice to have' fringe benefit than a necessary incentive. On this, preferences undoubtedly differ. Franz Schubert first composed a work for money at the age of 19. Ten to 12 years later he was still offering publishers his work for 'a moderate remuneration'.[29] Beethoven, on the other hand, was notoriously avaricious, playing publishers off against one another to gain desired publication fees for specific compositions and segmenting his markets geographically, with different publications for England, France, and the various German-speaking territories. Franz Joseph Haydn toiled most of his career on a handsome salary that provided no special incentive compensation for his prodigious creative output, but went for the big prizes when his patron Prince Nicolaus Esterhazy died and he was invited to London by impresario Johann Peter Salomon. Aaron Copland is said to have testified in a deposition on copyright royalties that he would pay people to listen to his music. As an author who has often purchased copies of my work at appreciable out-of-pocket cost and distributed them free to colleagues, I empathize. My daughter helps support

[29] FRANZ SCHUBERT'S LETTERS AND OTHER WRITINGS 30, 122–23, 135 (Otto Erich Deutsch ed., 1974).

her six children with the modest fees she receives playing bluegrass fiddle professionally, but dreams of the affluence a hit record would bring.

Although individual high-technology entrepreneurs may, like racetrack bettors and sweepstakes lottery players, be skewness-lovers, it seems less likely that decision-making in well-established corporate organizations conforms to the hypothesis. For the employed inventor, a particularly successful contribution can bring a sizeable bonus and promotion to a higher income bracket, but nothing remotely approximating the rewards populating the right-hand tail of the profit distributions Dietmar Harhoff and I have been surveying. For the research and development manager who drives an invention to commercial success, the fraction of benefits appropriated personally is similarly modest.[30] When corporate hierarchies must decide whether or not to invest substantial sums on the commercial development of an invention, risk aversion, not skewness affinity, is almost surely the behavioral norm. Such motivational differences may have much to do with the propensity observed during the first half of the twentieth century for a disproportionate share of the boldest technological innovations to originate outside the laboratories of large corporations,[31] and, more recently, for high-technology startups to be among the most prolific contributors to US technological dynamism.

To the extent that investments in technological and artistic creation are motivated by the longshot hope of a very large reward, intellectual property policies should sustain and reinforce that incentive system, not undermine it. This implies a role for strong patents and copyrights. But from that basic precept two more nuanced implications can be drawn.

First and most obvious, patents and copyrights ought not to be revoked or weakened simply because an innovator has made 'too much' money from his creation, for the prospect of a large reward is a crucial feature of the skewness-based incentive system. This does not mean that *any* weakening of intellectual property rights is necessarily counter-productive, for an expectation that patents or copyrights can be enforced fully has more relevance for some actors—notably, those who conform to the skewness-loving paradigm—than for others. There is compelling evidence that patent rights are not very important as a prerequisite for research and development investments in many (but not all) well-established industries.[32] The early market occupancy, reputational, and learning curve advantages of being a first mover are in

30 For an early statement of this managerial risk aversion hypothesis, see WILLIAM J. FELLNER, COMPETITION AMONG THE FEW 172–73 (1949).

31 See, eg, JOHN JEWKES ET AL., THE SOURCES OF INVENTION (1959); F. M. Scherer, *Schumpeter and Plausible Capitalism*, 30 J. ECON. LITERATURE 1416 (1992).

32 See Richard C Levin et al., *Appropriating the Returns from Industrial Research and Development*, Brookings Papers on Economic Activity (No. 3) 783–832 (1987); Wesley J. Cohen et al., *Appropriability Conditions and Why Firms Patent and Why They Do Not in the American Manufacturing Sector* (working paper), Carnegie-Mellon University (June 1997).

many instances much more important. Because of these non-patent advantages, the enforcement of compulsory patent licensing in more than 100 US antitrust settlements does not appear to have had much, if any, adverse impact on the target companies' investments in research and development.[33] Fine-tuning is needed in remedying antitrust abuses and avoiding impediments to further innovation caused when one inventor has a powerful position blocking the advances of others who could otherwise make significant improvement inventions. In particular, policymakers should be particularly cautious in avoiding the early erosion of rights for that class of innovators or creators for whom the uncertain hope of large rewards was a plausibly significant motivational factor.

A second implication follows. Although exceptions exist, there is reason to believe that the enforcement of intellectual property rights is biased in favor of large, well-established organizations, whose behavior conforms least well to the skewness paradigm, and against the independent innovators who conform most closely. David does occasionally slay Goliath in the courtroom, but more often than not, a well-heeled corporation can afford to persevere in costly litigation until the financially weaker party submits—if not in absolute defeat, then in a settlement that yields less than the spectacular reward that might otherwise have been achieved by an independent innovator.[34] If the skewness hypothesis is anywhere near correct, there ought to be policy adjustments to remedy the imbalance.

There may also be implications for international economic policy. Just as the larger payoffs available with statewide (and perhaps also nationwide) lotteries encourage more intensive participation than smaller city-based lotteries,[35] the larger rewards attainable by selling innovative products in a global marketplace probably strengthen incentives for both skewness-loving and risk-averse actors. There is much to be said therefore for continuing efforts to harmonize national intellectual property systems and, as under the single-filing-locus system endorsed by many European Union nations, to reduce the costs of securing protection outside one's home nation.

[33] See F M. SCHERER, THE ECONOMIC EFFECTS OF COMPULSORY PATENT LICENSING Ch. IV (1977).

[34] The nerd's prayer: 'Oh Lord, help me to capture the market with my killer app; but if Thou willst otherwise, let me at least sell out to Microsoft'.

[35] See Philip J Cook & Charles T. Clotfelter, *The Peculiar Scale Economies of Lotto*, 83 AM. ECON. REV. 634 (1993); *The Massachusetts State Lottery*, John F. Kennedy School of Government case study C16–91–1025.0 (1992).

Name index